Laser-Enhanced Ionization Spectrometry

CHEMICAL ANALYSIS

A SERIES OF MONOGRAPHS ON
ANALYTICAL CHEMISTRY AND ITS APPLICATIONS

Editor
J. D. WINEFORDNER

VOLUME 136

A WILEY-INTERSCIENCE PUBLICATION

JOHN WILEY & SONS, INC.

New York / Chichester / Brisbane / Toronto / Singapore

Laser-Enhanced Ionization Spectrometry

Edited by

JOHN C. TRAVIS AND GREGORY C. TURK

Analytical Chemistry Division
National Institute of Standards and Technology
Gaithersburg, Maryland

A WILEY-INTERSCIENCE PUBLICATION

JOHN WILEY & SONS, INC.

New York / Chichester / Brisbane / Toronto / Singapore

This text is printed on acid-free paper.

Copyright © 1996 by John Wiley & Sons, Inc.

All rights reserved. Published simultaneously in Canada.

Reproduction or translation of any part of this work beyond
that permitted by Section 107 or 108 of the 1976 United
States Copyright Act without the permission of the copyright
owner is unlawful. Requests for permission or further
information should be addressed to the Permissions Department,
John Wiley & Sons, Inc., 605 Third Avenue, New York, NY
10158-0012.

Library of Congress Cataloging in Publication Data:

Laser-enhanced ionization spectrometry / edited by
   John C. Travis and Gregory C. Turk.
         p. cm.—(Chemical analysis ; v. 136)
      "A Wiley-Interscience publication."
      Includes index.
      ISBN 0-471-57684-0 (alk. paper)
      1. Laser spectrometry. I. Travis, John C. II. Turk, Gregory
   Chester, 1951–   III. Series.
   QD96.L3L34   1996
   621.36′6—dc20   95-36614

Printed in the United States of America

10 9 8 7 6 5 4 3 2 1

CONTENTS

CONTRIBUTORS

Ove Axner, Department of Physics, Umeå University, Umeå, Sweden

Paul B. Farnsworth, Department of Chemistry, Brigham Young University, Provo, Utah

Robert B. Green, Associated Western Universities, Inc. (Northwest Division), Richland, Washington

Nicolò Omenetto, Environment Institute, Commission of the European Communities, Joint Research Centre, Ispra (Varese), Italy

Halina Rubinsztein-Dunlop, Department of Physics, The University of Queensland, Brisbane, Queensland, Australia

John C. Travis, Analytical Chemistry Division, National Institute of Standards and Technology, Gaithersburg, Maryland

Gregory C. Turk, Analytical Chemistry Division, National Institute of Standards and Technology, Gaithersburg, Maryland

Nikita B. Zorov, Department of Chemistry, Moscow State University, Mosow, Russia

PREFACE

Since its inception in 1976, laser-enhanced ionization (LEI) has received much attention in the field of analytical atomic spectrometry, spawning research groups around the world and an extensive body of research literature. The reason for this interest, which in many ways seems out of proportion to the actual implementation of LEI in working analytical laboratories, is the defining feature of LEI—the ability to use light to *selectively* ionize atoms. LEI is a variety of optical spectroscopy that does not use a light detector. The method is in fact an optical-to-electrical transducer and does not require some other transducer, such as a photomultiplier, to perform this task. The very first LEI paper refers to the method as the opto-galvanic effect.

Terminology has always been a source of confusion for LEI and related methods. The term *laser-enhanced ionization* would be more properly called laser-enhanced *collisional* ionization. Unlike related methods that use lasers to selectively induce ionization, the laser does not ionize in LEI. It only enhances the rate of an already existing source of collisional ionization. The heat of a flame or plasma, at atmospheric pressure, ionizes the laser-excited atoms. This combination has found a comfortable niche in analytical atomic spectrometry, where flames, plasmas, and furnaces are common tools of the trade.

Our goal in putting together this book was to summarize in a single work the present understanding of the details and capabilities of LEI, which are now spread throughout hundreds of original research papers. We have restricted the scope to the field of analytical chemistry.

The first two chapters set up the physical framework of the LEI process, dividing the task between the production of ions by LEI and the generation of signal by the field-induced motion of ions and electrons. Chapter 1 describes models for laser-enhanced ion production using either a rate-equation formalism or a density-matrix formalism. Chapter 2 deals with topics such as induced LEI currents and space charge effects. It derives a highly idealized functional form for the LEI electron and ion current signals and a somewhat more realistic approximate numerical model that tracks both the temporal behavior of the ion and electron distributions deposited by a laser and the resulting current.

In Chapters 3 and 4, the capabilities of LEI as an analytical method are described. Instrumentation, sources of noise, limits of detection, interferences, and applications are the major topics in this section of the book. Included in Chapter 3 is an extensive compilation of published LEI detection limits.

Most LEI measurements are done using a flame as the atom reservoir, and for this reason most of the book is devoted to flame LEI. However, considerable effort has been devoted to the development of nonflame technologies for LEI, and this is the topic of Chapter 5.

The final chapter is not devoted to pure LEI measurements, but rather to an interesting class of measurements that fall in an area of intersection between LEI and laser-induced fluorescence (LIF). LEI and LIF begin in the identical manner, with laser excitation of atoms. Chapter 6 covers a number of topics that can be seen as the effect of LEI on fluorescence and vice versa.

We are greatly indebted to the authors who have contributed to the understanding and development of LEI spectrometry through their work. We thank them also for their patience during preparation of this book.

JOHN C. TRAVIS
GREGORY C. TURK

Gaithersburg, Maryland
March 1996

CHEMICAL ANALYSIS

A SERIES OF MONOGRAPHS ON
ANALYTICAL CHEMISTRY AND ITS APPLICATIONS

J. D. Winefordner, *Series Editor*

Laser-Enhanced Ionization Spectrometry

FUNDAMENTAL MECHANISMS OF LASER-ENHANCED IONIZATION: THE PRODUCTION OF IONS

OVE AXNER

Department of Physics, Umeå University,
S-901 87 Umeå, Sweden

HALINA RUBINSZTEIN-DUNLOP

Department of Physics, The University of Queensland,
Brisbane, Queensland 4072, Australia

1.1. INTRODUCTION

The basic principle of the laser-enhanced ionization (LEI) technique is to enhance an existing ionization process in an atomic or molecular reservoir by optical means (laser illumination) and to detect the increased rate of ionization by some galvanic (electrical) means. The first experimental investigation in the mid-1970 using the LEI spectrometry technique therefore appeared under the name opto-galvanic spectrometry (OGS), a name now used exclusively for phenomena taking place in discharges.

The LEI technique has most often been used with flames as atomic reservoirs, although investigations in graphite furnaces, thermionic diodes, plasmas, and cells have also been carried out. Here the word *reservoir* signifies a medium that is able to both vaporize and atomize an introduced sample. The common denominator for these reservoirs is that they can be characterized as *weakly ionized plasmas*, implying that there exists a certain omnipresent thermal ionization.

In LEI the thermal ionization rate is enhanced by exciting atoms/molecules using resonant laser light tuned to a specific atomic/molecular transition. This atomic/molecular excitation process will then result in the enhancement of some existing ionization rate from within the plasma. By applying an electric

Laser-Enhanced Ionization Spectrometry, edited by John C. Travis and Gregory C. Turk.
Chemical Analysis Series, Vol. 136.
ISBN 0-471-57684-0 © 1996 John Wiley & Sons, Inc.

field across the volume of interaction, the created charges will separate and hence can be efficiently detected (1–5).

The first two chapters of this book will cover the basic theory of LEI spectrometry. The present chapter will focus upon how these charges are created, and Chapter 2 will consider the charge detection process in more detail.

In Section 1.2 we consider thermal ionization processes of atoms in a weakly ionized plasma, in general, and in a flame, in particular. Concepts such as thermal equilibrium and processes responsible for thermal ionization are discussed. An expression for the thermal ionization rate of atoms in flames is derived.

Section 1.3 is concerned with the representation of the atomic system exposed to resonant laser illumination and subsequent enhancement of the ionization rate. In Section 1.3.1 we first briefly consider various descriptions of the interaction between atoms and laser light. A discussion of the most suitable model describing the processes involved in laser-enhanced ion production is then presented. The bulk of the theoretical description is derived within the framework of the rate-equation formalism. Therefore, we derive the rate equations for a general three-level system exposed to two exciting laser fields in Section 1.3.2. In Section 1.3.3 the rate equations are solved under steady-state conditions, thereby giving the number of excited atoms as a consequence of laser illumination. Both general as well as more simplified expressions for one- and two-step excitation are given. The LEI signal has to exceed the level of the electric signal given by the thermal ionization. Section 1.3.4 is therefore devoted to the derivation of an expression for the ionization rate of the excited atoms/molecules for one-step excitation, in relation to the thermal ionization rate. Section 1.3.5 relaxes the coupling to thermal ionization (in order to simplify the theoretical treatment of the system) and presents expressions for the degree of ionization of the atoms/molecules in the interaction volume. In Section 1.3.6 we briefly comment upon the concept of LEI enhancement when two-step excitation is used instead of one-step LEI excitation. Finally, Section 1.3.7 briefly considers pulsed vs. continuous-wave (cw) excitation.

There has been a considerable discussion in the literature as to what sort of ionization mechanism caused by laser excitation is responsible for the most efficient ionization. Section 1.4 discusses the concept of collisional ionization vs. photoionization. This section also illuminates some of the advantages of the LEI technique with respect to photoionization techniques [e.g., resonance ionization spectroscopy (RIS) (5)].

Section 1.5 is concerned with the ionization efficiency of excited atoms in flames, a concept that is of crucial importance for the LEI technique, as it constitutes the basis for creation of the signal.

For any given element there are many possible excitation routes that can be employed to achieve an LEI signal. However, there always exists one specific transition that yields the most sensitive signal. In Section 1.6, an expression for the most sensitive one-step LEI transition under unsaturated conditions is derived. Comparisons with measurements performed on Li and Na are presented.

Section 1.7 covers the relation between the number of ionized atoms and the beam area (i.e., focusing of the laser beam). This concept is closely related to that of anomalous contributions from atoms outside of the interaction region, which is dealt with in more detail in Section 1.8.

Section 1.9 presents experimental evidence for the influence of two-photon excitation and dynamic Stark broadening and Stark splitting of peaks. In order to model these types of effects, Section 1.10 derives equations for one- and two-step excitation LEI signals using the density-matrix approach. Finally, Section 1.11 compares the experimental findings given in Section 1.9 with the theory presented in Section 1.10.

1.2. THERMAL IONIZATION

1.2.1. Thermal Equilibrium

Whenever a system (atomic or molecular) is in contact with a thermal bath at uniform temperature, it will attain a state of thermodynamic equilibrium. When such a system is disturbed from the outside by, for example, laser light illumination, it will very rapidly rearrange itself and assume a different state. Depending on various characteristics of the thermal bath and the laser excitation processes (in particular various timescales), this new state might or might not be characterized as a state of thermodynamic equilibrium. Before we study how the system's thermodynamic equilibrium state is altered by excitation processes within the atom or molecule, let us first look at the thermodynamic equilibrium state itself in more detail.

Consider an atomic or molecular system retained in a thermal bath. Assume furthermore, that this heat bath can be characterized as a weakly ionized plasma. Then, it also possesses a small but not always insignificant amount of thermal ionization (6). Modeling some processes in this plasma in the most simple way we can first define a thermal ionization rate contant, $k_{th.ion}$, for each specific process in the heat bath. One such process, for example, is the collisional ionization process between a species M and a collisional partner X:

$$[M][X] \xrightarrow{k_{th.ion.}} [M^+][e^-][X] \qquad (1)$$

where [M] is the concentration of the species M; [M$^+$] is the concentration of its positive ion; [X] is the concentration of the collisional partner X; and [e$^-$] is the concentration of electrons.

Similarly, there exists also a recombination rate constant, $k_{recomb.}$, for the reverse process, given by

$$[M^+][e^-][X] \xrightarrow{k_{recomb.}} [M][X] \tag{2}$$

When the system is in thermal equilibrium, these two rates balance exactly. This enables us to write an expression for the relation between the concentrations of the species M, its ions, and the electrons. This relation is known as the Saha equation and is stated in this case (7) as

$$[M^+][e^-] = K_{ion}[M] \tag{3}$$

where K_{ion} is the ionization constant, which is simply given by the ratio of the thermal ionization rate constant to the recombination rate constant, i.e.,

$$K_{ion} = \frac{k_{th.ion.}}{k_{recomb.}} \tag{4}$$

It is worthwhile to note that even though the ionization and recombination processes in themselves require a third body (as collision partner) and each of their rates depends upon the concentration of the collision partner, the ionization constant (and hence the degree of ionization of the atomic system) is in fact independent of the concentration of this third partner as long as thermal equilibrium conditions prevail.

As has already been mentioned and as will be discussed in more detail in Chapter 2, in the LEI technique an electric field is applied across the volume of interaction to separate the created charges and make detection of those charges possible. This applied electric field will remove most of the charges from the interaction region (in particular the electrons due to their high mobility). Consequently, the electric field will prevent recombination from occurring at any significant rate—the only remaining process will be the thermal ionization rate (8)—and the Saha equation is no longer valid (7). Since we then have reached a situation in which we have altered the conditions for thermal equilibrium, it is possible to measure this thermal ionization rate. In the same way, if we succeed in *enhancing* the ionization rate by exciting atoms/molecules using, for example, resonant laser light, *the electric field will also make possible the detection of the **increased** ionization rate*. This is, in short, the fundamental principle of the LEI technique (3).

1.2.2. Processes Responsible for Thermal Ionization

Let us now look at various processes responsible for thermal ionization of atoms/molecules in a weakly ionized plasma in more detail. In particular let us look more closely at the situation in a flame, since flames are the most commonly used atomic reservoirs for LEI spectrometry. A flame can be defined as a weakly ionized plasma since it exhibits most of the characteristics of such a medium, i.e., it has a certain production and recombination of charge carriers, collective motions of free charge carriers, and a Debye shielding distance (the distance over which any charge concentration is shielded by fast-moving electrons) that is much smaller than the typical distances involved (again in a state of thermal equilibrium). It is of particular interest that, according to the definition of a *weakly* ionized plasma, the energy involved in the ionization processes should only be a minor part of the whole available amount of energy in the system. This is indeed the case in a flame.

In a flame, a variety of different interactions between atoms, molecules, and light can take place. Considering for the moment only the *major* processes that directly or indirectly lead to ionization of an atom or a molecule in the flame, we find that a possible classification can be done by dividing them into *physical* and *chemical* ionization processes (6). Physical processes can, in turn, be divided into *collisional* and *radiative* ionization processes.

The *collisional physical* ionization processes are those that are most often considered to be responsible for ionization of atoms in flames. As an example, collisional ionization processes are expected to be responsible for the ionization of alkali elements in the flame, e.g.,

$$Na + X \rightarrow Na^+ + e^- + X \tag{5}$$

where X represents any flame molecule (6, 9).

The *radiative physical* ionization processes can be exemplified by the photoionization process that takes place when atoms and molecules are irradiated with laser light of sufficiently short wavelength (and strong enough intensity) so as to result in the production of an electron, e.g.,

$$Na + h\nu \rightarrow Na^+ + e^- \tag{6}$$

or by the interaction between blackbody radiation and (primarily excited) atoms/molecules. These processes can take place for almost any atom/molecule.

Chemical ionization processes, on the other hand, can be most easily characterized as those ionization processes in which the formation of a new chemical bond takes place. Most alkaline earth atoms are believed to ionize in such a way. The formation of the $CaOH^+$ ionic molecule from atomic Ca in the presence of the OH molecule represents such a process, often referred to as associative ionization:

$$Ca + OH \rightarrow CaOH^+ + e^- \tag{7}$$

There are also a number of charge redistribution processes that will not actively increase the number of charge carriers but rather transfer charges from one carrier to another. Although the number of charge carriers does not change, these types of processes might still be of importance, since they can constitute one reaction in a chain leading to a net ionization rate.

1.2.3. The Thermal Ionization Rate of an Atom in a Flame

Let us now derive an expression for the thermal ionization rate of an atom in a flame. We assume that we are considering an atom for which collisional ionization dominates other ionization processes. We assume further that the flame, with all its atomic and molecular species, is in a state of thermal equilibrium. Let us then estimate the thermal collisional ionization rate of atoms in such a flame.

1.2.3.1. Thermal Ionization of Ground State Atoms

An expression for the collisional ionization rate for ground state atoms can be written, according to a simple gas-kinetic theory, as a product of four different factors:

    a. The average relative velocity between two species in the flame
    b. The cross section of the collisional processes
    c. The concentration of the collisional species in the flame
    d. A factor involving the activation energy of the ionization process

Therefore, assuming that we can assign a cross section to the process of thermal ionization of ground state atoms in a flame by collisional ionization and that the activation energy for such a process is equal to that of the ionization limit, we can write the expression for the collisional ionization rate for ground state atoms, i.e., the state-specific ionization rate from state 1, $k_{1,ion}$:

$$k_{1,ion} = \left(\frac{8kT}{\pi\mu}\right)^{1/2} \sigma^X_{1,ion} \exp(-E_{ion}/kT)[X] \tag{8}$$

where the first square-root factor represents the average relative velocity of two species in the flame; k is the Boltzmann constant; T represents the temperature; $\sigma_{1,\text{ion}}^{X}$ is the cross section for ionization from the ground state (level 1) via a collision with the species X, where [X] again is the concentration of such species in the flame; μ is the reduced mass of the system [$\mu = m_{\text{atom}} m_{X}/(m_{\text{atom}} + m_{X})$, where m_{atom} and m_{X} are the masses of the atom under consideration and collision species X, reprectively]; and E_{ion} is the energy of the ionization limit.

Inserting appropriate values for the foregoing variables, assuming the major flame constituent to be the main quenching partner and a temperature of 2500 K, we obtain an average relative velocity of atoms in a flame of approximately 2×10^{3} m/s and a concentration of species in the flame of approximately 3×10^{24} m^{-3}. Furthermore, assuming that the cross section of a ground state atom is approximately equal to the geometric size of the atom (which in turn is approximately equal to a typical gas-kinetic value of the cross section), e.g., 10^{-16} cm^{2}, we obtain values for the ionization rate on the order of 10^{-3}–10^{-6} Hz for elements with ionization limits between 5 and 7 eV (i.e., 40,000–56,000 cm^{-1}). However, these values are in general several orders of magnitude *lower* than the measured ones. For example, experimental thermal ionization rates of alkali elements in an $H_2/O_2/N_2$ flame at 1930 K have previously been measured to be approximately $10^{2}, 10^{1}, 10^{-1}$, and 10^{-2} Hz for the elements Cs, K, Na, and Li, respectively. This big discrepancy between estimated and measured thermal ionization rates can be accounted for if we also consider that excited atoms can participate in the ionization process.

1.2.3.2. *Thermal Ionization of Excited State Atoms*

Let us again assume that the flame, with its species, is in a state of thermal equilibrium. This implies that the concept of detailed balance between each pair of atomic levels is valid. However, as will be seen from the subsequent discussion, since every atom has a virtually infinite number of excited levels, the assumption that we have a detailed balance between *all* pairs of levels leads to an absurd consequence, namely, a huge ionization rate. Hence, this model also fails to describe correctly the total thermal ionization rate, this time by overestimating it.

However, despite the problem of the model (that it predicts a significantly overestimated ionization rate), it still has a certain interest since it can illustrate very effectively some other important phenomena (e.g., the increase of the thermal ionization rate when laser light induces transitions between pairs of atomic levels, and in particular the role played, in such a case, by other closely lying levels—a matter of importance when a general atomic system is being reduced to a few-level system). Consequently, we will take a close look at this model for reasons that will be clarified shortly.

Let us start by assuming that we indeed have a detailed balance between all atomic levels, i.e., the atomic energy levels are populated according to Boltzmann's distribution law at thermal equilibrium:

$$n_i = \frac{g_i \exp(-E_i/kT)}{Z} n_{\text{atom}} \tag{9}$$

when n_i is the number density of atoms in state i (often referred to as the population of the level) $[\text{m}^{-3}]$; g_i, the degeneracy of the level; E_i, the energy of the state i; n_{atom}, the total number density of neutral atoms $[\text{m}^{-3}]$; and Z, the partition function, i.e.,

$$Z = \sum_i g_i \exp(-E_i/kT) \tag{10}$$

The partition function takes care of the condition that the sum of the population of all levels should be equal to n_{atom}, i.e.,

$$\sum_i n_i = n_{\text{atom}} \tag{11}$$

This implies that the fractional density of atoms in level i (or fractional thermal population) is simply given by

$$\frac{n_i}{n_{\text{atom}}} = \frac{g_i \exp(-E_i/kT)}{Z} \tag{12}$$

In addition, the condition of detailed balance, which requires any pair of atomic levels to be in equilibrium, is described by

$$n_i k_{ij} = n_j k_{ji} \tag{13}$$

where k_{ij} is the collisional transfer rate between the two states i and j $[\text{s}^{-1}]$. This relation can be used to relate any pair of collisional excitation and deexcitation rates to each other according to

$$\frac{k_{ij}}{k_{ji}} = \frac{n_j}{n_i} = \frac{g_j}{g_i} \exp(-\Delta E_{ji}/kT) \tag{14}$$

where ΔE_{ji} is the energy difference between the two states j and i, i.e., $\Delta E_{ji} = E_j - E_i$.

If we now assume that the ionization probability for the individual levels scales with an activation energy factor equal to the energy difference between the level and the ionization limit (i.e., in the same way as in Section 1.2.3.1, above), we can write a state-specific thermal ionization rate for an excited atom in state i in a flame, $k_{i,\text{ion}}$, as

$$k_{i,\text{ion}} = \left(\frac{8kT}{\pi\mu}\right)^{1/2} \sigma_{i,\text{ion}}^{X} \exp\left(-(E_{\text{ion}} - E_i)/kT\right)[X] \tag{15}$$

where we again have written the collisional ionization rate in terms of a cross section. This time $\sigma_{i,\text{ion}}^{X}$ represents the state-specific cross section for ionization, i.e., the cross section for ionization from the specific level i through a collision with the species X.

Hence, we find that each excited state possesses a certain collisional ionization rate, $k_{i,\text{ion}}$.

1.2.3.3.   Thermal Ionization of Multilevel Atoms

If we then take into account the actual thermal populations of each excited state, we can write the total thermal ionization rate for the atomic system (consisting of N atomic levels), dn_{ion}/dt [m^{-3}s^{-1}], as follows:

$$\frac{dn_{\text{ion}}}{dt} = \sum_{i=1}^{N} n_i k_{i,\text{ion}} = \sum_{i=1}^{N} \left(\frac{n_i}{n_{\text{atom}}}\right) k_{i,\text{ion}} n_{\text{atom}} \tag{16}$$

If we now insert the expressions for the state-specific fractional thermal population and the state-specific collision ionization rate from Eqs. (12) and (15), we find that the energy of the particular state i cancles out and that *each of the levels in the atom* (the ground state as well as the excited states) *will contribute approximately the same amount to the total collisional ionization rate* (only in proportion to their degeneracy factor, g_i, and state-specific cross section, $\sigma_{i,\text{ion}}^{X}$), namely,

$$\frac{n_i}{n_{\text{atom}}} k_{i,\text{ion}} = \left(\frac{8kT}{\pi\mu}\right)^{1/2} \sigma_{i,\text{ion}}^{X} \frac{g_i \exp\left(-E_{\text{ion}}/kT\right)}{Z}[X] \tag{17}$$

Hence, the total thermal ionization rate of an atom in a heat bath at temperature T can be written as

$$\frac{dn_{\text{ion}}}{dt} = \left(\frac{8kT}{\pi\mu}\right)^{1/2} \frac{\exp\left(-E_{\text{ion}}/kT\right)}{Z}[X] \sum_{i=1}^{N} g_i \sigma_{i,\text{ion}}^{X} n_{\text{atom}} \tag{18}$$

If we assume, for simplicity, that all ionization cross sections are approximately the same (which is not a necessary assumption, but simplifies the interpretation of the expression), we can write this expression as

$$\frac{dn_{ion}}{dt} \approx \left(\frac{8kT}{\pi\mu}\right)^{1/2} \sigma_{ion}^X \frac{\exp(-E_{ion}/kT)}{Z}[X] \sum_{i=1}^{N} g_i n_{atom} \qquad (19)$$

This implies, in turn, that the total thermal ionization rate, dn_{ion}/dt, can also rather easily be expressed in terms of the state-specific thermal ionization rate of the ground state, $k_{1,ion}$, [see Eq. (8)], namely,

$$\frac{dn_{ion}}{dt} = \frac{1}{Z} \sum_{i=1}^{N} g_i k_{1,ion} n_{atom} \qquad (20)$$

This expression illustrates that, as we sum over all states in an atom, the total thermal ionization rate can become exceedingly large (since there is virtually an infinite number of bound states, each with a finite cross section for ionization). In fact, as we let the number of states, N, increase drastically (i.e., include almost an infinite number of states), the total ionization rate approaches the limiting value

$$\frac{dn_{ion}}{dt} = \frac{1}{Z} \sum_{i=1}^{N} g_i k_{1,ion} n_{atom} \xrightarrow[N \to \infty]{} \exp(E_{ion}/kT)k_{1,ion} n_{atom} \qquad (21)$$

This implies that the activation energy dependence of the collision process, $\exp(-E_{ion}/kT)$, which is given implicit in the expression for $k_{1,ion}$ [see Eq. (8)], effectively cancels out. Hence, as the ionization rate for the ground state *underestimates* the actual total thermal ionization rate for an atom (as was discussed in Section 1.2.3.1, above), one finds that the estimated total thermal ionization rate is *overestimated* when we include *all* excited states in the description of the atom.

The normal way to avoid this problem is to assume that the condition of detailed balance is not valid for the highest lying states in the atom. We can then, somewhat arbitrarily, assume that this condition is only valid for states up to a certain level, either expressed in terms of energy or in terms of principal quantum number (e.g., assuming that the mean radius of the electronic orbit should be a certain fraction of the mean free path in the heat bath). We can then write

$$\frac{dn_{ion}}{dt} \approx \left(\frac{8kT}{\pi\mu}\right)^{1/2} \sigma_{ion}^X \frac{\exp(-E_{ion}/kT)}{Z}[X] \sum_{i=1}^{i_{cutoff}} g_i n_{atom} = \frac{1}{Z} \sum_{i=1}^{i_{cutoff}} g_i k_{1,ion} n_{atom}$$

$$(22)$$

Since the sum of the degeneracy factors of all levels up to (and including) a certain cutoff principal quantum number, n_{cutoff}, is given (for a single valence electron atom) by the expression

$$\sum_{i=1}^{i_{cutoff}} g_i = \frac{2n_{cutoff}^3 + 3n_{cutoff}^2 + n_{cutoff}}{3} \tag{23}$$

we find that this implies that the total ionization rate of an atomic system will be approximately 3–4 orders of magnitude larger than the thermal ionization rate of ground state atoms for n_{cutoff} values between 10 and 30.

Even though this simplification might not be found fully satisfactory (since the choice of the cutoff position strongly affects the thermal ionization rate), what is of importance in this case is that the relationship between the *enhanced* ionization rate (due to laser illumination) and the *thermal* ionization rate is rather insensitive to the specific choice of this cutoff. Let us therefore assume that Eq. (22) is sufficiently accurate to use in the derivation of the relationship between laser-enhanced ionization rates and thermal ionization rates.

1.3. LASER-ENHANCED IONIZATION IN THE RATE-EQUATION FORMALISM

1.3.1. Representation of the Interaction Between Atoms and Laser Light

In order to model the processes involved in LEI, let us start with the concept of excitation of atoms/molecules by laser light and its relationship to the enhanced ionization rate. In the following presentation, emphasis has been placed on presenting the description of the excitation and deexcitation processes so as to illustrate the *basic* physical principles (such as the relation between thermal ionization and laser-enhanced ionization rates, the number of excited atoms as a consequence of laser-illumination, or the degree of ionization of the atoms/molecules in the interaction volume), sometimes at the cost of full stringency of the theoretical treatment.

When deriving expressions for the number of excited atoms in various levels due to laser excitation, it is common to treat the atoms as though they consist of only a few bound states. Of special interest are, of course, the states between which laser light can induce transitions, but, as we will show, other non-laser-connected levels might also be of importance.

In what follows, we will study more specifically the population of the states of a simplified three-level system, since it serves several purposes.

First, it can be used for the case when the atoms are being excited by a one-step excitation process, i.e., between the ground state and one of the two excited states in the model. The three-level model can, in such a case, predict the populations of the two laser-connected levels, as well as the non-laser-connected level, which can represent either a multitude of states lying higher than the uppermost laser-connected level or any of the levels that might exist between the two laser-connected levels (6, 10, 11). Hence, the non-laser-connected levels can represent the influence of higher lying or intermediate levels upon the two laser-connected levels.

The three-level model can also be used for predicting the populations in a three-level system exposed to two exciting laser light fields (a two-step excitation process). However, in such a case, the three-level system is not able to include any of the influences of other higher lying or intermediate states upon the laser-connected levels. Such an influence can be accounted for in a phenomenological way, if it is felt necessary, by the use of results obtained in the study of one-step excitation upon three-level systems. The three-level system can also not describe the simultaneous influence of higher lying states and intermediate states upon the one-step excitation case (a situation which, however, is not of highest importance for a full understanding of the general processes above).

The proper treatment of the interaction of laser light and a reduced atomic system can only be performed by solving the time-dependent Schrödinger equation for a few-level atomic system exposed to laser light. The resulting wave function can be represented, for example, by a density-matrix formalism. In such a description several properties of the laser light have to be described in detail, for example, whether it consists of one or several modes as well as its bandwidth. This quasi-monochromatic light field can then induce transitions between levels, with strengths primarily described by the Rabi frequency of each transition, i.e., the frequency at which the atomic population alternates periodically between the states involved.

However, when studying atom–light interactions in collisionally dominated environments, as is the case for in flames, the density-matrix formalism is usually simplified to a rate-equation approach (12), justified mainly by considerations of the high collision rates in flames and the large bandwidth of the laser light. In the rate-equation formalism, all coherence effects induced by the finite bandwidth of the light are neglected, since the light is assumed to be significantly broader than the atomic transitions and can be characterized by only a spectral irradiance.

The simplification of the density-matrix formalism to the rate-equation approach is a good approximation particularly for one-step excitation or for excitations in general with light significantly broader than the atomic transitions (13, 14). However, in the case of two-step excitation LEI when the laser light is not necessarily significantly broader than the transitions, not all physical properties of the combined atomic and light system will be correctly represented by the rate-equation formalism, which has been shown recently (15). Effects that are not properly taken care of are primarily those caused by intense light fields, such as two-photon excitation and dynamic Stark effects (broadening and splitting of transitions). However, because of the simplicity of the rate-equation formalism, extensive derivations and calculations of laser-induced atomic excitation (16, 17) as well as the LEI signal strengths using the rate-equation formalism have frequently occurred in the literature (4, 9, 18–20).

Below we give a simplified derivation of expressions for the one- and two-step LEI signal strengths based on the rate-equation formalism. This will serve as a basis for an illustrative description of the various physical phenomena that influence the potential of LEI as a powerful technique for trace element analysis. This is done in the following way. In Section 1.3.2 the rate equations for a general three-level system exposed to one or two exciting laser light fields are presented. We will then derive expressions for the fraction of atoms excited in the interaction region due to the laser excitation (Section 1.3.3). Section 1.3.4 will be devoted to the derivation of an expression for the one-step excitation LEI ionization rate of the excited atoms/molecules and its relation to the thermal ionization rate. Next, in Section 1.3.5, the degree of ionization of the atoms/molecules in the interaction volume is calculated. Section 1.3.6 will then briefly comment on LEI enhancement, i.e., the relation between the two-step and the one-step LEI signals. Finally, Section 1.3.7 discusses pulsed vs. cw excitation, all in terms of the rate-equation approach.

The rate-equation formalism derived in Section 1.3 is then used as a tool in some of the subsequent sections: Section 1.4 compares collisional ionization LEI with one-step excitation followed by photoionization; Sections 1.5, 1.6, and 1.7 describe the concepts of ionization efficiency of excited atoms in flames, the most sensitive one-step transition in LEI, and the number of ions produced vs. beam area (focusing), respectively, all based upon the rate-equation formalism; Section 1.8 discusses anomalous contributions to the signal from excitation of atoms outside the geometrically confined interaction region.

The last three sections in this chapter address the needs and possibilities of using the density-matrix formalism to describe anomalous features occurring in the LEI technique such as anomalous lineshapes, two-photon contributions, and dynamic Stark effects (broadening, splitting, and shifts): Section 1.9 describes some experimental findings that cannot be explained by the rate-equation formalism; Section 1.10 gives a simplified derivation of the density-matrix formalism for two-step excitation of atoms in highly collisional media; Section 1.11 compares the experimental findings with predictions from the density-matrix formalism.

1.3.2. A Three-Level System Exposed to One- or Two-Step Excitation

Let us now derive an expression for the number density of excited atoms as a consequence of laser illumination within the rate-equation formalism (12, 21). We assume, for simplicity, that the atoms can be treated as consisting of only three bound levels (the levels 1, 2, and 3; see Fig. 1.1) among which the laser excitation takes place in one or two steps (at one or two different wavelengths), weakly coupled to an ionization continuum. Such a three-level system can be described by the following rate-equation system:

$$\frac{dn_3}{dt} = k_{13}n_1 + (k_{23} + B_{23}I_{23}^v)n_2 - (k_{31} + k_{32} + A_{32} + B_{32}I_{32}^v + k_{3,\text{ion}})n_3 \quad (24)$$

$$\frac{dn_2}{dt} = (k_{12} + B_{12}I_{12}^v)n_1 - (k_{21} + A_{21} + B_{21}I_{21}^v + k_{23} + B_{23}I_{23}^v + k_{2,\text{ion}})n_2$$
$$+ (k_{32} + A_{32} + B_{32}I_{32}^v)n_3 \quad (25)$$

and

$$n_{\text{atom}} = n_1 + n_2 + n_3 \quad (26)$$

where n_1, n_2, and n_3 are the number densities of atoms in the three states 1, 2, and 3, respectively [m^{-3}]; n_{atom} is the total number density of neutral atoms

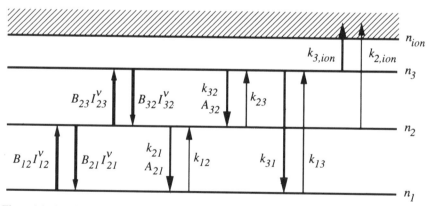

Figure 1.1. A schematic representation of various excitation and deexcitation processes in a three-level atom: n_1, n_2, and n_3 are the number densities of the three levels, respectively; k_{21} is the sum of the collisional deexcitation and spontaneous emission rates between levels 2 and 1; k_{12} is the collisional excitation rate between levels 1 and 2; $k_{3,\text{ion}}$ ($k_{2,\text{ion}}$) is the collisional ionization rate from level 3 (2); $B_{12}I_{12}^v$ ($B_{23}I_{23}^v$) and $B_{21}I_{21}^v$ ($B_{32}I_{32}^v$) are the absorption and stimulated emission rates and I_{21}^v ($=I_{12}^v$) and I_{23}^v ($=I_{32}^v$) are the spectral irradiances of the laser light.

$[m^{-3}]$; k_{21} (k_{32} and k_{31}) is the collisional deexcitation rate and A_{21} (A_{32}) is the spontaneous emission rate between levels 2 and 1 (3 and 2, and 3 and 1, respectively) $[s^{-1}]$; k_{12} (k_{23} and k_{13}) is the collisional excitation rate between the levels 1 and 2 (2 and 3, and 1 and 3, respectively) $[s^{-1}]$; $k_{3,ion}$ ($k_{2,ion}$) is the collisional ionization rate from level 3(2) $[s^{-1}]$; $B_{12}I_{12}^v$ ($B_{23}I_{23}^v$) and $B_{21}I_{21}^v$ ($B_{32}I_{32}^v$) are the absorption and stimulated emission rates $[s^{-1}]$; and I_{21}^v ($= I_{12}^v$) and I_{23}^v ($= I_{32}^v$) are the spectral irradiances of the laser light, defined as the intensity per unit (absolute) frequency bandwidth $[W/(m^2\ Hz)]$. The relations between the Einstein A_{21}, B_{21}, and B_{12} factors for spontaneous and stimulated emission and absorption are given in this notation by

$$B_{12} = \frac{g_2}{g_1}B_{21} = \frac{g_2}{g_1}\frac{\lambda^3}{8\pi hc}A_{21} \tag{27}$$

where h is Planck's constant $[6.62 \times 10^{-34} J/Hz]$, and c is the speed of light $[3.0 \times 10^8\ m/s]$.

In the foregoing description, the three-level system has been coupled to the ionization continuum by the state-specific collisional ionization rates $k_{3,ion}$ and $k_{2,ion}$ for completeness. Laser-induced chemical processes (molecular formation as well as chemical ionization) have been omitted since they are generally not of major importance in comparison with collisional processes (22–24).

1.3.3. The Number of Excited Atoms

The main objective of this subsection is to solve the rate equations for the number of atoms excited during the laser pulse as a consequence of laser light illumination. More specifically, let us express the number of excited atoms in relation to the total number of neutral atoms in the interaction region. We call this quantity here the fraction of neutral atoms excited (and will denote it with C; see Section 1.3.3.1, below). The general expressions will then be simplified for the two special cases of one- and two-step excitation, respectively. Let us first, however, derive the general (steady-state) solution of the system of rate-equations.

1.3.3.1. General Solution of the System of Rate Equations

This system of equations can be most easily solved by assuming that an equilibrium is established among the three bound levels within the first part (nanosecond) of the laser pulse. In addition, since the ionization collision rates are, in general, much smaller than deexcitation rates within the three-level system, the ionization rates can, to a first order, be neglected in comparison with the other rates in the treatment of the three-level system (25). Hence, the rate equations [(Eqs. (24)–(25)] can be solved for steady-state conditions.

However, when a significant fraction of atoms is ionized within the duration of the laser pulse, it is important also to include the effect of depletion of neutral atoms in the interaction region (due to ionization) in the solution of the rate equations. Depletion of neutral atoms limits the total number of ionized atoms to the total number of species under study and introduces a time dependence in the expression for the number of excited atoms despite the steady-state treatment of the rate equations. It is therefore of importance to account for depletion properly already at this early stage. This can most easily be done by relating the number densities of atoms in the excited states, $n_i(t)$, to the total number density of (remaining) *neutral* atoms within the system at each specific time, $n_{atom}(t)$, rather than to the total number density of species under study, n_{tot}, which is a time-independent quantity. By using the concept of the fraction of neutral atoms in an excited state, as is done below, the derived expressions will be *time independent* and thus valid throughout the entire laser pulse, irrespective of the fraction of atoms being ionized (i.e., the degree of depletion of the atomic system) (20). (This also implies that the steady-state condition only applies to the *relative* distribution of excited and ground-state atoms within the three-level neutral atomic system; there is no steady-state condition related to the number density of neutral atoms, ions, or the ionization rate.)

Let us define the fractions of neutral atoms that end up in state 2 and state 3 as a consequence of the laser illumination as C_2 and C_3, respectively, such that

$$C_2 = \frac{n_2(t)}{n_{atom}(t)} \tag{28}$$

and

$$C_3 = \frac{n_3(t)}{n_{atom}(t)} \tag{29}$$

We should emphasize that even though both $n_i(t)$ and $n_{atom}(t)$ in the expressions above are time dependent, the fractions of excited atoms (i.e., C_2 and C_3) are time independent throughout the laser pulse (except for a very rapid transient phenomenon in the beginning of the excitation pulse, before the steady-state condition has been established).

Whenever the system is exposed to laser light and excitation takes place, the rate equations can be solved under steady-state conditions, giving the ratios of the number density of excited atoms in state 2 and state 3 to the total number of neutral atoms. The general expressions for the fractions of neutral atoms excited are then given by

$$C_2 = \frac{(k_{12} + B_{12}I_{12}^{\nu})(k_{31} + k_{32} + B_{32}I_{32}^{\nu}) + k_{13}(k_{32} + B_{32}I_{32}^{\nu})}{(k_{21} + B_{12}I_{12}^{\nu} + B_{21}I_{21}^{\nu})(k_{31} + k_{32} + B_{32}I_{32}^{\nu}) + (k_{31} + B_{12}I_{12}^{\nu})(k_{23} + B_{23}I_{23}^{\nu})}$$

(30)

and

$$C_3 = \frac{(k_{12} + B_{12}I_{12}^{\nu})(k_{23} + B_{23}I_{23}^{\nu}) + k_{13}(k_{21} + k_{23} + B_{21}I_{21}^{\nu} + B_{23}I_{23}^{\nu})}{(k_{21} + B_{12}I_{12}^{\nu} + B_{21}I_{21}^{\nu})(k_{31} + k_{32} + B_{32}I_{32}^{\nu}) + (k_{31} + B_{12}I_{12}^{\nu})(k_{23} + B_{23}I_{23}^{\nu})}$$

(31)

where, for simplicity, spontaneous emission and some exciting collisions have been neglected, as these are generally inferior to deexciting collisions (i.e., $A_{32}, A_{21}, k_{12}, k_{23}$, and $k_{13} \ll k_{21}, k_{31}$, and k_{32}).

The behavior of the system exposed to either one- or two-step laser excitation is of particular interest. The foregoing expressions can then be rearranged and simplified to give the most proper description of the system for these two specific conditions, i.e., with or without the second-step laser illumination ($I_{23}^{\nu} = I_{32}^{\nu} \neq 0$ and $I_{23}^{\nu} = I_{32}^{\nu} = 0$, respectively). In the following discussion we will refer to the two cases when the second-step laser is *on* or *off* by the superscript on or off, respectively.

1.3.3.2. Simplified Solution of the System of Rate Equations for One-Step Excitation

For one-step laser excitation, the two expressions for the fraction of neutral atoms in states 2 and 3, C_2 and C_3, reduce to C_2^{off} and C_3^{off}, given by

$$C_2^{\text{off}} = (C_2^{\text{off}})_{\text{sat}} \frac{B_{12}I_{12}^{\nu}}{B_{12}I_{12}^{\nu} + (C_2^{\text{off}})_{\text{sat}}k_{21}}$$

(32)

and

$$C_3^{\text{off}} = (C_3^{\text{off}})_{\text{sat}} \frac{B_{12}I_{12}^{\nu}}{B_{12}I_{12}^{\nu} + (C_2^{\text{off}})_{\text{sat}}k_{21}}$$

(33)

where the $(C_2^{\text{off}})_{\text{sat}}$ and the $(C_3^{\text{off}})_{\text{sat}}$ factors are the population density fractions of the excited levels at full saturation of the first-step transition with the second step turned off. Expressions for the $(C_2^{\text{off}})_{\text{sat}}$ and the $(C_3^{\text{off}})_{\text{sat}}$ factors are given in Table 1.1.[1] In order to arrive at these equations, some spontaneous emission

---

[1] Expressions for a number of variables in the equations in Sections 1.3.3.2–1.3.3.5 are all collected in Table 1.1.

Table 1.1. Variables Appearing in the Expressions for the One- and Two-Step Fractions of Excited Atoms, C_2^{off}, C_2^{on}, and C_3^{on}

Saturated population density fractions for one-step excitations:

$$(C_2^{off})_{sat} = \frac{g_2}{g_1 + g_2}$$

$$(C_2^{off})_{sat}^* = \frac{(C_2^{off})_{sat}}{1 + \zeta (C_2^{off})_{sat}}$$

$$(C_3^{off})_{sat} = \zeta (C_2^{off})_{sat}$$

$$(C_3^{off})_{sat}^* = \zeta (C_2^{off})_{sat}^*$$

Saturated population density fractions for two-step excitations:

$$(C_2^{on})_{sat} = \frac{g_2}{g_1 + g_2 + g_3}$$

$$(C_3^{on})_{sat} = \frac{g_3}{g_1 + g_2 + g_3}$$

Branching ratios:

$$\zeta = Y_{32} \frac{g_3}{g_2} \exp(-\Delta E_{32}/kT)$$

$$Y_{31} = \frac{k_{31}}{k_{31} + k_{32}}$$

$$Y_{32} = \frac{k_{32}}{k_{31} + k_{32}}$$

Constants in the expression for the population density fractions for two-step excitations:

$$a = (C_3^{on})_{sat} k_{31} + (C_2^{on})_{sat} k_{21}$$

$$b = \frac{(C_3^{on})_{sat}}{(C_2^{off})_{sat}} (k_{31} + k_{32})$$

$$c = [(C_3^{on})_{sat}]^2 (k_{31} + k_{32}) \left(k_{21} - \frac{k_{31}}{(C_2^{off})_{sat}} \right)$$

and exciting collisions have again been neglected in comparison with deexciting collisions or laser excitation i.e., $k_{13} \ll (A_{32}, A_{21}, k_{12}$, and/or $k_{23}) \ll (k_{21}, k_{31}$, and/or $k_{32})$ and $k_{12} \ll B_{12} I_{12}^\nu$. Furthermore, we have used the relation between the Einstein coefficients for absorption and stimulated emission, i.e., $g_1 B_{12} = g_2 B_{21}$.

These expressions describe the well-known feature of optical saturation for excitation of atoms by laser light (i.e., that the fraction of atoms excited increases linearly with laser light intensity for low intensities and less than linearly for high laser intensities). Since both practical and theoretical aspects of optical saturation have been discussed previously in several papers in the literature (26–30), we will not comment upon this issue in detail here.

1.3.3.3. Simplified Solution of the System of Rate Equations under Two-Step Excitation

The expression for the fraction of atoms excited (to level 3) under two-step excitation (neglecting all other non-laser-connected levels) looks like

$$C_3^{on} = (C_3^{on})_{sat} \frac{B_{12}I_{12}^v \, B_{23}I_{23}^v}{(B_{12}I_{12}^v + a)(B_{23}I_{23}^v + b) + c} \tag{34}$$

where, as earlier, spontaneous emission and exciting collisions have been neglected in comparison with deexciting collisions or laser excitation, i.e., $k_{13} \ll (A_{31}, A_{32}, A_{21}, k_{12},$ and/or $k_{23}) \ll (k_{21}, k_{31},$ and/or $k_{32})$; $k_{12} \ll B_{12}I_{12}^v$; and $k_{23} \ll B_{23}I_{23}^v$ and where the expressions for the a, b, and c variables in the foregoing equation can be found in Table 1.1.

In the derivation of the expressions for the fraction of neutral atoms excited we have completely neglected the influence of other high-lying levels. However, since all levels are coupled to each other by collisions (in one way or another), it is important also to study the influence of other excited levels on this excitation process.

1.3.3.4. Simplified Solution of the System of Rate Equations under One-Step Excitation: The Influence of Higher Lying States

If we now want to investigate to what extent other non-laser-connected levels will influence the excitation process, we can simply include the coupling to other excited states by studying how the existence of level 3 will influence the population of level 2 for the one-step excitation case.

As can be seen from the expressions for the fractions of atoms excited in the one-step excitation case [Eqs. (30) and (31) together with Table 1.1], the population of the upper state (level 3) (if treated as typical of any high-lying non-laser-connected level) will be directly proportional to that of the upper laser-connected level (level 2) for all laser irradiances

$$C_3^{off} = \xi C_2^{off} \tag{35}$$

where again ξ is given in Table 1.1.

In the derivations of C_2^{off} and C_3^{off} above [Eqs. (32) and (33)] we have assumed that exciting collisions to other nearby (higher lying) states are smaller than typical quenching rates. This assumption is, of course, not valid if the energy separation between the different states is small enough.

If we now relax the condition that the exciting collisions to other nearby (higher lying) states are smaller than typical quenching rates we can write the expressions for the fraction of neutral atoms excited as

$$C_2^{\text{off}} = (C_2^{\text{off}})_{\text{sat}}^* \frac{B_{12}I_{12}^{\nu}}{B_{12}I_{12}^{\nu} + (C_2^{\text{off}})_{\text{sat}}^*(k_{21} + k_{23}Y_{31})} \tag{36}$$

and

$$C_3^{\text{off}} = (C_3^{\text{off}})_{\text{sat}}^* \frac{B_{12}I_{12}^{\nu}}{B_{12}I_{12}^{\nu} + (C_2^{\text{off}})_{\text{sat}}^*(k_{21} + k_{23}Y_{31})} \tag{37}$$

The expressions with the superscript asterisk (*) thus represent the cases when the couplings to other excited states are explicitly taken into account. The Y_{31} and Y_{32} are the branching ratios from level 3 down to level 1 and level 2, respectively. Expressions for the various variables in the foregoing equations are, as before, collected in Table 1.1.

In the above derivation, we have assumed that the exciting and deexciting collisional rates between level 2 and 3 are related by the condition of detailed balance, i.e.,

$$k_{23} = k_{32} \frac{g_3}{g_2} \exp\left(-\Delta E_{32}/kT\right) \tag{38}$$

What can be found from Eqs. (36) and (37) is, first of all, that the upper non-laser-connected levels (here represented by level 3 and C_3^{off}) are still replicas of that of the upper laser-connected level, C_2^{off}, related by the factor ζ (since $(C_3^{\text{off}})_{\text{sat}}^* = \zeta(C_2^{\text{off}})_{\text{sat}}^*$). This implies that Eq. (35) is still valid and hence that the population of the highest level in this description (level 3) will be related to the upper laser transition almost in a Boltzmann-like manner:

$$\frac{C_3^{\text{off}}}{C_2^{\text{off}}} = \frac{(C_3^{\text{off}})_{\text{sat}}^*}{(C_2^{\text{off}})_{\text{sat}}^*} = \zeta = Y_{32} \frac{g_3}{g_2} \exp\left(-\Delta E_{32}/kT\right) \tag{39}$$

The deviation from Boltzmann distribution is mainly given by the deexcitation branching ratio factor (Y_{32}), i.e., the ratio of the collision rate back to the laser-connected level to the sum of all collision rates (back to the laser-connected level as well as those down to lower levels from which the direct collisional ionization rate can be assumed to be insignificant in comparison with that from the uppermost state).

Secondly, we find that the fraction of neutral atoms excited to state 2 decreases the higher are the collisional rates to other high-lying states (k_{23}). However, since the excited state population will be redistributed among these higher states, a larger fraction of all neutral atoms will be transferred to excited states when the coupling between the upper levels is strong than in the case when other high-lying states are not being considered. In addition, we can also see that the requirements for saturation change somewhat with the degree of coupling of the upper state of other excited states (however, not as drastically as one first might expect). This saturation excitation rate, $(C_2^{\text{off}})_{\text{sat}}^*(k_{21} + k_{23}Y_{31})$, varies only between the values $(C_2^{\text{off}})_{\text{sat}}^* k_{21}$ (for $E_3 \gg E_2$, so that $k_{23} \ll k_{21}$) and $(C_2^{\text{off}})_{\text{sat}}^* k_{21} + (C_3^{\text{off}})_{\text{sat}}^* k_{31}$ (for $E_3 \approx E_2$, so that $k_{23} \gg k_{21}$, i.e., fast mixing).

As will be discussed below, this more extensive expression for the fraction of neutral atoms excited [Eqs. (36) and (37)] must be used when, for example, the laser-enhanced ionization rate is to be related to the thermal ionization rate. In descriptions where this relation is not of any major importance, the simpler expressions [Eqs. (32)–(33)] can often be used for estimates of the fraction of neutral atoms excited.

1.3.3.5. Simplified Solution of the System of Rate Equations under One-Step Excitation: The Influence of Intermediate States

If we now want to investigate to what extent lower lying excited states (here denoted intermediate states) influence the population of the laser-connected levels, we can proceed in a way similar to that just described except that we treat the third level as an intermediate level (i.e., $E_3 < E_2$).

In this case, similar equations can be derived for the fractions of neutral atoms excited as those above [Eqs. (36) and (37)]. Note, however, that in this case k_{23} is a deexciting collision, which might therefore be somewhat larger than an exciting one. This implies that the actual fraction of neutral atoms excited to the upper laser-connected state decreases with an increasing number of non-laser-connected states in a way similar to the treatment of high-lying states above (10, 19). This time, however, a certain part of the excited atoms will find their way to lower lying excited states (intermediate states). The distinction between higher lying and lower lying states is of importance for LEI since atoms in intermediate states generally have lower ionization probabilities than atoms in higher ones (this is discussed further below). Consequently, this implies that the number of excited atoms (to the upper laser-connected level), and hence the LEI signal of an element with numerous intermediate states can be lower than that of an element with few intermediate levels.

In addition, the existence of intermediate levels implies that the requirement for saturation will change with the number of quenching channels in a way similar to the case with high-lying states described in the previous subsection.

Here it is also worth mentioning specifically that the solutions of the rate equations presented above all are derived under steady-state conditions (see detailed discussion in Section 1.3.3.1). This formalism can thus only be used for those (intermediate) states that are in close contact with the laser-connected levels [i.e., for which Eq. (38) in valid] but not for long-lived intermediate states that have k_{31} or $k_{32} \ll$ (laser pulse length)$^{-1}$, i.e., metastable states. Metastable states will act as a trap of atoms and can only be correctly treated by a full temporal solution of the system of rate equations (10, 19). It has recently been shown that ordinary metastable states have lifetimes ranging from a few nanoseconds (which thus do not really act as a metastable state for laser excitations in the 5–30 ns range) to some microseconds in an air/acetylene flame (31–34)—the longer lifetimes for those states that are forbidden to decay by parity (the same parity as the ground state) and shorter for those that do not decay for other reasons [spin conservation of J quantum number restrictions (32)]. In addition, the lifetimes of some metastable states (in Pb and Au) have been found to vary by an order of magnitude, depending on the actual stoichiometry of the interaction region (32–34).

1.3.4.   The Ionization Rate for One-Step Excitation and Its Relation to Thermal Ionization

When deriving and solving an equation for the production rate of ions from the atoms under consideration let us first consider the case of one-step excitation, i.e., when $I_{23}^{v} = I_{32}^{v} = 0$. Let us start by deriving a simple expression for the ionization rate of atoms that follows a one-step LEI excitation of atoms in flames and relate it to that for thermal ionization.

As previously mentioned, many excited atoms ionize primarily by collisions with flame molecules. It is only when very high light intensities are used for excitation (e.g., using focused laser beams in the flame) or when the upper laser-connected state is far from the ionization level (several electron volts) that photoionization can compete with collisional ionization (as discussed in more detail below). Therefore, in this subsection we will also assume that the atoms ionize mainly by collisional ionization.

In our derivation of the one-step ionization rate of atoms in flames, let us first recall (from Section 1.2) that under thermal equilibrium conditions each excited level contributes approximately equally to the ionization rate. Since we now represent the atom by a three-level system, we have simplified our description of the atomic/molecular system significantly. Consequently, we must investigate in more detail how this simplification affects that actual laser-enhanced ionization rate.

Prior to determination of the relationship that relates the laser-enhanced ionization rate to the thermal ionization rate of atoms in flames (9), let us investigate how the existence of other high-lying as well as low-lying states will influence the laser-enhanced *ionization rate* for one-step excitation in a few-level atomic description (similar to the way we investigated the extent to which other non-laser-connected levels influenced the fraction of neutral atoms excited in Sections 1.3.3.4 and 1.3.3.5).

1.3.4.1. The Influence of Higher Lying States upon the Ionization Rate for a Few-Level System

If we, as already mentioned, first assume that collisional ionization is the dominant ionization route, the laser-induced ionization rate for a three-level system can be written in accordance with that for the thermal case in Eq. (16)[2]:

$$\frac{dn_{ion}}{dt} = k_{2,ion}n_2 + k_{3,ion}n_3 \tag{40}$$

where n_{ion} is the number density of ions, related to the total number density of the analyte species (neutral atoms and ions), n_{tot}, by

$$n_{tot} = n_{ion} + n_{atom} \tag{41}$$

and where $k_{2,ion}$ and $k_{3,ion}$ are the state-specific ionization rates for the levels 2 and 3, respectively.

Since we know that each level (except perhaps those that lie above the artificially imposed cutoff limit) contributes approximately the same to the ionization rate when the system is in thermal equilibrium, let us, for simplicity, relate the two ionization rates to each other in the most simple way, i.e., by the Boltzmann factor:

$$k_{2,ion} = k_{3,ion} \exp\left(-\Delta E_{32}/kT\right) \tag{42}$$

Since the population of level 3 is related to that of level 2 almost in a Boltzmann-like manner [Eq. (39)], i.e.,

$$\frac{C_3^{off}}{C_2^{off}} = \frac{(C_3^{off})^*_{sat}}{(C_2^{off})^*_{sat}} = \xi = Y_{32}\frac{g_3}{g_2}\exp\left(-\Delta E_{32}/kT\right) \tag{43}$$

---

[2] We have here neglected the ionization from the ground state since it only gives rise to a thermal ionization.

we find that the rate of production of ions in the one-step case of the three-level system can be written

$$\frac{dn_{ion}}{dt} = C_2^{off}\left[1 + \frac{g_3}{g_2}Y_{32}\right]k_{2,ion}\,n_{atom} \tag{44}$$

where the first factor of unity within the brackets originates from the ionization directly from level 2, while the second factor $(g_3/g_2)Y_{32}$ represents the ionization which goes through the higher lying level, level 3.

What we can see from this treatment is that all high-lying levels still contribute to the total ionization rate, although no longer exactly to the same extent. The ionization rate from the upper level (level 3) is reduced by a factor $(g_3/g_2)Y_{32}$, which is a product of the ratio of degeneracies of the two states and a branching ratio (the ratio of the collision rate back to the laser-connected level to the sum of all collision rates, i.e., back to the laser-connected level as well as to lower levels).

If we also include a multitude of upper states [for example, by assuming that the right-hand side of Eq. (40) constitutes of a summation over a large number of states], we will find that they *all* contribute to the total ionization rate in proportion to the population of the directly laser-pumped level (level 2 in this case), and still reduced by a similar branching ratio factor. This implies that the total ionization rate in this case can be written approximately as

$$\begin{aligned}
\frac{dn_{ion}}{dt} &= C_2^{off}\frac{k_{2,ion}}{g_2}\sum_{i=\text{upper laser level}}^{i_{cutoff}} Y_{i2}g_i n_{atom} \\
&\approx C_2^{off}Y_{i'2}k_{2,ion}\sum_{i=\text{upper laser level}}^{i_{cutoff}} (g_i/g_2)n_{atom}
\end{aligned} \tag{45}$$

where in the last step we have introduced an effective (mean, or typical) branching ratio, $Y_{i'2}$ (under the assumption that the various high-lying excited state branching ratios do not differ considerably from each other), which in turn is unity for the case of ionization from the upper laser-connected level.

It seems here, at first glance, that the one-step laser-enhanced ionization rate will be very dependent upon the number of higher lying states that are included [all contribute to the sum in Eq. (45)]. However, this is not really the case. For each higher lying state (which thus will contribute to an enhancement of the ionization rate by constituting a term in the sum above) a similar term should be added to the denominator of the expression for $Y_{i'2}$. The result of this operation is that these contributions more or less cancel out. Consequently, the one-step laser-enhanced ionization rate is not extremely sensitive to the number of higher lying states included in the model.

As an example, for a two-level system (with equal degrees of degeneracy of the two levels), 50% of all neutral atoms will be excited and hence contribute to ionization under saturated conditions, whereas approximately 90% of all atoms will be in excited states if we allow the atom to have 10 additional high-lying states (still under saturated conditions). Consequently, as shown by this example, the fraction of atoms excited increases less than 100% for an 10-fold increase in the number of excited states in the model of the atom.

1.3.4.2. The Influence of Lower Lying States upon the Ionization Rate for a Few-Level System

We can study the influence of intermediate levels in a similar way. However, what can be found from such a study is that each intermediate level will have a population far below that which would exist at that particular level if a detailed balance at thermal equilibrium existed between the upper laser-connected level and that particular level. Consequently, all the intermediate levels will contribute significantly less to the total ionization rate than all the higher lying states. Therefore, since excited atoms redistribute rather rapidly among all excited levels in highly collisional media such as flames, the lower lying states will efficiently reduce the effective ionization rate.

1.3.4.3. The Influence of all Non-Laser-Connected Levels upon the Ionization Rate for a Few-Level System

The conclusion to be drawn from the above discussion [Eq. (45)] is that the ionization rate of the laser-connected level has been increased (when laser excitation is applied) by an amount approximately equal to the C_2^{off} factor times a ratio of degeneracies, g_1/g_2, and the inverse Boltzmann factor between the ground state and the upper laser-connected level (i.e., $k_{2,ion}/k_{1,ion}$), as compared to the thermal case [one term of the sum in Eq. (22)]. Each of the higher lying non-laser-connected states, here represented by level 3, will contribute to the increase in total ionization rate by an amount proportional to that of the laser-connected level, reduced by a branching ratio factor, Y_{32}.

The lower lying levels will all be considerably underpopulated (in comparison with a Boltzmann distribution originating from the laser-connected level) and thus do not contribute much at all to the total ionization rate. Since the number of levels that contribute to the total ionization rate has been limited to those below a certain cutoff level, the relation between the total laser-enhanced ionization rate [Eq. (45)] and the total thermal ionization rate

[Eq. (22)] is given by

$$
\frac{(dn_{ion}/dt)_{laser\ excitation}}{(dn_{ion}/dt)_{thermal\ excitation}} = C_2^{off}\ Y_{i'2} \frac{\displaystyle\sum_{i=upper\ laser\ level}^{i_{cutoff}} (g_i/g_2)}{\dfrac{1}{Z}\displaystyle\sum_{i=1}^{i_{cutoff}} g_i} \frac{k_{2,ion}}{k_{1,ion}}
$$

$$
\approx C_2^{off}\ Y_{i'2} \frac{\displaystyle\sum_{i=upper\ laser\ level}^{i_{cutoff}} (g_i/g_2)}{\displaystyle\sum_{i=1}^{i_{cutoff}} (g_i/Z)} \exp\left(\Delta E_{21}/kT\right)
\tag{46}
$$

where we have neglected the ionization from all lower lying levels. This implies that the sum in the numerator runs over all states whose energy is approximately equal to or larger than that of the upper laser-connected level. In addition, we have assumed that the two state-specific collisional ionization rates are related to each other in a similar manner to that in Eq. (42), i.e., by the (inverse) Boltzmann factor. In addition, for all elements whose ground state is the only low-lying state, we can replace the partition function, Z, with the first sum in its series, i.e., the degeneracy of the ground state, g_1.

As can be seen from Eq. (46), the ionization rate of the system will be increased (as compared to the thermal ionization rate) by the product of four different factors:

a. The fraction of neutral atoms that are actually excited to the upper state (per degree of degeneracy), C_2^{off}

b. The fraction of the excited atoms that will quench back to the upper laser-connected level, $Y_{i'2}$

c. The ratio of the number of levels over which the laser-excited population is being redistributed (in comparison to the degeneracy of the upper laser-connected level), and which thus will contribute significantly to the ionization rate (i.e., in principle, those that have an energy higher than that of the upper laser-connected level, u.l.l.), to the total number of levels (in comparison to the partition function, or, quite often, the degeneracy of the lower laser-connected level) that contribute to the thermal ionization,

$$
\sum_{i_{u.l.l.}}^{i_{cutoff}} (g_i/g_2) \Bigg/ \sum_{i=1}^{i_{cutoff}} (g_i/Z)
$$

d. The inverse Boltzmann factor corresponding to the energy difference between the two laser-connected levels (i.e., the activation energy gained by the laser excitation), $\exp(\Delta E_{21}/kT)$

A similar kind of discussion can be carried out for the two-step excitation case, although a derivation following the same lines as given above would be rather tedious. However, in many cases the foregoing derivation can be considerably simplified, which makes modeling of both the one-step and two-step excitation processes possible.

1.3.5. The Degree of Ionization for One- and Two-Step Excitation

Let us now derive expressions for the degree of ionization of atoms for one- and two-step excitation (the degree of ionization is here defined as the fraction of atoms in the interaction volume that ionize during the laser pulse). The derivation of the ionization rate above focused upon the relation between the thermal ionization and the enhanced ionization due to laser radiation. However, if we now relax the coupling between these two types of ionization processes[3] (since we have related the one-step enhanced ionization rate to the thermal ionization rate), we can considerably simplify the description by assigning an *effective* collisional ionization rate from the uppermost laser-coupled level, $k_{2,\text{ion}}^{\text{eff}}$. Such an effective ionization rate should then include all the redistribution processes of excited atoms among other excited states. This implies that $k_{2,\text{ion}}^{\text{eff}}$ then roughly corresponds to the factors $Y_{i'2} \sum (g_i/g_2)k_{2,\text{ion}}$ in Eq. (45), above. This assumption will then simplify the description of the ionization process, leading to the following experessions

$$\frac{dn_{\text{ion}}}{dt} = \begin{cases} k_{2,\text{ion}}^{\text{eff}} n_2 & \text{for one-step excitation} \\ k_{3,\text{ion}}^{\text{eff}} n_3 & \text{for two-step excitation} \end{cases} \tag{47}$$

We still assume that the two effective collisional ionization rates are related according to

$$k_{2,\text{ion}}^{\text{eff}} = k_{3,\text{ion}}^{\text{eff}} \exp\left(-\Delta E_{32}/kT\right) \tag{48}$$

(which might not be a very accurate assumption, as will be discussed in Section 1.5 in more detail, but proves to be a good enough approximation for our purposes here).

---

[3] There are several reasons why we should not necessarily focus upon the relation between the laser-induced ionization rate and the thermal ionization rate but instead study only the laser-enhanced ionization rate. One such reason is that the thermal ionization rate takes place in the entire flame whereas the laser-enhanced ionization rate only takes place in the interaction region defined by the overlap between the laser light and the flame. Hence, any relation between the two must take this volume effect into consideration. Another effect that must be considered is that the charge detection efficiency most probably will not be the same over the entire flame (as will be discussed in more detail in Chapter 2).

When solving the equation for production of ions, we get two different expressions for the one- and two-step cases, i.e., with and without the second-step laser applied, respectively:

$$\frac{dn_{ion}^{off}}{dt} = C_2^{off} k_{2,ion}^{eff}(n_{tot} - n_{ion}) \tag{49}$$

$$\frac{dn_{ion}^{on}}{dt} = C_3^{on} k_{3,ion}^{eff}(n_{tot} - n_{ion}) \tag{50}$$

where the expressions for the fractions of neutral atoms in the excited states are given in Section 1.3.3, above.

The time-dependent solutions to these equations, determining the number density of ions and electrons produced after an interaction time τ of laser light illumination, can then be readily expressed as

$$n_{ion}^{off}(\tau) = \Phi^{off}(\tau)n_{tot} \tag{51}$$

$$n_{ion}^{on}(\tau) = \Phi^{on}(\tau)n_{tot} \tag{52}$$

where Φ^{off} and Φ^{on} represent the degrees of ionization of the atomic system when the second exciting step is turned off and on, respectively, and which are given by

$$\Phi^{off}(\tau) = 1 - \exp(-C_2^{off} k_{2,ion}^{eff}\tau) \approx C_2^{off} k_{2,ion}^{eff}\tau \tag{53}$$

$$\Phi^{on}(\tau) = 1 - \exp(-C_3^{on} k_{3,ion}^{eff}\tau) \approx C_3^{on} k_{3,ion}^{eff}\tau \tag{54}$$

The last step in these two expressions is valid only when the fraction of ionized atoms is significantly smaller than unity.

Equations (53) and (54) are very useful when we want to gain some insight into how various physical parameters affect the degree of ionization in the system. Of particular interest are, on the one hand, the relations between the degrees of ionization and the laser intensity and, on the other, the relations between the degrees of ionization and the product of the effective collisional ionization rate and the laser pulse duration.

Let us for simplicity study a one-step excitation case for which the degree of ionization is expressed by Eq. (53). When plotting the degree of ionization vs. the laser intensity, as is done in Fig. 1.2, we find two different types of curves that look very similar but describe basically different phenomena. The five curves a, b, c, d, and e in the figure represent situations for different products of the ionization rates and laser pulse lengths ($k_{2,ion}^{eff}\tau$). The five curves correspond to cases when the product $k_{2,ion}^{eff}\tau$ is 10^{-6}, 10^{-4}, 10^{-2}, 10^0, and 10^2, respectively. The typical situations these five cases represent are as follows:

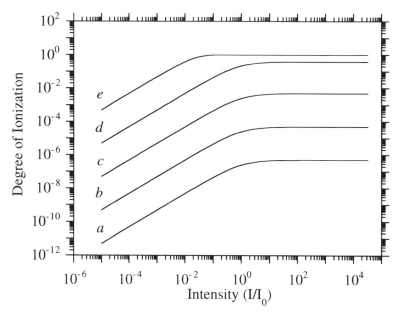

Figure 1.2. Degree of ionization vs. laser intensity for a one-step excitation case followed by collisional ionization [Eqs. (32), (48), and (53)]. Curves a, b, c, d, and e represent situations for different products of ionization rates and laser pulse lengths ($k_{2,ion}^{eff} \tau$). The five curves correspond to cases when the product $k_{2,ion}^{eff} \tau$ is 10^{-6}, 10^{-4}, 10^{-2}, 10^{0}, and 10^{2}, respectively; I_0 is the intensity for which the two terms in the denominator of Eq. (32) are equal.

a. The atoms are excited by excimer, Nd:YAG, or N_2 pulsed laser light (in the 5–30 ns region) to really low-lying states from which the collisional ionization rate is very small (around 10^2 Hz).

b. The atoms are excited by a similar type of laser to states somewhat closer to the ionization limit—however, from which the ionization rate is still rather small (around 10^1 kHz).

c. The atoms are excited by a similar type of laser to states even closer to the ionization limit from which the ionization rate is significant (on the order of megahertz).

d. The atoms are excited either by a similar type of laser to states with very high ionization rates (around 10^2 MHz) or by a flashlamp-pumped laser system (with pulse lengths in the microsecond range) to excited states from which the ionization rate is significant (on the order of megahertz).

e. The atoms are excited either by flashlamp-pumped laser system to states close to the ionization limit (with collisional ionization rates around 10^2 MHz) or by cw light to a low-lying state from which the thermal ionization rate is again rather small (10^2 kHz).

Since the collisional ionization rate from lower lying states is normally significantly lower than the inverse of the pulse duration from excimer, Nd:YAG, and N_2 laser systems, the product $k_{2,\text{ion}}^{\text{eff}}\tau$ is significantly smaller than unity. This is the situation in the cases a, b, and c in Fig. 1.2. We find that the series expansion of the expression of the degrees of ionization as done above in Eq. (53) is adequate. Consequently, the entire intensity dependence of the degrees of ionization in these cases is given by the intensity dependence of the expression for the fraction of atoms excited, i.e., C_2^{off}. We therefore conclude that the degree of ionization increases linearly with light intensity until optical saturation is reached, i.e., the main limiting factor of the degree of ionization (for higher laser intensities) is optical saturation of the transition. This is what is seen in the three curves a, b, and c. The maximum degree of ionization is then given by the product of the effective collisional ionization rate, $k_{2,\text{ion}}^{\text{eff}}$, the duration of the light–matter interaction, τ, and the saturated fraction of atoms excited, $(C_2^{\text{off}})_{\text{sat}}$.

In case e, when the atoms are excited to states from which the effective inonization rate is much faster then the inverse of the interaction time, we find that optical saturation is no longer the limiting process for the degree of ionization. In this case we find that a substantial degree of ionization can be obtained long before optical saturation becomes of significance. The maximum degree of ionization in this case is, of course, close to unity. Although the process of neutral atom depletion is completely different from that of optical saturation, it is often very difficult to distinguish between the two processes since the two curves behave quite similarly with respect to intensity. However, what is of importance is that the apparent saturation intensity, defined as the intensity for which the two low- and high-intensity asymptotes meet, is shifted compared to the true saturation intensity when atom depletion sets in.

Finally, case d is the situation where the degree of ionization is both limited by optical saturation and a high degree of ionization, which thus constitutes of an intermediate case.

The conclusion about the degree of ionization as a function of laser light intensity is that both atomic depletion as well as ordinary optical saturation might limit the signal for high laser intensities. It is, in general, rather difficult to distinguish between the two processes (since the actual shapes of the mathematical functions describing the two features are very similar) and so other means of investigation must often be used in order to draw any correct conclusions about the true limiting process in the prevailing situation.

1.3.6.   The LEI Enhancement: A Relation Between the Two-Step and One-Step LEI Signals

As we have already discussed, the two-step LEI technique will be generally superior to one-step LEI as far as sensitivity is concerned. It is therefore of interest to define an LEI enhancement, χ, as the ratio of the degree of ionization for two-step excitation to that for one-step excitation. The LEI enhancement is thus a measure of how much the degree of ionization, and hence the LEI signal, is expected to increase when a second exciting step is applied:

$$\chi = \frac{\Phi^{on}}{\Phi^{off}} \tag{55}$$

Using the previously derived expressions for the degree of ionization [Eqs. (53) and (54)], we arrive at an important and interesting expression for the LEI enhancement, which is independent of several experimental parameters. This expression is

$$\chi = \frac{\Phi^{on}(\tau)}{\Phi^{off}(\tau)} = \frac{1 - \exp(-C_3^{on} k_{3,ion}^{eff} \tau)}{1 - \exp(-C_2^{off} k_{2,ion}^{eff} \tau)} \approx \frac{1 - \exp(-C_3^{on} k_{3,ion}^{eff} \tau)}{C_2^{off} k_{2,ion}^{eff} \tau} \tag{56}$$

where in the last step we have assumed that the degree of ionization for one-step excitation is significantly smaller than unity.

Since the effective collisional ionization rate from level 3 is significantly larger than that from level 2, a considerable signal enhancement is expected. This has indeed been found in many cases.

If we first assume that the ratio of the two effective collisional ionization rates roughly scale by the Boltzmann factor, i.e., by the factor $\exp(-hc/\lambda_{23} kT)$, where λ_{23} represents the wavelength of the second-step excitation, we than find, for a typical second-step excitation (wavelengths of the second step are normally between 500 and 670 nm), that the effective collisional ionization rate from level 3 should be a factor of 5×10^3 and 10^5 larger than from level 2.

Experimentally, LEI enhancements of up to 3000 have been found (35). This seems to be in good agreement with the foregoing estimate. However, a number of experiments have also shown LEI enhancement factors that are only in the range between 2 and 1000, which is not really in agreement with the estimates based upon Eq. (56).

One way to explain such small enhancement factors is to assume that there is already a significant degree of ionization from the first step. This assumption, however, seems to be untenable, as proved by several of the studied cases. The results show that the one-step signal for these type of elements is not as strong as from the elements that exhibit the most sensitive signals; moreover, the first laser excitation only excites the atoms to states a few electron volts below the ionization limit (from which the collisional ionization rate is not expected to be of significant strength).

Another explanation for such low enhancement factors is that the previous assumption concerned with the two collisional ionization rates scaling as the Boltzmann factor is not fully correct. Such a situation could indicate that collisional ionization is not the main process responsible for ionization for these types of elements. Any other kind of ionization process, for example, associative ionization, will of course scale in a completely different manner. It is plausible for the elements exhibiting the lowest enhancement that associative ionization is in fact the main contributor to the overall ionization increase that follows the one-step excitation process. However, even for those elements for which collisional ionization is believed to be the main ionizing process, the LEI enhancement numbers found do not really follow the estimates made from the expression above.

It has been shown by Axner and Berglind (36) that the ionization efficiency of excited atoms in flames (defined as the probability that an excited atom ends up as a positive ion and an electron rather than decays down to the ground state) does not really follow a simple Boltzmann-like type of expression. In that study, it was found that the ratio of the ionization efficiencies of excitations of Na and Li to states approximately 1 eV (8000 cm^{-1}) below the ionization limit to those of excitations to states significantly closer to the ionization limit was 4–7 times larger than that given by Boltzmann factor. (This is discussed further in Section 1.5, below.)

If we now generalize this finding to other elements and if we also assume, somewhat arbitrarily, that the actual ratio of $k_{2,\text{ion}}^{\text{eff}}$ to $k_{3,\text{ion}}^{\text{eff}}$ in Eq. (56) is approximately 1 order of magnitude larger than that given by the ordinary Boltzmann relation, we find that the expected LEI enhancements should be roughly between 2 and 4 orders of magnitude. This is indeed in closer agreement with experimental findings.

1.3.7. Pulsed vs. Continuous-Wave Excitation

A comparison between the efficiency of pulsed vs. continuous-wave (cw) excitation in LEI can now be performed using some of the expressions just derived. This can most easily be done by estimating the fraction of all atoms entering the flame that will be ionized under typical conditions with the two types of excitation modes—pulsed and cw, respectively.

Let us define the *total* ionization efficiency of the LEI technique, Ξ, as the ratio of the average number of ions produced per second (as a consequence of laser illumination), $\bar{\dot{N}}_{\text{ion}}$, to the total number of atoms entering the flame per second, $\dot{N}_{\text{tot}}$. Hence, $\bar{\dot{N}}_{\text{ion}}$ is the average ion production rate (averaged over a number of pulses) [s^{-1}], while $\dot{N}_{\text{tot}}$ is the flux of species under investigation entering the flame [s^{-1}].

When deriving an expression for the total ionization efficiency of the LEI technique, let us first assume that the flame has a certain flame rise velocity, v, a given length, L, and a width, w. This implies that all the species entering the flame under a given period of time, Δt, and that atomize in the flame (assuming a degree of atomization of η) will be distributed in a volume given by $vLw\,\Delta t$. Hence, the total (average) number density of atomized analyte species, $n_{tot}[\mathrm{m}^{-3}]$, is given by

$$n_{tot} = \eta\frac{\dot{N}_{tot}\,\Delta t}{vLw\,\Delta t} = \eta\frac{\dot{N}_{tot}}{vLw} \tag{57}$$

Let us now determine two quantities of special interest, namely, the fraction of atoms illuminated, φ (i.e., the fraction of *all* atomized species under investigation entering the flame that is temporally and spatially coincident with the laser beam) and the atomic illumination time, τ_{atom} (i.e., the actual time the atoms are exposed to laser light) for the pulsed and cw cases, respectively, and study them separately in the cases of pulsed and cw excitation.

1.3.7.1. Fraction of Atoms Illuminated and Atomic Illumination Time for Pulsed Excitation

Let us now first look at the pulsed situation. We assume that the laser light has a diameter of d and a pulse repetition rate of R_p. The fraction of atoms illuminated when pulsed laser excitation is used, φ_{pulsed}, can then be written

$$\varphi_{pulsed} = \frac{(\pi d^2/4)\,Ln_{tot}}{\dot{N}_{tot}}R_p = \eta\frac{\pi d^2}{4vw}R_p \tag{58}$$

where we have used the relation between n_{tot} and $\dot{N}_{tot}$ in Eq. (57) in the last step. This expression thus holds for the case where the repetition rate is so low that the atoms in the interaction region can be fully replaced by new atoms between two consecutive laser pulses, i.e., when:

$$R_p \leqslant \frac{v}{d} \tag{59}$$

Assuming a flame rise velocity of 10 m/s and a laser beam diameter of 0.5 cm, we find that the condition for expression (58) to hold is that the repetition rate is smaller than or roughly equal to 2 kHz.

In addition, expression (58) assumes that the laser pulse duration under pulsed excitation, τ_{laser}^{pulsed}, is so short that the atoms can be considered to be "frozen" in space during the excitation procedure ($\tau_{laser}^{pulsed} \ll d/v = 500\,\mu s$). This, in turn, implies that the atomic illumination time for pulsed excitation, τ_{atom}^{pulsed}, is equal to the laser pulse duration:

$$\tau_{atom}^{pulsed} = \tau_{laser}^{pulsed} \tag{60}$$

1.3.7.2. Fraction of Atoms Illuminated and Atomic Illumination Time for cw Excitation

The most significant difference for cw excitation is that the atomic illumination time no longer is necessarily equal to the laser pulse duration, but rather is related to the atomic transit time across the laser beam. In order to monitor an enhanced ionization rate and eliminate drifts and low-frequency fluctuations, cw laser light is normally chopped. The atomic illumination time therefore depends on the cw chopping frequency, R_{cw}, as well as its duty cycle, γ.

For sufficiently *low* chopping frequencies, most illuminated atoms will have time to drift in as well as out of the laser beam during one laser illumination period. Hence, the atomic illumination time for low-chopping-frequency cw excitation, $\tau_{atom}^{cw,low}$, is simply equal to the drift time across the laser beam (d/v):

$$\tau_{atom}^{cw,low} = \frac{d}{v} \tag{61}$$

The condition for this to hold is thus that the laser illumination time, τ_{laser}^{cw}, given by

$$\tau_{laser}^{cw} = \frac{\gamma}{R_{cw}} \tag{62}$$

is longer than the drift time across the laser beam (d/v). This is equivalent to

$$R_{cw} \leqslant \gamma \frac{v}{d} \tag{63}$$

Since a low chopping frequency is tantamount to a long laser illumination time and therefore most often also a long interillumination time, only a certain fraction of all atomized species under investigation entering the flame will be illuminated. For low cw chopping frequencies the fraction of atoms illuminated, φ_{cw}^{low}, can therefore be written

$$\varphi_{cw}^{low} = \frac{dv\tau_{laser}^{cw} \, Ln_{tot}}{\dot{N}_{tot}} R_{cw} = \eta\gamma\frac{d}{w} \tag{64}$$

where in the last step we have used Eqs. (57) and (62).

Thus, the fraction of atoms illuminated is given by three factors: the degree of atomization, the duty cycle of the light, and the relative width of the laser beam with respect to that of the flame. It is interesting to note that the fraction of atoms illuminated is independent of the chopping frequency, depending only on the duty cycle and diameter of the laser light.

For sufficiently *high* chopping frequencies, on the other hand, the laser illumination time τ_{laser}^{cw} is *shorter* than the drift time across the laser beam (d/v). This implies that the conditions for this case to be valid is that

$$R_{cw} > \gamma \frac{v}{d} \tag{65}$$

This implies further that the atoms will experience a number of laser illumination periods on their journey across the interaction region. Hence, the atoms are exposed to a train of laser pulses, each with a duration of τ_{laser}^{cw}. The number of illuminations experienced by the atoms, M, is given by the ratio of the drift time across the interaction region to the chopping period, i.e., $(d/v)/(1/R_{cw})$. The total atomic illumination time for high-chopping-frequency cw excitation, $\tau_{atom}^{cw,high}$, which is given by the product of M and τ_{laser}^{cw}, can simply be written as a product of the laser light duty cycle and the drift time across the interaction region:

$$\tau_{atom}^{cw,high} = M\tau_{laser}^{cw} = \gamma \frac{d}{v} \tag{66}$$

The fraction of atoms illuminated for high chopping frequencies, φ_{cw}^{high}, is independent of the duty cycle of the laser light (since all atoms in the "corridor" under the laser beam will interact with the laser light):

$$\varphi_{cw}^{high} = \eta \frac{d}{w} \tag{67}$$

Hence, in this case the fraction of atoms illuminated is independent of both the laser repetition rate as well as the duty cycle of the light.

It is now possible to get some estimates of the atomic illumination times and the fraction of atoms illuminated for the three cases of pulsed excitation, low-frequency cw excitation, and high-frequency cw excitation, respectively, by inserting some typical values into the appropriate expressions above [Eqs. (58), (60), (61), (64), (66), and (67)].

Let us assume a pulse duration of 10 ns for the pulsed excitation (as a compromise between the 5 ns long Nd:YAG pulses and 20–25 ns long excimer pulses that are frequently used in various laboratories) and a duty cycle of 50% for the cw excitation. In addition, let us again assume a flame rise velocity of 10 m/s and a laser beam diameter of 0.5 cm. We then get the following atomic illumination times:

$$\tau_{atom}^{pulsed} = 10 \text{ ns}; \qquad \tau_{atom}^{cw,low} = 500 \text{ μs}; \qquad \tau_{atom}^{cw,high} = 250 \text{ μs}$$

We can thus see that the atomic illumination times are 4–5 orders of magnitude *longer* for cw excitation than for pulsed excitation.

Assuming furthermore a flame width of 1 cm and a degree of atomization of unity, we find that the fraction of atoms illuminated for the cw cases are quite high:

$$\varphi_{cw}^{low} = 25\% \qquad \text{and} \qquad \varphi_{cw}^{high} = 50\%$$

For the pulsed case, the fraction of atoms illuminated is strongly dependent on the repetition rate of the laser. For a 10 Hz repetition rate we find that the fraction of atoms illuminated is less than a percent, whereas for a 50 Hz repetition rate it is around 1%:

$$\varphi_{pulsed}^{10Hz} = 0.2\% \qquad \text{and} \qquad \varphi_{pulsed}^{50Hz} = 1\%$$

With knowledge of the atomic illumination times (τ_{atom}^{pulsed}, $\tau_{atom}^{cw,low}$, and $\tau_{atom}^{cw,high}$) and the fraction of atoms illuminated (φ_{pulsed}, φ_{cw}^{low}, and φ_{cw}^{high}) for the three cases of pulsed excitation, low-frequency chopped cw excitation, and high-frequency chopped cw excitation, respectively, expressions for the total ion production rate and the total ionization efficiency of the LEI technique can now be derived.

1.3.7.3.   *Total Ionization Ēciency of the LEI Technique*

Irrespective of whether we are dealing with pulsed or cw excitation, the average ion production rate (averaged over a number of pulses), $\bar{N}_{ion}[\text{s}^{-1}]$, can be written as a product of the flux of species entering the flame, $\dot{N}_{tot}[\text{s}^{-1}]$, the fraction of atoms illuminated, φ, and the degree of ionization of the illuminated atomic system, $\Phi(\tau_{atom})$

$$\dot{\bar{N}}_{ion} = \varphi\Phi(\tau_{atom})\dot{N}_{tot} \tag{68}$$

This implies that the total ionization efficiency of the LEI technique, Ξ, defined as the ratio of the total ion production rate, $\dot{\bar{N}}_{ion}$, to the total flux of atoms entering the flame, $\dot{N}_{\text{tot}}$, can be simply written as a product of the fraction of atoms illuminated and the degree of ionization of the illuminated atomic system,

$$\Xi = \frac{\dot{N}_{\text{ion}}}{\dot{N}_{\text{tot}}} = \varphi\Phi(\tau_{\text{atom}}) \tag{69}$$

The fraction of atoms illuminated for the three cases of pulsed excitation (φ_{pulsed}), low-frequency cw excitation ($\varphi_{\text{cw}}^{\text{low}}$), and high-frequency cw ($\varphi_{\text{cw}}^{\text{high}}$) excitation are given above by Eqs. (58), (64), and (67), respectively, while the degree of ionization of the illuminated atoms are given by the Eqs. (53) or (54) for one- and two-step excitation, respectively, with the atomic illumination time τ_{atom} equal to $\tau_{\text{atom}}^{\text{pulsed}}$, $\tau_{\text{atom}}^{\text{cw,low}}$, or $\tau_{\text{atom}}^{\text{cw,high}}$ [from Eqs. (60), (61), and (66)] for the three cases of pulsed excitation, low-frequency cw excitation, and high-frequency cw excitation, respectively.

It is worth noting here that the degree of ionization of the illuminated atoms for the high-chopping-frequency cw case is simply equal to $\Phi(\tau_{\text{atom}}^{\text{cw,high}})$, where the argument of the expression of the degree of ionization is the total illumination time, despite the fact that the atoms are illuminated by a series of shorter light pulses. The reasons for this are simply that the time for establishing a steady-state condition (normally on the order of nanoseconds) can be neglected in comparison with the pulse length of the "on" cycles (during which a steady-state condition prevails) and that recombination is generally slower than the inverse of the pulse length of the "off" periods of the light (so that the degree of ionization remains the same throughout the entire "off" cycle). Hence, although the atoms experience the "off" periods of the light, they do not react to them. This implies furthermore that the atoms experience a degree of ionization roughly equal to a situation in which they were illuminated by a continuous light pulse of duration $\tau_{\text{atom}}^{\text{cw,high}}$. This is illustrated mathematically in the Appendix.

It is now possible to compare the total ionization efficiency of the LEI technique for pulsed and cw excitation on a quantitative basis. In doing so, let us first simplify our conditions somewhat. Since it can be shown that high-chopping-frequency cw excitation is preferable to low-frequency cw excitation (the former making better use of the atoms as well as having better reduction of flame noise), let us compare the total ionization efficiency of the LEI technique for high-chopping-frequency cw excitation with that of pulsed excitation. We can then write

$$\frac{\Xi_{\text{cw}}}{\Xi_p} = \frac{\varphi_{\text{cw}}^{\text{high}}}{\varphi_{\text{pulsed}}} \frac{\Phi(\tau_{\text{atom}}^{\text{cw,high}})}{\Phi(\tau_{\text{atom}}^{\text{pulsed}})} \tag{70}$$

The ratio of the fraction of atoms illuminated for the cw case to that of the pulsed case can simply be estimated according to

$$\frac{\varphi_{cw}^{high}}{\varphi_{pulsed}} = \frac{\eta(d/w)}{\eta(\pi d^2/4vw)R_p} = \frac{4v}{\pi d R_p} \tag{71}$$

Using the same assumption about flame and laser parameters as above, we find that the ratio of the fraction of atoms illuminated for the cw case to that of the pulsed case is 250 for a 10 Hz repetition rate pulsed laser system and 50 for a 50 Hz system. Hence, we find that the cw excitations can interact with significantly more atoms (by approximately 2 orders of magnitude) than can most excimer- or Nd:YAG-based pulsed laser systems.

For the ratio of the degrees of ionization, we find a situation that is somewhat more complicated. The degree of ionization of the illuminated atoms, for both the pulsed and the cw cases, can be written for the one- and two-step excitation cases, in accordance with Eqs. (53) and (54), as

$$\varphi^{off}(\tau_{atom}) = 1 - \exp(-C_2^{off}k_{2,ion}^{eff}\tau_{atom}) \tag{72}$$

and

$$\varphi^{on}(\tau_{atom}) = 1 - \exp(-C_3^{on}k_{3,ion}^{eff}\tau_{atom}) \tag{73}$$

When comparing the total ionization efficiencies of pulsed vs. cw excitations, one must take into account the differences in intensity, atomic illumination time, and spectral width of the two light sources.

In general, it has been found that optical transitions are saturated whenever pulsed excitation is used. This implies that the expressions for the fraction of atoms excited in Eqs. (72) and (73); C_2^{off} and C_3^{on}, therefore attain their saturated values, $(C_2^{off})_{sat}$ and $(C_3^{on})_{sat}$, under pulsed conditions. In a similar manner, it is possible to conclude that cw excitation in general does not saturate the transitions. Hence, the expressions for the fraction of atoms excited. Eqs. (32) and (34), take their low intensity values in the cw case.

The ratio of the degree of ionization for the case of cw to that of pulsed excitation for one-step excitation then becomes

$$\frac{\Phi_{cw}^{off}(\tau_{atom}^{cw,high})}{\Phi_p^{off}(\tau_{atom}^{pulsed})} = \frac{1 - \exp[-(B_{12}(I_{12}^v)^{cw}/k_{21})k_{2,ion}^{eff}\tau_{atom}^{cw,high}]}{1 - \exp[-(C_2^{off})_{sat}k_{2,ion}^{eff}\tau_{atom}^{pulsed}]} \tag{74}$$

It is now possible to get some quantitative estimates of this ratio.

For the cw case, let us assume a dye-laser high output of 100 mW, an atomic width of 4 GHz, an A_{21} factor of 0.5×10^8 Hz, equal degeneracy of the two laser-connected states, a collision deexcitation rate of 3×10^9 Hz, a wavelength of 500 nm, and a beam diameter of 5 mm. We then find that the ratio of the excitation rate to the deexcitation rate $(B_{12}I_{12}^\nu/k_{21})$ in the cw case is roughly 0.5×10^{-3} whereas the saturated value for the fraction of atoms excited, $(C_2^{off})_{sat}$, is 3 orders of magnitude larger, i.e., 0.5.

In addition, although the laser light from a cw laser might be considerably narrower than the atoms in the flame [so that the assumption that the atoms are illuminated by a broadband light source (which is needed for the rate-equation formalism) is not fully correct], let us here, for simplicity, assume that the fast collisions will smooth out the atomic response of the system so that we can assess an effective laser bandwidth equal to the value of the collisional broadening in flames.

Inserting these values together with the values of the atomic illumination times from the previous subsection, we find that the ratio of the degree of ionization for the case of cw to that of pulsed excitation for one-step excitation can be written

$$\frac{\Phi_{cw}^{off}(\tau_{atom}^{cw,high})}{\Phi_p^{off}(\tau_{atom}^{pulsed})} = \frac{1 - \exp\left[-k_{2,ion}^{eff}/8 \times 10^6\right]}{1 - \exp\left[-k_{2,ion}^{eff}/2 \times 10^8\right]} \tag{75}$$

A significant depletion ($>50\%$) of the atoms in the interaction region will then occur in the cw case if the collisional ionization rate is larger than 10 MHz and in the pulsed case if the collisional ionization rate is larger than 200 MHz.

This implies that the degree of ionization is close to unity for both the cw and the pulsed case for excitation to states with collisional ionization rates faster than a few hundred megahertz. Although not fully established in the literature, it is generally believed that ionization rates for high-lying atomic states, so-called Rydberg states, are of this order of magnitude. This implies that the ratio of the total ionization efficiency of the LEI technique for high-chopping-frequency cw excitation to that of pulsed excitation for one-step excitation to high-lying states simply is equal to the ratio of the fraction of atoms illuminated for the cw case to that of the pulsed case, i.e., 250 for a 10 Hz repetition rate pulsed layer system and 50 for a 50 Hz system.

Whenever the one-step excitation only excites atoms to states several electron volts from the ionization level, the collisional ionization rate can be significantly smaller. When the collisional ionization rate is smaller than a few megahertz, both the exponentials in the Eq. (74) can be expanded, which thus significantly simplifies the expression as well as the interpretation:

$$\frac{\Phi_{cw}^{off}(\tau_{atom}^{cw,high})}{\Phi_p^{off}(\tau_{atom}^{pulsed})} = \frac{1 - \exp\left[-k_{2,ion}^{eff}/8 \times 10^6\right]}{1 - \exp\left[-k_{2,ion}^{eff}/2 \times 10^8\right]} \approx \frac{k_{2,ion}^{eff}/8 \times 10^6}{k_{2,ion}^{eff}/2 \times 10^8} = 25 \tag{76}$$

We then find that the ratio of the total ionization efficiency of the LEI technique for high-chopping-frequency cw excitation to that of pulsed excitation for *one-step excitation* to low-lying states is equal to 6250 for a 10 Hz repetition rate pulsed laser system and 1250 for a 50 Hz system.

For collisional ionization rates in between these to extremes, i.e., for $k_{2,ion}^{eff}$ in the range between a few and a few hundred megahertz, the ratio of the total ionization efficiency of the LEI technique for high-chopping-frequency cw excitation to that of pulsed excitation for one-step excitation takes values in between these, i.e., between 250 and 6250 for a 10 Hz repetition rate pulsed laser system and between 50 and 1250 for a 50 Hz system.

This implies that up to 3.5 orders of magnitude more atoms can be ionized in the cw case than in the pulsed case for one-step excitation. However, as can easily be seen from the assumptions made above, the estimated values depend directly upon the actual values of several parameters included (the A_{21} factor, the k_{21} value, the laser light intensity in the cw case, the existence of photoionization from the upper laser-connected level in the pulsed case, etc.). Hence, the actual ratio of the two total ionization efficiencies of the LEI technique for a given situation might differ considerably from that estimated above. It can generally be concluded, though, that the total ionization efficiency is significantly smaller for pulsed excitation than for cw excitation in the one-step excitation case.

When the atoms have such a high ionization limit that two-step excitation has to be used, the situation will be different. In such a case, the cw signal will be decreased considerably in comparison with the pulsed signal since the expression for the fraction of neutral atoms excited in the two-step case, Eq. (34), consists, in the unsaturated case, mainly of a product of *two* ratios of the laser excited rates and the quenching rates—one for each transition. This implies that the number of excited atoms in the upper laser-connected level will decrease approximately 3 orders of magnitude in the cw case $[(B_{12}I_{12}^v/k_{21}) \approx (B_{23}I_{23}^v/(k_{31} \text{ or } k_{32})) \approx 0.5 \times 10^{-3}]$ whereas both transitions can easily be saturated in the pulsed case, giving rise to a saturated value for the fraction of atoms excited, $(C_3^{on})_{sat}$, of 1/3. This leads to

$$\frac{\Phi_{cw}^{on}(\tau_{atom}^{cw,high})}{\Phi_p^{on}(\tau_{atom}^{pulsed})} = \frac{1 - \exp\left[-k_{3,ion}^{eff}/16 \times 10^9\right]}{1 - \exp\left[-k_{3,ion}^{eff}/3 \times 10^8\right]} \tag{77}$$

As can be seen from this expression, the degree of depletion for the cw case is, in general, rather low (since the collisional ionization rate in flames presumably is below 16 GHz). For those states for which the collisional ionization rate is also below 300 MHz, we find that both exponents can be expanded, yielding a ratio of degrees of ionization for the case of cw to that of pulsed excitation for two-step excitation of 2%:

$$\frac{\Phi_{cw}^{on}(\tau_{atom}^{cw,high})}{\Phi_{p}^{on}(\tau_{atom}^{pulsed})} = \frac{1 - \exp\left[-k_{3,ion}^{eff}/16 \times 10^9\right]}{1 - \exp\left[-k_{3,ion}^{eff}/3 \times 10^8\right]} \approx \frac{k_{3,ion}^{eff}/16 \times 10^9}{k_{3,ion}^{eff}/3 \times 10^8} = 0.02 \quad (78)$$

This implies that the ratio of the total ionization efficiency of the LEI technique for high-chopping-frequency cw excitation to that of pulsed excitation for *two-step excitation* to states with collisional ionization rates approximately below 100 MHz is 5 (250 times 0.02) for a 10 Hz repetition rate pulsed laser system and roughly unity (50 times 0.02) for a 50 Hz system.

This suggests that the two modes of operation will then give approximately the same *total* ionization efficiency of LEI.

When an actual experiment is being performed, however, it is important to keep in mind that what is of importance is the signal-to-noise level. Although the cw case might produce a signal up to 3 orders of magnitude larger, as shown above, the produced charges and hence the signal will be distributed over approximately 6 orders of magnitude longer time (a typical signal integration time in the pulsed case is in the microsecond range). If the noise from the background consists of flicker noise (which is directly proportional to the time of measurement) the signal-to-noise ratio can be approximately 3 orders of magnitude *lower* in the cw case than in the pulsed case. If the background noise can instead be considered short-noise limited, which, however, is probably not the case for cw measurements, the noise will increase only with the square root of the detection time whereby the two modes of operation at most can give similar signal-to-noise ratios.

It has been found experimentally by Havrilla et al. (37) that the detection powers (that is, the signal-to-noise ratio) for pulsed excitation (in the case of microsecond-long pulses) are approximately equal to the cw case under the conditions prevailing in that study.

In summary we can conclude that in practice, although significantly higher total ionization efficiencies can be obtained using cw excitation, pulsed excitation will always be preferable to cw excitation since the most important limitation of the latter is in fact not the *total ionization efficiency of LEI* but rather the larger amounts of noise collected with the longer duty cycle present in cw excitation.

1.4. COMPARISON BETWEEN COLLISIONAL IONIZATION AND PHOTOIONIZATION

It is also of interest to compare the collisional LEI technique with the case where the excited atoms are mainly being ionized by photoionization. Since photoionization is one of the major techniques for ionization of atoms in low-pressure environments in methods such as resonance ionization spectroscopy (RIS) or resonance ionization mass spectrometry (RIMS) (38–41) and since it is virtually atomic-reservoir-independent, one might be tempted to try to increase the degree of ionization in the LEI technique by making use of photoionization.

The use of photoionization in LEI has been discussed by, for example, Curran et al. (42) and Axner and Sjöström (43). However, as will be shown below, photoionization of excited atoms can only compete with collisional ionization of atoms when either the excited state is far from the ionization limit (and hence the collisional ionization rate is very small) or very high laser light intensities are used (43).

In order to compare the one- and two-step (collisional ionization) LEI cases with the case of one-step excitation followed by photoionization from the excited state by a second photon, let us first derive an expression for the degree of ionization in the photoionization case, which can be easily related to the expressions for the one- and two-step excitation LEI cases given in Eqs. (53) and (54), above.

The degree of ionization of an atomic system that is exposed to an ordinary one-step excitation between the levels 1 and 2 and subsequently photoionized from the excited level (level 2) into the ionization continuum can be written

$$\Phi^{\text{P.I.}}(\tau) = 1 - \exp\left(-C_2^{\text{P.I.}} \cdot R_{2,\text{ion}}^{\text{P.I.}} \tau\right) \approx C_2^{\text{P.I.}} \cdot R_{2,\text{ion}}^{\text{P.I.}} \tau \tag{79}$$

where $R_{2,\text{ion}}^{\text{P.I.}}$ is the photoionization rate given by the simple expression

$$R_{2,\text{ion}}^{\text{P.I.}} = I_{2,\text{ion}}^{\text{P.I.}} \frac{\sigma_{2,\text{ion}}}{h\nu_{2,\text{ion}}} \tag{80}$$

and where $I_{2,\text{ion}}^{\text{P.I.}}$ is the irradiance of the photoionizing laser light [W/m²]; $\sigma_{2,\text{ion}}$, the cross-section for photoionization from the excited state 2 to the ionization continuum; $\nu_{2,\text{ion}}$, the frequency of the light; and $C_2^{\text{P.I.}}$, the fraction of atoms excited on level 2, this time roughly given by

$$C_2^{\text{P.I.}} = (C_2^{\text{off}})_{\text{sat}} \frac{B_{12} I_{12}^{\nu}}{B_{12} I_{12}^{\nu} + (C_2^{\text{off}})_{\text{sat}}(k_{21} + R_{2,\text{ion}}^{\text{P.I.}})} \tag{81}$$

Let us now equalize the expression for the degree of ionization in the interaction volume for the photoionization case with the two LEI cases in order to find out the one and two-step collisional-ionization-equivalent photoionization intensities, $(I_{2,\text{ion}}^{\text{P.I.}})_{\text{equiv.}}^{1-\text{step}}$ and $(I_{2,\text{ion}}^{\text{P.I.}})_{\text{equiv.}}^{2-\text{step}}$, respectively, i.e., the intensity needed in the photoionization step in each case in order to obtain the same degree of ionization for the photoionized atoms as for the collisionally ionized ones.

Equalizing the two expressions for the one- and two-step LEI cases from above [Eqs. (53) and (54)] with that for one-step excitation followed by photoionization [Eqs. (79)–(81)], we obtain

$$C_2^{\text{P.I.}}(I_{2,\text{ion}}^{\text{P.I.}})_{\text{equiv.}}^{1-\text{step}} \frac{\sigma_{2,\text{ion}}}{h\nu_{2,\text{ion}}} = C_2^{\text{off}} k_{2,\text{ion}}^{\text{eff}} \tag{82}$$

$$C_2^{\text{P.I.}}(I_{2,\text{ion}}^{\text{P.I.}})_{\text{equiv.}}^{2-\text{step}} \frac{\sigma_{2,\text{ion}}}{h\nu_{2,\text{ion}}} = C_3^{\text{on}} k_{3,\text{ion}}^{\text{eff}} \tag{83}$$

where the expressions of the C_2^{off} and C_3^{on} are as previously given (Table 1.1).

Assuming now for convenience that the first step is saturating the transition strongly, i.e., $B_{12} I_{12}^{\nu} \gg a$ and $(k_{21} + R_{2,\text{ion}}^{\text{P.I.}})$, that the second-step excitation is barely saturating the upper transition in the two-step case, i.e., $B_{23} I_{23}^{\nu} \approx b$, and that c is much less than $B_{12} I_{12}^{\nu}$, we can relate the intensities needed in the photoionization case to achieve equally strong ionization rates in the LEI and photoionization cases:

$$\frac{(I_{2,\text{ion}}^{\text{P.I.}})_{\text{equiv.}}^{2-\text{step}}}{I_{23}^{\text{LEI}}} = \frac{(I_{2,\text{ion}}^{\text{P.I.}})_{\text{equiv.}}^{2-\text{step}}}{I_{23}^{\nu} \Delta\nu_L} = \frac{B_{23}}{\Delta\nu_L} \frac{h\nu_{2,\text{ion}}}{2\sigma_{2,\text{ion}}} \frac{k_{3,\text{ion}}^{\text{eff}}}{k_{31} + k_{32}}$$

$$= \frac{g_3}{g_2} \frac{A_{32}}{\Delta\nu_L} \frac{(\lambda_{23})^3}{16\pi\sigma_{2,\text{ion}}\lambda_{2,\text{ion}}} \frac{k_{3,\text{ion}}^{\text{eff}}}{k_{31} + k_{32}} \tag{84}$$

$$\frac{(I_{2,\text{ion}}^{\text{P.I.}})_{\text{equiv.}}^{1-\text{step}}}{(I_{12}^{\text{sat}})^{\text{LEI}}} = \frac{B_{12}}{\Delta\nu_L} \frac{hc}{\lambda_{2,\text{ion}}\sigma_{2,\text{ion}}} \frac{k_{2,\text{ion}}^{\text{eff}}}{k_{21}} = \frac{g_2}{g_1} \frac{A_{21}}{\Delta\nu_L} \frac{(\lambda_{12})^3}{8\pi\sigma_{2,\text{ion}}\lambda_{2,\text{ion}}} \frac{k_{2,\text{ion}}^{\text{eff}}}{k_{21}} \tag{85}$$

where we have used the fact that $B_{23} I_{23}^{\nu} = b$ in the first equation for a comparison of the photoionization situation with the two-step collisional ionization LEI case and where we have related the one-step collisional-ionization-equivalent photoionization intensities to the saturation intensity of the first step in the second expression for a comparison of the photoionization situation with the one-step LEI case. In addition, we have here denoted the bandwidth of the laser light $\Delta\nu_L$, and we have used the relation between the Boltzmann A and B factors given by Eq. (27).

The one- and two-step collisional-ionization-equivalent photoionization intensities can now be estimated if we insert some typical value ($g_3 = 10$; $g_2 = 6$; $g_1 = 2$; $A_{32} = 0.1 \times 10^8\,\text{s}^{-1}$; $A_{21} = 1 \times 10^8\,\text{s}^{-1}$; $\Delta\nu_L = 10\,\text{GHz}$; $\lambda_{12} = \lambda_{23} = \lambda_{2,\text{ion}} = 500\,\text{nm}$; $\sigma_{2,\text{ion}} = 10^{-18}\,\text{cm}^2$; $k_{3,\text{ion}}^{\text{eff}} = k_{31} + k_{32}$). Furthermore, in the one-step case, we assume, that the ratio of the $k_{2,\text{ion}}^{\text{eff}}$ to k_{21} represents the probability that an excited atom will ionize rather than return to the ground state. Keeping the ratio of $k_{2,\text{ion}}^{\text{eff}}$ to k_{21} approximately 1 order of magnitude larger than the ordinary Boltzmann relation [and the energy deficit of the first excited state approximately equal to the energy carried by the first-step photon, i.e., $k_{2,\text{ion}}^{\text{eff}} \approx 10k_{21} \exp(-hc/\lambda_{12}kT)$], we find

$$\frac{(I_{2,\text{ion}}^{\text{P.I.}})_{\text{equiv.}}^{2-\text{step}}}{I_{23}^{\text{LEI}}} \approx 8 \times 10^4 \approx 10^5 \tag{86}$$

and

$$\frac{(I_{2,\text{ion}}^{\text{P.I.}})_{\text{equiv.}}^{1-\text{step}}}{(I_{12}^{\text{sat}})^{\text{LEI}}} \approx 300 \tag{87}$$

Hence, we find in this example that in order to obtain the same degree of ionization of the atoms, an almost 5 orders of magnitude higher intensity is required in the photoionization step in the case of one-step excitation followed by photoionization than in the uppermost excitation step in the two-step LEI collisional ionization case. Consequently, the scheme of two-step excitation followed by collisional ionization is *several orders of magnitude more sensitive* than the scheme of one-step resonant excitation followed by photoionization in the flame (43)! In a similar way, it can be shown that the two-step excitation followed by photoionization can in general not rival the ordinary two-step LEI excitation followed by collisional ionization due to the high collisional ionization rate from high-lying states.

Thus, in general, the two-step LEI technique followed by collisional ionization is several orders of magnitude more sensitive (when one is comparing the number of created charges for a given laser intensity) than ionization techniques using one-step excitation followed by photoionization of the excited atoms.

The reasons for the superiority of the two-step LEI technique to the technique utilizing one-step excitation scheme followed by photoionization are (i) the transition probabilities for excitations to high-lying bound states are often orders of magnitude larger than the transition probabilities for photoionization and (ii) the collisional ionization efficiency is reasonably high for many excited states (often in the range of 0.1–1.0). What this also implies is that the *selectivity* of the two-step LEI technique is superior to the photoionization techniques. As will be discussed in Chapter 3, any excess amount of light will increase the possibilities of creating background signals by wing excitations of various atoms (primarily alkali elements). Therefore, 4–5 orders of magnitude stronger light in the second step can in many cases give rise to strong background signals. In such a case the signals from the analyte under study will be completely drowned in the background signals from concomitant elements in the sample.

On the other hand, when comparing the one-step (collisional) LEI with photoionization, we find that there is only a 2–3 order of magnitude difference between the intensity needed for saturation of the first-step excitation and that which will give a considerable photoionization rate.

Although many of the aforementioned estimates of "typical" numbers are supposed to reflect real situations, the actual numbers can of course vary significantly with respect to these estimates in any particular situation. Consequently, it is plausible that although photoionization will not rival collisional ionization in most one-step excitation cases, there are some situations when this will take place (in situations when the photoionization cross sections are considerably higher than in the foregoing assumption, when the laser beams are strongly focused or when high laser light intensities are being used).

In addition, what might also be of importance is that photoionization will only play a significant role when pulsed excitation is used; whenever cw excitation is performed the laser light intensities used are far too small for photoionization processes to ever rival collisional ionization.

It is therefore of interest to study the relation between the degree of ionization and the intensity in the one-step excitation case, as was done above, under the assumption that photoionization indeed can supersede collisional ionization. If we assume that the photoionization rate from the excited state is a factor of 300 weaker than collisional ionization at the intensity corresponding to optical saturation, the relation between the degree of ionization and intensity is represented by the curves in Fig. 1.3: here, curves a–e represent the same situations as in Fig. 1.2, where the atoms were exposed to collisional ionization solely.

From Fig. 1.3 it can easily be seen that the existence of photoionization changes the behavior considerably for high intensities. For the situations when the collisional ionization is smaller than 1% of the inverse interaction time (i.e., curves a–c) (less than 1 MHz collisional ionization rate for most pulsed laser system) photoionization will prevent the signal from leveling off for high intensities. The existence of photoionization can in fact completely disguise the onset of optical saturation, as is the case in the situation shown by curve b, so that information about the true physical processes will be very difficult to extract in the system studied. A more thorough investigation of the influence of photoionization on the LEI signal can be found in Omenetto et al. (44).

To recapitulate what was concluded above, photoionization will not, in general, be significant in comparison with collisional ionized two-step LEI. Neither will photoionization give any significant contribution to the signal

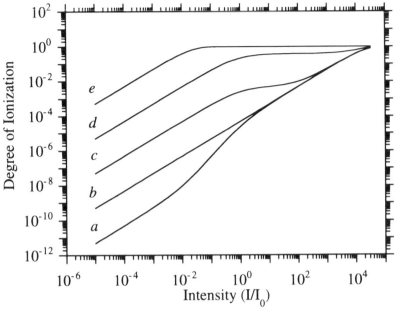

Figure 1.3. Degree of ionization vs. laser intensity for a one-step excitation case exposed to both collisional ionization and photoionization according to Eqs. (79)–(81). The five curves represent the same situations as in Fig. 1.2, i.e., different products of ionization rates and laser pulse length ($k_{2,ion}^{eff} \tau = 10^{-6}$, 10^{-4}, 10^{-2}, 10^0, and 10^2 for cases a, b, c, d, and e, respectively). The photoionization rate from the excite state has here been assumed to be a factor of 300 weaker than collisional ionization at the intensity corresponding to optical saturation.

when the first-step excitation brings the atoms close enough to the ionization limit so that they can benefit from a high collisional ionization rate.

1.5. THE IONIZATION EFFICIENCY OF EXCITED ATOMS IN FLAMES

In general, when an atom has absorbed a photon and is left in an excited state, several processes might affect its future fate: it can undergo spontaneous emission, light-induced radiative transitions, or collisional transitions to other bound or unbound states. Even though we do not know all the processes influencing excited atoms in flames in detail, we do know that atoms exposed to collisions have a certain probability of becoming ionized in the subsequent collisional process. This means that we can define a probability of ionization, or ionization yield, Y^i, from an excited state i, simply as the probability that an excited atom will become ionized by the subsequent process (collisional, radiative, or chemical), in a similar way to the definition of, for example, the fluorescence yield.[4] In practice, this then implies that Y^i is simply the ratio of the rate of the ionizing process to the total rate of those processes which transfer the atom to another state, i.e.

$$Y^i = \frac{\text{ionization rate from state } i}{\text{total collisional, radiative, and chemical transfer rate from state } i}$$

(88)

Such a summation is of course practically impossible to perform, however, since it includes such a large number of unknown rates. In fact, it is even almost impossible to measure to *direct* (single-step) ionization yield, since what any measurement might detect is the *effective* ionization rate that automatically takes into account possible branchings/transfers to other states prior to ionization.

However, since the atom can also be transferred to other excited states, it can undergo an arbitrary number of collisional transfers before becoming ionized. Consequently, in each of these states, the atoms experience a given ionization yield. Summing up all possible transfer yields and ionization yields finally gives the true (or effective) ionization yield of an excited atom in the flame as defined above.

An alternative approach to dealing with direct ionization yields is to define an ionization efficiency of an atom excited to a state i, ζ^i, as the probability that an excitation of an atom to a state i by absorption of a photon will lead to a subsequent ionization of the atom before it finds its way down to the ground state again. This definition can then be formulated for an atomic system as

$$\zeta^i = \frac{\text{the number of ions formed}}{\text{the number of atomic excitations to state } i}$$

(89)

All the individual ionization efficiencies (for each excited state) are coupled to each other through

$$\zeta^i = \frac{k_{i,\text{ion}} + \sum_{j \neq i}(k_{ij} + A_{ij})\zeta_j}{k_{i,\text{ion}} + \sum_{j' \neq i}(k_{ij'} + A_{ij'})}$$

(90)

where the summations are performed over all bound states other than the laser-populated state, and where the spontaneous emission factors should be included only in the cases of dipole allowed transitions.

---

[4] It is woth noting that the term *ionization yield* has previously occurred in the literature in a paper by Smith et al. (45), although with a different meaning than that used here. The definition of the term in that paper is identical to that of *degree of ionization* in this chapter.

Although this summation is again virtually impossible to carry out, the concept of ionization *efficiency* has a significant advantage as compared to the concept of ionization *yield* since it corresponds exactly to the experimentally measurable quantities—the number of ions produced as a function of the number of photons absorbed.

It is possible to estimate the expected behavior of the ionization efficiency of highly excited atoms exposed to collisional ionization as a function of principle quantum number. Such an estimate was done by Axner and Berglind (36). It was found that the ionization efficiency should decrease monotonically from the value of unity at the ionization limit. However, it was found that the decrease of ionization efficiency is not as fast as the decrease of the Boltzmann factor despite the fact that each separate pair of exciting and deexciting collisional rates scales as the Boltzmann factor, i.e., $\exp(-\Delta E/kT)$, where ΔE is the energy between the Rydberg level and the ionization limit.

In the literature, there are only a few measurements of the ionization efficiency of excited atoms in flames. However, measurements of the ionization efficiency of a number of excited np states in Na and Li in an acetylene/air flame have been performed by Axner and Berglind (36). The ionization efficiency of virtually the entire series of np states in Li and Na was measured (Li, np states, $n \geqslant 4$; Na, np states, $n \geqslant 5$). In that study, it was found that the ionization efficiency of excited Li and Na atoms in an acetylene/air flame is close to unity for atoms excited to states close to the ionization limit and that it decreases as a function of energy deficit to the ionization limit. The decrease, however, was in general found to be not as fast as the Boltzmann factor, i.e., $\exp(-\Delta E/kT)$. This is also in agreement with what a simple theory of the estimated ionization efficiency for collisionally ionized excited atoms in flames predicts (36). Therefore, the ionization efficiency can be expressed as

$$\zeta^i = f(\Delta E)\exp(-\Delta E/kT) \qquad (91)$$

where $f(\Delta E)$ in the following discussion will be referred to as the Boltzmann-compensated ionization efficiency. Consequently, this factor takes care of any deviation of the ionization efficiency from a pure Boltzmann-type behavior.

The measured ionization efficiencies and the Boltzmann-compensated ionization efficiency values for the np series of Li and Na are presented in Figs. 1.4 and 1.5, respectively. However, due to contradictory values of the transition probability factors for the $2s-np$ (Li) and the $3s-np$ (Na) transitions in the literature (46–48), the evaluation of the measured data gives two somewhat different sets of values for the ionization effeciencies. Therefore, there are two sets of data related to each element in these two figures.

From Fig. 1.4, displaying the ionization efficiencies, one can conclude that the ionization efficiencies are close to unity for states with $\Delta E < kT(kT =$

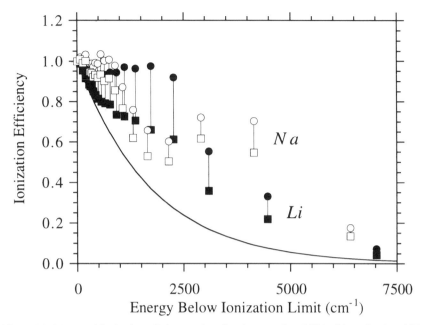

Figure 1.4. Measured ionization efficiency values for the np series of Li (solid markers) and Na (white markers). For each of the elements two sets of data points are presented, representing evaluations of the measured data from either transition probability A-factor values from Wiese et al. (46, 47) (circles) or from Lindgård and Nielsen (48) (squares). The curve represents an ionization efficiency equal to the Boltzmann factor, $\exp(-\Delta E/kT)$, where $kT = 1742\,\mathrm{cm}^{-1}$, corresponding to a temperature of 2500 K.

$1742\,\mathrm{cm}^{-1}$ for $T = 2500\,\mathrm{K}$) and approximately 50% for states with $\Delta E \approx 2.5\,kT\,(np \approx 6p$ in both elements).

In Fig. 1.5, displaying the Boltzmann-compensated ionization efficiencies, we have also included three different fits: (i) one fit for each element (curve a for Na and curve b for Li), calculated as the third-order polynomial that best fits all data for that particular element, are displayed; (ii) a similar fit based upon all data points (from both elements) is included (curve c); (iii) for comparison, curve d represents a Boltzmann-compensated ionization efficiency equal to unity. Due to the lack of known ionization efficiencies of other elements in flames, the fit shown as curve c will serve as an estimate of the ionization rate for other elements ionizing by collisional ionization. The parameters of the fits are presented in Table 1.2.

Consequently, we can conclude that excited atoms in flames in general experience a significant ionization efficiency already for rather low principal quantum numbers (e.g., $\zeta^i = 0.5$ for the $6p$ states of Li and Na), a few times larger than the Boltzmann factor (4–7 times larger) for states a few kT below the ionization limit (e.g., for the $4p$ and $5p$ states).

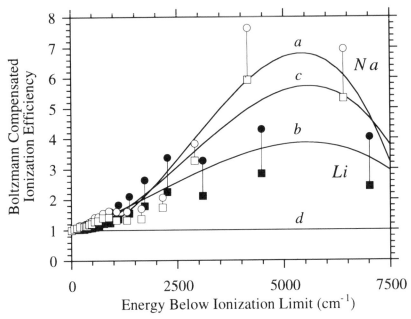

Figure 1.5. Measured Boltzmann-compensated ionization efficiency values for the *np* series of Li (solid markers) and Na (white markers). For each of the elements two sets of data points are presented, representing evaluations of the measured data from either transition probability *A*-factor values from Wiese et al. (46, 47) (circles) or from Lindgård and Nielsen (48) (squares). The curves represent three fits (*a–c*) together with a curve for a unity Boltzmann-compensated ionization efficiency (*d*). The three fits are as follows: *a*, the third-order polynomial that best fits the measured Boltzmann-compensated ionization efficiency values for Na; *b*, the third-order polynomial that best fits the values for Li; and *c*, the third-order polynomial that best fits all the data points (for both Na and Li).

Table 1.2. Parameters for the Fits to the Boltzmann-Compensated Ionization Yields for Li and Na in an Acetylene/Air Flame Using the Following Fitting Function[a]

$$1 + \varepsilon_1 \left(\frac{\Delta E}{kT} \right) + \varepsilon_2 \left(\frac{\Delta E}{kT} \right)^2 + \varepsilon_3 \left(\frac{\Delta E}{kT} \right)^3$$

Element	Li	Na	Overall Average
ε_1	−0.62	0.80	0.18
ε_2	2.20	0.35	1.27
ε_3	−0.45	−0.10	−0.27

[a] In this fit kT has been taken as $1742 \, \text{cm}^{-1}$, corresponding to $T = 2500 \, \text{K}$.

1.6.   THE MOST SENSITIVE TRANSITION FOR ONE-STEP LEI: SIGNAL STRENGTH VS. PRINCIPAL QUANTUM NUMBER

By using some of the relations derived earlier concerning laser excitation, excited state populations, and ionization rates in flames, it is now possible to estimate the optimum transitions for one-step LEI. We refer here to the optimum transition as the most sensitive one, i.e., the transition yielding the highest sensitivity in LEI.

It is obvious that whenever there exist a number of transitions that can all be saturated by the laser light, the best choice (for the strongest signal) is to use the one that excites the atoms to the highest level possible, i.e. as close to the ionization limit as possible, since then the atom can benefit from the highest possible ionization yield. However, since the transition probability (and hence the laser-induced excitation rate) decreases very fast as a function of principal quantum number of the upper state, n^*, transitions to highly excited levels ($n^* > 10$) will not, in general, be saturated by laser light from normal pulsed dye lasers. Then the question arieses, which is the most sensitive transition in one-step LEI for nonsaturated transitions?

We will now derive an expression for the relation between the LEI signal strength and the principal quantum number of the upper state for a general type of atom exposed to collisional ionization. In this derivation we will make use of a few of the equations derived earlier.

Let us assume first that the degree of ionization of atoms (exposed to one exciting laser step) in the interaction region can be described by Eq. (53), i.e.,

$$\Phi^{\text{off}}(t) = 1 - \exp(-C_2^{\text{off}} k_{2,\text{ion}}^{\text{eff}} t) \approx C_2^{\text{off}} k_{2,\text{ion}}^{\text{eff}} t \tag{92}$$

and for simplicity let us assume that the fraction of atoms ionized is low (less than approximately 10%), so that the last approximation is valid (this is not a necessary assumption—it merely simplifies the interpretation of the result). Then, let us use the expression for the fraction of atoms excited in level 2:

$$C_2^{\text{off}} = (C_2^{\text{off}})_{\text{sat}}^* \frac{B_{12} I_{12}^\nu}{B_{12} I_{12}^\nu + (C_2^{\text{off}})_{\text{sat}}^* (k_{21} + k_{23} Y_{31})} \approx \frac{B_{12} I_{12}^\nu}{k_{21} + k_{23} Y_{31}} \tag{93}$$

where the last approximation is valid if we assume that the transition in unsaturated.

By defining $\zeta^{(2)}$, the ionization efficiency from level 2, as

$$\zeta^{(2)} = \frac{k_{2,\text{ion}}^{\text{eff}}}{k_{21} + k_{23} Y_{31}} \tag{94}$$

and by using the relation between the Einstein A_{21} and B_{12} factors for spontaneous and stimulated absorption, defined previously [Eq. (27)], we can, using the above equations, write an expression for the degree of ionization of atoms as

$$\Phi^{\text{off}}(\tau) = \frac{g_2}{g_1} \frac{(\lambda_{12})^3}{8\pi hc} A_{21} \zeta^{(2)} I_{12}^{\nu} \tau \tag{95}$$

where τ represents the duration of the laser pulse. As can be seen from the expression, the one-step LEI signal is determined by the amount of energy per unit frequency-bandwidth and per unit area to which the atoms are exposed, i.e., $I_{12}^{\nu} \tau$ (and not solely the spectral intensity). The quantity of energy per unit area is normally referred to as the *fluence* (or exposure). Let us therefore define the spectral fluence, F_{ν}, as this quantity:

$$F_{\nu} = I_{12}^{\nu} \tau \tag{96}$$

We can then write the sensitivity for one-step (unsaturated) LEI, $\mathfrak{R}$, defined as the degree of ionization normalized to the spectral fluence, as

$$\mathfrak{R} = \frac{\Phi^{\text{off}}(\tau)}{F_{\nu}} = \frac{g_2}{g_1} \frac{(\lambda_{12})^3}{8\pi hc} A_{21} \zeta^{(2)} \tag{97}$$

Let us now make use of the relation between the A factor for a transition and the (effective) principal quantum number of the upper state, n^*:

$$A_{21} \approx A_0 \frac{1}{(n^*)^3} \tag{98}$$

which mainly is valid for unperturbed series of Rydberg states, especially in elements which can be treated as one-electron atoms. Here A_0 is an element-specific constant.

Let us then relate the principal quantum number, n^*, to the energy deficit of the excited state, $\Delta E(= E_{\text{ion}} - E_2)$, by the well-known Bohr expression for Rydberg states,

$$\Delta E = \frac{hcR}{(n^*)^2} \tag{99}$$

where R is the Rydberg constant (109, 737 cm^{-1}), and let us also express the wavelength for a certain transition in terms of this energy deficit:

$$\lambda_{12} = \frac{hc}{E_2 - E_1} = \frac{hc}{E_2} = \frac{hc}{E_{\text{ion}} - \Delta E} \tag{100}$$

If we again express the ionization efficiency, $\zeta^{(2)}$, as a product of a Boltzmann-compensated factor times an exponential factor, i.e.,

$$\zeta^{(2)} = f(\Delta E)\exp(-\Delta E/kT) \tag{101}$$

we can write the expression for the sensitivity [Eq. (97)] (49) as

$$\Re = \Re_0 \frac{(\Delta E)^{3/2}}{(E_{ion} - \Delta E)^3} f(\Delta E)\exp(-\Delta E/kT) \tag{102}$$

where $\Re_0$ is a constant containing a number of physical and experimental constants:

$$\Re_0 = A_0 \frac{g_2}{g_1} \frac{(hc)^{1/2}}{8\pi R^{3/2}} \tag{103}$$

This expression for the degree of ionization of atoms thus depends upon only one variable, the energy between the Rydberg level and the ionization limit, ΔE. We know that the degree of ionization has a mazimum for some given energy deficit (the principal quantum number) since the signal is small for lower states due to a small Boltzmann factor, rather small for higher states due to a small transition probability factor, and thus larger in the region in between.

It is possible to determine the position of this maximum by, for example, simply equating the derivative of the function with respect to the energy deficit to zero. Let us first, for simplicity, take the Boltzmann-compensated factor to be equal to unity. Then, one finds that the maximum signal is obtained for an energy deficit, $(\Delta E)_{opt}$, given by

$$(\Delta E)_{opt} = \left(\frac{3}{2} + \frac{9}{2}\frac{kT}{E_{ion}}\right)kT \tag{104}$$

Since E_{ion} normally is approximately $20kT$–$30kT$ for most elements in many flames, this simplified expression indicates that the highest one-step LEI sensitivity, $\Re^{max}$, should be found for excitations to states that are positioned approximately $1.7kT$ below the ionization limit, i.e., $(\Delta E)_{opt} \approx 1.7kT$. In an acetylene/air flame (with an estimated temperature of 2500 K) this corresponds roughly to $3000\ cm^{-1}$. This very simple estimate then predicts that the most sensitive one-step LEI transitions for Li and Na (under unsaturated conditions) are those that excite the atoms to the $6p$ state in Li (at $3096\ cm^{-1}$ below the ionization limit) and the $7p$ state in Na (at $2909\ cm^{-1}$ below the ionization limit).

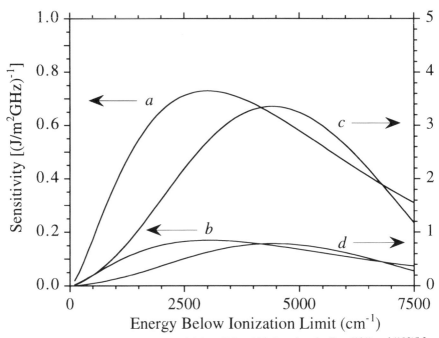

Figure 1.6. Predicted one-step LEI sensitivity of Li and Na [as given by Eqs. (101) and (102)] for the two cases of a Boltzmann-compensated factor, $f(\Delta E)$, equal to unity (Li, curve a; Na, curve b—left axis) and equal to the third-order polynomial expression that represents the best fit to the data presented in Fig. 1.5 (Table 1.2) (Li, curve c; Na, curve d—right axis).

The expression for the sensitivity for one-step excitation above, Eq. (102), is also plotted in Fig. 1.6 for clarity. Here, sensitivities for Li and Na are plotted both for the case of the Boltzmann-compensated factor being equal to unity (Li, curve a; Na, curve b) and for a Boltzmann-compensated factor that is taken as the third-order polynomial expression best fitting the average of the measured Boltzmann-compensated factor for Li and Na, as was discussed in Section 1.5 (see Fig. 1.5 and Table 1.2) (curves c and d). The sensitivity is given in units of $[J/(m^2\ GHz)^{-1}]$. This implied that the degree of ionization becomes 30% for a sensitivity of 3 $[J/(m^2\ GHz)^{-1}]$ for a typical experimental situation with 10 µJ of laser light with a spectral bandwidth of 10 GHz across an area of 10 mm².

As can be seen from curves c and d in Fig. 1.6, the full expressions predict that the maximum sensitivity will occur for excitations to states approximately 4000–4500 cm^{-1} below the ionization limit. This estimate (which takes into account the actual deviation of the degree of ionization from the pure Boltzmann-like behavior) thus predicts that the most sensitive one-step LEI transitions for Li and Na (under unsaturated conditions) are those that excite the atoms to the $5p$ state in Li (at 4471 cm^{-1} below the ionization limit) and the $6p$ state in Na (at 4152 cm^{-1} below the ionization limit).

It should be borne in mind, however that this simplified derivation is built upon some general assumptions. As was mentioned previously in Eq. (98), the n-dependence of the A factor is mainly valid for unperturbed series of Rydberg states, especially in elements that can be treated as one-electron atoms. Since the atomic reservoir in this case has a rather high temperature (2500 K), it is found that the predicted optimum transitions excite atoms to states that are not necessarily correctly described as pure Rydberg states. Consequently, for this type of hot medium, the simplified expression above [Eq. (102)] is expected to describe properly only one electron types of elements. In order to estimate the most sensitive transition for other elements, the expression making use of the actual A factors, Eq. (97), should preferably be used (whenever individual transition probability factors are known).

In order to check the consistency of the simplified theory, Eq. (102), with that of the more extensive one, Eq. (97), the most sensitive transitions for Li and Na were also calculated by the use of Eq. (97) with the actual A factors (and not by using the A-vs.-n dependence). This more accurate calculation predicts that the $3s$–$6p$ transition in Na indeed should be the most sensitive

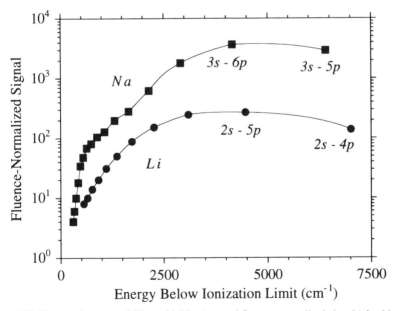

Figure 1.7. Measured one-step LEI sensitivities (spectral fluence normalized signals) for Li and Na as a function of energy deficit of the upper laser level.

one, approximately 30% more sensitive than the $3s-5p$ state (49). In the case of Li, on the other hand, the more extensive theory predicts that all the three transitions, $2s-5p$, $2s-6p$, and $2s-7p$ should have virtually the same sensitivity (within 10%).

Figure 1.7 displays the actual experimental findings for Li and Na; the spectral-fluence-normalized signal (degree of ionization) is plotted (in arbitray units) vs. the energy deficit of the upper state (to the ionization limit). As can be seen from the measurements, the most sensitive transitions in Li and Na are $2s-5p$ and $3s-6p$, respectively. This is in fairly good agreement with the aforementioned theoretical predictions (49, 50).

Since most elements look rather similar when being excited, we can assume that the above findings (that the most sensitive transitions in one-step LEI are those that excite the atoms to states approximately $2kT-3kT$ below the ionization limit) also are valid for other elements subjected to collisional ionization in the flame.

1.7.   THE NUMBER OF IONS PRODUCED VS. THE AREA OF THE LASER BEAM

It is also of importance to study the relation between the number of ions produced and the area of the laser beam (for a given laser pulse energy) in order to find the optimum focusing conditions of LEI. By using some of the relations derived earlier concerning laser excitation rates, excited state populations, and ionization rates in flames it is now possible to derive an expression for the number of ions produced in the flame (as a consequence of laser illumination) as a function of the beam area for one-step LEI.

Recalling some useful expressions for the degree of ionization [Eq. (79)], the fraction of atoms excited [Eq. (81)] and the photoionization rate [Eq. (80)], we can write the number of ions produced (N_i) for a given interaction region (given by L times A, where L is the length of the interaction region and A is the area of the laser beam) as

$$N_i = \Phi^{\text{off}}(\tau) n_{\text{tot}} L A \tag{105}$$

where

$$\Phi^{\text{off}}(\tau) = 1 - \exp[-C_2^{\text{P.I.}}(k_{2,\text{ion}}^{\text{eff}} + R_{2,\text{ion}}^{\text{P.I.}})\tau] \tag{106}$$

$$C_2^{\text{P.I.}} = (C_2^{\text{off}})_{\text{sat}}^* \frac{B_{12} I_{12}^v}{B_{12} I_{12}^v + (C_2^{\text{off}})_{\text{sat}}(k_{21} + R_{2,\text{ion}}^{\text{P.I.}})} \tag{107}$$

and

$$R^{\text{P.I.}}_{2,\text{ion}} = I^{\text{P.I.}}_{2,\text{ion}} \frac{\sigma_{2,\text{ion}}}{h\nu_{12}} \tag{108}$$

we have here assumed that the energy of the photon is high enough to induce photoionization directly from the intermediate level.

Let us now write the spectral irradiance, I^ν_{12}, and the irradiance of the photoionizing light, $I^{\text{P.I.}}_{2,\text{ion}}$, in terms of a pulse energy, E_p; a pulse length, τ; a spectral width, $\Delta\nu$; and a laser beam area, A. This then implies that we can write an expression for the one-step LEI signal (for those elements and transitions for which the energy of the second photon is enough in order to photoionize the excited atom) as a function of A:

$$N_i = c_1 A \left\{ 1 - \exp\left[-\frac{c_2(A + c_3)}{A(A + c_4)} \right] \right\} \tag{109}$$

where

$$c_1 = n_{\text{tot}} L \tag{110}$$

$$c_2 = \frac{B_{12} E_p}{\Delta\nu} \frac{k^{\text{eff}}_{2,\text{ion}}}{k_{21}} \tag{111}$$

$$c_3 = \frac{E_p}{h\nu_{12}} \frac{\sigma_{2,\text{ion}}}{\tau k^{\text{eff}}_{2,\text{ion}}} \tag{112}$$

and

$$c_4 = \frac{E_p}{h\nu_{12}} \frac{\sigma_{2,\text{ion}}}{\tau k^{\text{eff}}_{2,\text{ion}}} + \frac{B_{12} E_p}{\tau \Delta\nu k_{21}(C^{\text{off}}_2)_{\text{sat}}} \approx \frac{B_{12} E_p}{\tau \Delta\nu k_{21}(C^{\text{off}}_2)_{\text{sat}}} \tag{113}$$

The relation between number of ions produced, N_i, and beam area is plotted for four different pulse energies in Fig. 1.8 (1000, 100, 10, and 1 μJ for curves a, b, c, and d, respectively). In this plot the following "typical" values have been used: $A_{21} = 10^8 \, \text{s}^{-1}$; $g_2 = 2g_1$; $\lambda_{12} = 250 \, \text{nm}$; $\sigma_{2,\text{ion}} = 10^{-16} \, \text{cm}^2$; $\tau = 10 \, \text{ns}$; $\Delta\nu = 10 \, \text{GHz}$; $k_{21} = 10^9 \, \text{s}^{-1}$; $k^{\text{eff}}_{2,\text{ion}} = 10^7 \, \text{s}^{-1}$; $L = 10 \, \text{cm}$; and $n_{\text{tot}} = 10^{10} \, \text{cm}^{-3}$ (roughly corresponding to a concentration of 1 ppm of the element under investigation in the solution, elemental mass 50 u, 5% nebulizer efficiency, unity degree of atomization, aspiration rate 5 mL/min, and a flame rise velocity of 5 m/s). All this implies that the four constants c_1 to c_4 take values of $c_1 = 10^9 \, \text{mm}^{-2}$; $c_2 = 310 \, E_p(\text{mJ}) \, \text{mm}^2$; $c_3 = 130 \, E_p(\text{mJ}) \, \text{mm}^2$; and

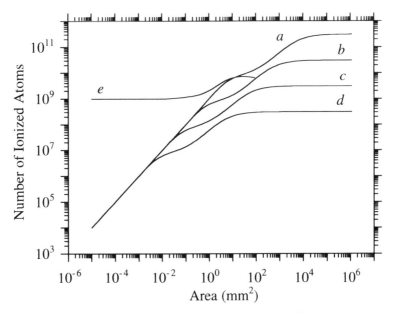

Figure 1.8. The relation between the number of ions produced, N_i, and beam area, A, plotted for four different pulse energies (1000, 100, 10, and 1 µJ for curves a, b, c, and d, respectively), together with a "typical" experimental curve (e), constructed by roughly averaging several experimental curves. Note that the experimental curve has been inserted with no absolute relation to the theoretical ones.

$c_4 = 4650\, E_p(\text{mJ})\,\text{mm}^2$. Here $E_p(\text{mJ})$ is the laser pulse energy given in millijoules, and the laser beam area, A, is given in square millimeters.

As can be seen from the theoretically predicted curves in Fig. 1.8, they all consist of four different regions.

In region 1, for the smallest beam areas, all curves merge into a single curve, which is directly proportional to the beam area. This displays the situation when the beam area is so small that *all atoms in the interaction region are being ionized* (with 100% efficiency). This takes place for such small areas that photoionization dominates the collisional ionization processes (the photoionization rate is larger than the inverse of the pulse duration). The linear dependence of the signal with area (a slope of 1 in the log-log diagram) then only arises from the increase in the number of atoms in the interaction region.

Eventually, all four curves bend slightly (although for different beam areas). This indicates the start of region 2, which consists of the beam areas for which neutral atom depletion no longer limits the signal. However, all atoms are still exposed to a situation where the lower transition is optically saturated and photoionization dominates collisional ionization.

For larger beam areas, the intensity has dropped so that photoionization no longer rivals collisional ionization (region 3). The curve then retains its slope of 1 in the log-log diagram (the signal is directly proportional to the number of illuminated atoms, all of which are optically saturated and exposed solely to collisional ionization).

For very large beam areas, finally, the signal is independent of the beam area (region 4). This situation occurs for such large areas that the first transition is no longer saturated. That means that for any change in area, the changes in excitation rate of the atoms and the number of atoms in the interaction region exactly cancels out.

The four curves in Fig. 1.8 illustrate typical situations predicted by the theory. It is worth noting that the exact positions of the onset of all these different regions depend on a number of different parameters and the four curves in the figure therefore should only be taken as examples of some situations that the theory predicts.

It is obvious that curves of this kind have much in common with normal saturation curves, i.e., signal vs. laser irradiance. Such experimental curves, which have appeared on several occasions in the literature, can be used for assessing values of saturation parameters and photoionization cross sections. The aforementioned curves [number of atoms ionized (or signal) vs. beam area], however, have not yet appeared in the literature. The main reason is probably that the experimental curves do not have much in common with the predicted ones.

The exact form of an experimental curve depends on a number of parameters, the atoms under investigation, the transition used, the spatial homogeneity of the beam, the laser pulse energy, etc. Despite this, a "typical" experimental curve is displayed in Fig. 1.8 for comparison (curve e). Curve e has been constructed out of a rough averaging of several experimental curves and serves therefore only as an example of an experimental curve (51). This implies that there is no direct relation between the absolute scales of the theoretical and "typical" experimental curves in Fig. 1.8. Only the shapes of the curves with respect to each other are of importance in this comparison.

As can be seen from the figure, the agreement between the theoretical and the "typical" experimental curves is quite bad. The reasons for this are as follows.

First of all, the theory assumes an unlimited, homogeneous distribution of atoms in space. This is not the case in a flame. For example, the atomic density is higher in the central part of the flame than in the outer parts. A similar argument is valid for the charge detection efficiency (see Chapter 2). Therefore, whenever the beam area is made large enough (in an actual experiment) the experimental signal will decrease. These two effects thus limit the experimental signal for larger beam areas (for areas roughly larger than some tens of square millimeters, a number which, however, depends very much on the actual element under investigation, its atomization temperature, and the flame used).

For smaller beam areas (for some elements/transitions, smaller than a few square millimeters; for others, smaller than some tens of square millimeters), it is found, in fact, that the experimental signals do not decrease with decreasing beam area as predicted by theory. The reason for this is that when the laser beam is made small (by focusing or by using apertures) the actual area of the interaction region (defined as the volume in space that the laser beam traces out in the atomic vapor) constitutes only a small part of the entire atomic cloud in the flame. This implies that if any amount of diffracted or scattered light from the laser beam (due to diffraction from apertures or laser light scattered from particles/molecules in the flame) or any kind of emitted fluorescence from the geometrically defined interaction region were to excite some atoms in other parts of the flame, the signal from the small geometrically defined interaction region would no longer necessarily dominate the total measured signal. The interaction volume has thus been made so small that the main part of the measured signal originates from atoms *outside* the actual geometrically defined interaction region. This concept has not yet been studied very extensively (although it is of considerable importance for a proper understanding of the physical principles that govern the total number of ions produced, and thus the signal, in flame LEI). However, some results from the published experimental investigation of this phenomena are briefly presented and discussed in the following section.

1.8. ANOMALOUS CONTRIBUTIONS FROM ATOMS OUTSIDE THE INTERACTION REGION

1.8.1. Introduction

We have just seen that the measured number of produced ions as a function of interaction area does not agree with the theoretical predictions, in particular for smaller areas. This discrepancy can be discussed in terms of the influence that "scattered, resonant light" might have on the atoms situated outside the actual, geometrically confined interaction region (the region defined by the laser light and the atomic vapor), here called the *true interaction region*. The term *scattered, resonant light* in this context is used in its broadest meaning, i.e., it refers to laser light diverted from its original path by diffraction or scattering from apertures, particles, or molecules in the flame as well as by fluorescence from atomic species in the true interaction region.

In general, the interference effects which are associated with excitation of atoms outside the true excitation region by scattered, resonant light can occur in all kinds of techniques in which the detection volume is significantly larger than the geometrically defined interaction region. In LEI there is a substantial charge-detecting electric field in a large part of the flame.

Furthermore, the presence of scattered, resonant light outside of the geometrically confined interaction region implies that erroneous conclusions may be drawn when, for example, number densities, ionization yields, signal enhancements, and saturation effects are studied by all such types of techniques unless special precautions are taken.

There are as yet very few experimental results that illustrate the behavior of the signal caused by excitation of atoms outside of the true interaction region by scattered, resonant light in LEI spectrometry (52–55). Effects of this kind are therefore not yet fully understood, nor is their extent well known. We will here examine this phenomenon by briefly describing a study on anomalous contributions to the LEI signal from Sr atoms positioned outside the interaction region (53).

1.8.2. Some Experimental Examples: The Sr Anomaly

1.8.2.1. One-Step Excitation

In the first set of experiments Sr atoms were excited by one-step excitation from the ground state, $5s^2\,{}^1S_0$, to the $5s5p\,{}^1P_1$ state (a transition at 460.733 nm). The laser wavelength was scanned across this resonance (from 460.4 to 461.2 nm) for various beam areas and laser pulse energies (and hence intensities). Due to a high transition probability the transition was optically saturated for all atoms in the true interaction region in all of the measurements performed.

The results from a set of one-step experiments in which the output from the dye laser was scanned across the Sr resonance for different pulse energies (12–550 µJ) and for different beam diameters (2, 4, and 8 mm) are displayed in the Figs. 1.9–1.11.

From the first set of curves (Fig. 1.9), obtained using a rather small beam diameter (2 mm), one can easily see a distinct peak that is clearly visible in all of the four scans. Worth noticing here is that even though the intensity of the light was increased by a factor of 45 between the lowest and highest intensities, the peak signal increased only by a factor of 2 (compare the 12 and 550 kW/cm^2 curves in Fig. 1.9).

This finding can be easily interpreted as a manifestation of the normal optical saturation phenomenon. However, what is not in agreement with the normal behavior of optical saturation is that the measured signal in such a case should also become broader the more the transition is saturated.

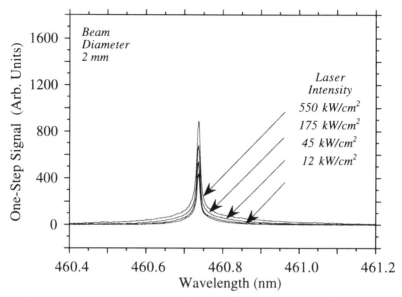

Figure 1.9. A set of one-step scans between 460.2 and 461.2 nm in 3 ppm Sr. The beam diameter was 2 mm. The intensities of the light in the four scans were (from bottom to top) 12, 45, 175, and 550 kW/cm².

A simple theory for the behavior of saturated transitions predicts that the width of the measured peaks should increase with the light intensity (as the square root of the intensity if the laser bandwidth is more narrow than the induced transition widths), as illustrated in Fig. 1.12. In this figure a Lorentzian type of broadening curve describes a two-level system exposed to both collisional ionization and photoionization. This curve is fitted to the peak values of the experimental curves in Fig. 1.9. Such a fit can reproduce the measured peak signals rather accurately. However, as can be seen from a comparison between Figs 1.9 and 1.12, the theory cannot reproduce the width of the curves. Hence, *optical saturation **cannot** be responsible for the weak dependence between LEI signal strength and laser light intensity in this case,* since the two set of curves show completely different broadening behavior.

When the beam area and the laser pulse energies were increased four times so that the beam intensities [measured in units of W/m²] were the same i.e., doubling the beam diameter, the curves presented in Fig. 1.10 resulted. A fourfold larger interaction region normally implies that an approximately fourfold larger signal is to be expected. The signals were, however, only 20–40% larger than when the smaller aperture was used (compare Fig. 1.9). Hence, this finding is also in contrast with the expected behavior of ordinary LEI signals. Most of the other behaviors of the signals were similar to those

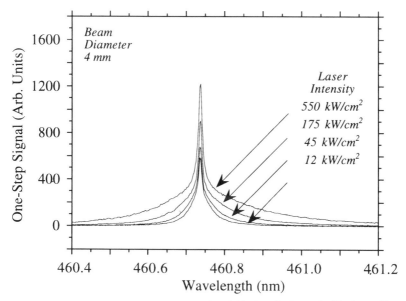

Figure 1.10. A set of one-step scans between 460.2 and 461.2 nm in 3 ppm Sr. The beam diameter was 4 mm. The intensities of the light in the four scans were (from bottom to top) 12, 45, 175, and 550 kW/cm² . The y-axis scale is identical to than in Fig. 1.9.

found in Fig. 1.9, i.e., a very small signal increase as a function of laser light intensity together with rather narrow peak widths. This time, however, especially for the higher intensities, the signals begin to look as though they are composed of a narrow peak positioned on top of a broader feature, i.e., a curve that is very wide and has pronounced wings.

When the diameter of the aperture was increased to 8 mm (with a corresponding increase in pulse energy so that approximately the same light intensities were still used) the signals (Fig. 1.11) showed the same type of basic behavior as was seen in the Fig. 1.10, i.e., a narrow peak on top of broad wings, although now the latter were somewhat more pronounced. Also this time, the signals (for a given intensity) were only marginally larger than with the previously used aperture (approximately 40–50%). A significant broadening of the peaks is clearly visible, in particular at the lower parts of the high-intensity peaks, i.e., in those curves taken at 175 and 550 kW/cm² .

From these experimental curves it was possible to gain some insight into the ture physical origin of the peaks and hence also to offer a possible explanation of their somewhat strange and unexpected behavior. The assumption was made that the "broadened wings" in Fig. 1.11 (the 175 and 550 kW/cm² curves) originated from atoms in the true interaction region, i.e., the region defined by

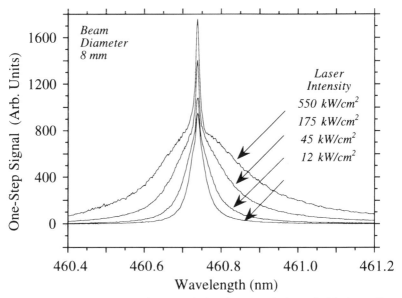

Figure 1.11. A set of one-step scans between 460.2 and 461.2 nm in 3 ppm Sr. The beam diameter was 8 mm. The intensities of the light in the four scans were (from bottom to top) 12, 45, 175, and 550 kW/cm². The y-axis scale is identical to those in Figs. 1.9 and 1.10.

the laser beam (in this case a volume with a circular area with a diameter of 8 mm and a length of 5 cm), whereas the sharp, narrowband peak on top came from atoms positioned outside the true interaction region and excited by scattered, resonant light. This assumption is supported by the fact that the broad, shoulder-like signals increase in width with increasing intensity of the light, in agreement with the prediction of the theory of optical saturation.

If this assumption is correct, then it should be possible to model the curves as comprising two different and independent parts. One such attempt is shown in Fig.1.13, which shows that an experimental one-step LEI curve can successfully be decomposed into two different Lorentzian curves. The experimental curve (curve *a*) has been fitted to a sum of two different Lorentzian curves, shown separately as curves *b* and *c*. Curve *d* is the difference between the experimental curve and the fit. As can clearly be seen from the figure, curve *c* models the narrowband peak very nicely while curve *b* describes the broad shoulders accurately. Hence, this supports the assumption that the signal is indeed composed of two different parts.

Furthermore, when comparing the three scans for the highest intensity, i.e., the 550 kW/cm² curves in the Figs 1.9, 1.10, and 1.11, one can clearly see that the saturation-broadened signal, i.e., the contribution from the atoms in the

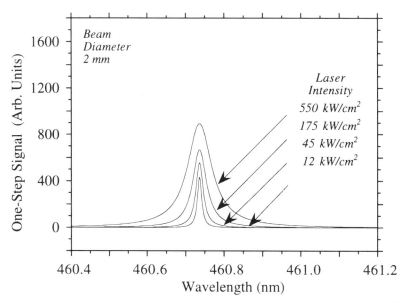

Figure 1.12. A set of theoretically calculated one-step scans fitted to the peak values of the experimental data in Fig. 1.9. The theoretical function used is based upon the steady-state solution of the rate equations for a two-level atom exposed to both collisional ionization and photoionization.

true interaction region, is also decreasing in magnitude when smaller apertures are used. This is again in agreement with the assumption about the origin of the two different parts of the signal (fewer atoms contribute to this wide signal). The signal from the atoms positioned outside of the true interaction region, however, is of comparable size in the three scans. Hence, it is found that the relative part of the detected signal that originates from atoms in the true interaction region is decreasing significantly with the decrease in the interaction region. In this case, already when the largest aperture was used (i.e., that with a diameter of 8 mm), the contribution from the atoms outside the interaction region was of equal magnitude to that from the true interaction region, as can clearly be seen from the 550 kW/cm² curve in Fig. 1.11. For the 4 mm diameter aperture case, the signal from atoms outside the interaction region constitutes more than 70% of the signal, as can be seen from the 550 kW/cm² curve in Fig. 1.10. This fraction is even higher for the experiment when a 2 mm aperture is used (see the 550 kW/cm² curve in Fig. 1.9). Hence, when high laser light intensities were used in conjunction with small apertures *the signal from the true interaction region was almost insignificant compared to that from atoms outside the true interaction region.*

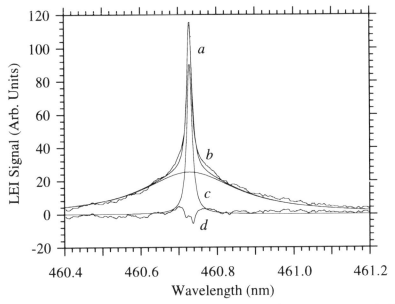

Figure 1.13. A one-step LEI scan between 460.4 and 461.2 nm in 3 ppm Sr (curve *a*). This experimental curve has been fitted to the sum of two Lorentzian curves shown separately (*b* and *c*). Curve *d* is the difference between the experimental curve and the fit.

1.8.2.2.   Two-Step Excitation

In order to verify the assumption of existence of scattered, resonant light outside of the true interaction region in the flame, an experiment was performed in which the atoms were excited by two laser beams, i.e., two-step excitation. In this experiment the laser beams were *temporally* but *not spatially* overlapping in the flame. The first excitation step was directed through the flame at a height of 20 mm above the burner head, while the second laser beam was positioned at a height of 24 mm. The diameters of the beams in this experiment were 2 mm. Hence, there was an unilluminated distance of 2 mm between the two beams in the flame. The second-step laser light had a wavelength of 554.333 nm and excited Sr atoms from the $5s5p\ ^1P_1$ state further to a higher lying $5s6d\ ^1D_2$ state.

Figure 1.14 displays the signal obtained from a solution of 3 ppm of Sr when only the first laser beam was on as curve *a*. In this case a signal similar to those presented in the previous subsection was obtained, i.e., a signal composed of one narrow and one broad peak. When the second laser was turned on, the signal increased considerably (approximately 1 order of magnitude) dispite the

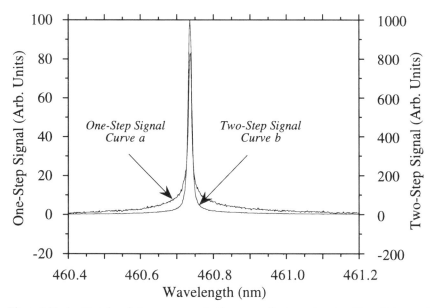

Figure 1.14. A comparison between a concentration-normalized one-step signal from Sr (curve *a*) and a concentration-normalized two-step signal (curve *b*) when the two beams are not overlapping in the flame (the first-step laser is scanned between 460.4 and 461.2 nm). Note the difference in scales. The intensity of the first step was $400\,kW/cm^2$ (corresponding to 380 µJ), and the intensity of the second step was $23\,kW/cm^2$. The beam diameter was 2 mm, and the distance between the two beams was 2 mm.

nonoverlapping geometry, as can be seen in curve *b*. Note that the two curves have different absolute scales in the figure in order to show more clearly the differences in shapes. It can clearly be seen that only the narrowband part of the one-step signal was enhanced by the presence of the second step. There is no longer any evidence of broadened wings in the two-step signal (since those are not increased by the second laser and thus have become insignificant due to the rescaling). This provides further evidence that the narrowband peak originates from excited, unsaturated atoms outside of the true interaction region.

Similar results have been obtained in a study of both Sr and Na by Seltzer and Green (52). They found that the total ionization signal produced by two parallel, nonoverlapping laser beams (separated by a distance of 2–3 cm and for both vertically and horizontally displaced beams) was significantly larger than the sum of the ionization signals recorded separately using a 460.7 nm first-step transition followed by a 355 nm photoionizing second-step excitation in Sr and a 589 nm first-step excitation followed by the same photoionization second-step excitation in Na, respectively.

1.8.2.3. Consideration of the Physical Processes Responsible for the "False" Signal in LEI

As already mentioned, the exact mechanism for the existence of scattered, resonant light in other parts of the flame is not known. However, at least three possible mechanisms might be responsible for the light that excites the atoms positioned in other areas of the flame than the true interaction region:

   a. The light might be diffracted from an aperture positioned outside the flame.

   b. The light might originate from fluorescence emitted from the excited analyte atoms in the true interaction region.

   c. The light might originate from laser light scattering from particles or flame molecules in the heavily illuminated interaction region in the flame.

From the previously described experiments of Axner and Sjöström (53) the conclusion can be drawn that the main cause of the scattered, resonant light cannot be diffraction from the aperture (point a). The reason for this is that *the signal from the scattered light did not increase significantly when the laser light intensity was increased substantially.* (If laser light diffraction from the aperture was to be the origin of the scattered, resonant light, the narrow signal, originating from unsaturated transitions in atoms, would increase linearly with the laser light intensity.)

In order to determine whether the scattered, resonant light outside the true interaction region originated mainly from emitted fluorescence from Sr atoms in the interaction region (point b, above), a one-step experiment was performed in which the number density of Sr atoms in the flame was increased successively. Although the shape of a log-log calibration curve of growth can vary significantly as a function of the Sr concentration, laser bandwidth, self-absorption, and postfilter effects, it is believed that in the range of concentrations studied (1–10 ppm) a slope close to 2 should have been obtained if the scattered, resonant light was due to emitted fluorescence from Sr atoms. (This is because both the fluorescence and the absorption process leading to ionization increase linearly with the number density of Sr while self-absorption still is small, therefore resulting in an overall quadratic dependence upon concentration.) It was found that the LEI signal was very close to *linear* with concentration in the range between 1 and 10 ppm. Therefore, there is no direct evidence that fluorescent light emitted from the excited analyte atoms in the true interaction region is mainly responsible for the narrowband LEI signal.

Another experiment was performed by Axner and Sjöström (53) in which the narrowband signal was investigated as a function of different flame compositions. This experiment was carried out to clarify the influence of scattered, resonant light from particles and molecules in the flame [point c, above]. It was found that the narrowband part of the LEI signal increased significantly (approximately by a factor of 3) with increasing richness of the flame. This indicates that the light-scattering process might originate from molecules or particles in the flame that increase in number density when the flame is operating fuel-rich, even though this mechanism also requires a signal from the atoms outside the true interaction region that increases linearly with the laser light intensity.

In order to investigate the origin of the anomalous signals in their LEI experiment, Seltzer and Green (52) inserted a ceramic plate into the flame to act as a light barrier, to create two regions of the flame that were separately illuminated by the two different laser beams. They found that the total signal recorded for simultaneous illumination by both laser beams was again larger than the sum of the two individual signals. This time, however, the total signal was smaller than what was observed without the ceramic plate in the flame. These results suggest that emitted fluorescent light or scattered, resonant light from particles and molecules in the flame could be partly but not solely responsible for the anomalous signals resulting from excitation by nonoverlapping beams.

Further experiments in order to investigate the origin of the anomalous signals in LEI have been performed by Turk (55). He obtained some degree of spatial selectivity to the LEI detection process in the flame by controlling the voltage applied to the detection electrodes. He could thereby detect with some selectivity the LEI signal originating from atoms only outside the true interaction region and compare the shape of the LEI signals with that from the entire flame. His results support the assumption discussed above that the narrow peak originates from regions in the flame outside the true interaction region and therefore only exposed to low, nonsaturating light intensities, while the broad "pedestal" comes from strongly saturated atoms inside the interaction volume.

1.8.2.4. Conclusion

When high laser intensities are used the atoms present in the true interaction region are strongly saturated. If then only a very small fraction of all the accessible resonant light finds its way to other parts of the flame, a substantial number of atoms in that region might be excited by this light. Many of the atoms outside of the true interaction region will therefore also contribute to the detected LEI signal since the detection process is not especially localized in the flame (all atoms in regions of the flame where the charge detection efficiency is noticeable will contribute to the signal). Furthermore, these atoms will not become saturated with the kind of intensities used in this experiment. Hence, their contribution to the signal will be a narrow peak (not affected by saturation broadening) that will add to the signal from the atoms in the true interaction region.

The exact mechanism for this process has not yet been determined. However, we can still conclude that since the total volume of the flame in which there is a considerable charge detection efficiency is often significantly larger than the true interaction region, the narrowband contribution from the atoms outside the interaction region might well be significant and even dominate the signal from the true interaction region, especially when smaller interaction volumes are used in conjunction with high laser light intensities.

1.9. ANOMALOUS CONTRIBUTIONS FROM TWO-PHOTON TRANSITIONS AND DYNAMIC STARK EFFECTS

1.9.1.   Introduction

Other types of anomalous lineshapes and signal strengths in two-color LEI spectrometry in flames have also been encountered (15, 56–58). It has been found, for example, that both lineshapes and signal strengths from atoms in flames can be significantly affected by the influence of various types of coherent contributions in the excitation process. These types of effects consist of, for example, two-photon excitations (i.e, when the atoms simultaneously absorb two photons, one from each laser) (15, 56–58) and dynamic Strak effects (i.e., interactions between the laser light fields and the atoms that cause rapid Rabi oscillations between atomic energy levels, giving rise to broadening, splitting, and/or shifts of transitions) (15, 59–61). In general, these types of effects are most pronounced when excitations with narrowband light of high intensities take place in weakly collisional media (62). However, it has been found that they can also have profound effects on excitations of atoms in strongly collisional flames when rather moderate laser intensities are used, obtainable in unfocused pulsed laser beams from modern dye laser systems (15, 59, 63).

1.9.2.   A Short Description of Two-Photon vs. Two-Step Excitation and Dynamic Stark Effects of Atoms in Flames

1.9.2.1.   Two-Photon vs. Two-Step Excitation

In general, it is possible to obtain two kinds of excitation of atoms in two-color LEI in flames.

a. The type when the atoms are first excited to the intermediate level by absorption of one photon and then, a short time later, excited to the uppermost level by absorption of a second photon—generally termed *two-step excitation*

b. The type when the atoms absorb two photons simultaneously and consequently are excited directly from the ground state to the uppermost state—termed *two-photon excitation*

The characteristic feature of the two-step signal is that it always occurs at the resonance wavelength of a scanned laser (even though the other laser might be slightly detuned from the resonance wavelength). The two-photon signal, on the other hand, appears at a wavelength for which the sum of the energy of the two photons corresponds to the energy difference between the uppermost and the lowermost levels. Hence, the position of this two-photon peak, when one laser is scanned, depends on the detuning of the other laser.

This implies that if one laser is slightly detuned from resonance and the other is scanned across the transition, *two* peaks appear—one corresponding to a two-step excitation at the resonant wavelength of the scanned laser, and one corresponding to a two-photon transition slightly shifted from the resonant wavelength. When both lasers are at resonance, on the other hand, both types of excitations occur simultaneously. Therefore, in general, when both lasers are at resonance, the signal is composed of both a two-step and a two-photon contribution—a fact that has often been neglected. The latter might add both positive as well as negative contributions to the conventional two-step signal (as will be discussed in more detail below).

1.9.2.2. Dynamic Stark Effects

Other phenomena that might influence the shape of LEI signals are dynamic Stark effects. Effects of this type occur as a consequence of the laser light field affecting and disturbing the actual atomic energy levels. These effects show up as (i) broadening, (ii) splitting, and/or (iii) shifting of the transitions and take place at sufficiently strong laser intensities (excitation and deexcitation rates).

In general, dynamic Stark broadening/splitting is obtained when a strong laser field induces a high excitation rate between two levels (e.g., the ground state and a excited state) so as to broaden and/or split the levels. Basically, whenever the exciting laser is broader (frequency-wise) than the transition, the levels become broadened, whereas the levels become split whenever the laser bandwidth is narrower than the transition. This dynamic Stark broadening can, for example, be detected by scanning a weak probe laser across a transition which emanates from the excited level. In this case, and as long as the level becomes broadened (and not split), this phenomenon is often referred to as power broadening.

When the exciting laser light is narrower than the transition (counting all widths, i.e., induced, collisional, and natural), each level will split into two levels, separated by the Rabi flop frequency of the system. The Rabi flop frequency is defined here as the the rapid and periodic oscillation of the atomic population in the two-level atomic system in the presence of a strong, narrowband laser field.

Whether a laser field is narrow- or broadband is determined by its relative width as compared to other widths in the system (the total width of the transition). Thus, if a certain laser field causes power broadening at a given laser intensity (the laser is slightly broader than the transition), it can turn out that, by increasing the laser intensity, the transition broadens so much that the width of the transition becomes larger than the bandwidth of the laser. Then, Strak splitting of the upper levels starts to occur. Hence, Stark broadening/power broadening can successively transform into Stark splitting as the intensity of the laser light is increased.

Furthermore, the width of a power-broadened peak is proportional to the inverse of the lifetime of the state. Consequently, it is proportional to the intensity of the light. The Stark splitting, on the other hand, is proportional to the square root of the intensity (59, 62).

Both the power broadening and the dynamic Stark splitting are, however, in many cases difficult to observe directly in flames. One reason is that the broadening and splitting are normally smaller than (or in the same order of magnitude as) other widths in the flame. Other reasons that prevent their direct observation are that spatial and temporal inhomogeneities smear out their structure.

Despite the small amount of broadening and splitting, these types of effects can affect the LEI signals significantly, in particular their shapes. The reason for this is that the detectable effects, which might give rise to very broad or split peaks (up to several hundred picometers, or gigahertz) from ordinary transitions in flames, are *indirect* consequences of the dynamic Stark splitting phenomenon rather than being the Rabi flop splittings themselves.

Features significantly broader than the actual power broadening or Stark splitting can occur, for example, when the strong laser (giving rise to the splitting of the intermediate level) is scanned and the weak narrowband, second-step probe is fixed at its resonance wavelength. The situation has much in common with that of the relation between power broadening and saturation broadening, which basically is the following. As already mentioned, power broadening takes place when a narrowband (weak) laser is scanned across a transition that emanates from a level strongly coupled to another level (e.g., the ground state) by intense laser light. The measured width of the peak is then directly related to the lifetime of the intermediate state (or the laser deexcitation rate). Saturation broadening, on the other hand, is observed when one scans a strong laser field (that strongly saturates the transition at resonance) across the absorption line profile. As a result, the transition is not only saturated at the line center but also (partly) saturated at the wings of the transition so that an appreciable signal is obtained even when the laser is slightly detuned from resonance. Saturation-broadened peaks are therefore significantly broader than power-broadened ones.

For the same reason that saturation-broadened peaks are much wider than power-broadened peaks, it is possible to get widely split curves in LEI (with significantly wider splits than that of the Rabi frequency) by scanning the strong first-step laser while holding the second one fixed at its resonance wavelength. This phenomenon is described in more detail in Section 1.9.3.2, below.

Finally, the dynamic Stark shift can in most cases be neglected as far as LEI in flames is concerned since it is an effect that gives rise to shifts of levels (and hence transitions) approximately 1 order of magnitude smaller than the size of the dynamic Stark splittings. Furthermore, shifts will only occur for two-photon transitions; the ordinary two-step transitions will not experience any shift.

1.9.3. Experimental Evidence of Two-Photon vs. Two-Step Excitation and Dynamic Stark Effects for Atoms in Flames

Two-step and two-photon excitation processes of atoms in flames have been documented qualitatively in the literature (15, 64–66). Turk et al. (65) scanned in turn the two dye laser light outputs in wavelengths around the resonance wavelengths. As a result, they were able to detect both the ordinary two-step signals (together with the one-step signals when one of the lasers was scanned far from resonance) and a two-photon signal (obtained when the detuning of one of the dye lasers is exactly the same but of reversed sign as the detuning of the other dye laser). Similar effects were found by Havrilla and Carter (66), who made three-dimensional plots of Cu, and Goldsmith (64), who studied hydrogen. Theoretically, excitation phenomena (including lineshape profiles) in the presence of various collisions have been studied by a number of authors using the density-matrix formalism (62,67–75). However, until recently, there were only two publications in which the excitation processes in atoms in such strongly collisionally dominated media as flames have been described, Lau (75) and Lin et al. (74). The paper by Lau treats only one specific experimental case and is therefore not particularly illustrative. Lin et al. treat a one-step excitation followed by either collisional ionization or nonresonant photoionization and is therefore also slightly beyond our present purview.

The first paper pointing out the importance of two-photon vs. two-step excitation as well as dynamic Stark effects in stepwise LEI in flames was published by Axner and Sjöström in 1992 (15). The authors made a thorough experimental as well as a first theoretical investigation based upon the density-matrix formalism of these phenomena. Additional theoretical studies have thereafter been performed by Axner and Ljungberg (59, 60), Ljungberg et al. (61), and Boudreau et al. (63). Some of the experimental results of Axner and Sjöström (15) will be presented here for subsequent comparison with theoretical predictions based on the density-matrix approach.

The experiments described in the following subsections are all done on the $5s^2 \, ^1S_0 - 5s5p \, ^1P_1 - 5s6d \, ^1D_2$ transitions in Sr atoms in an air/acetylene flame. The terminology adopted below uses the expression *resonance wavelengths* (λ_1^0 and λ_2^0) for the wavelengths needed for excitation of an atom (at rest) from one level to another level (from level 1 to level 2, and from level 2 to level 3, respectively) in the medium used at the center of the transition. Consequently, $\lambda_1^0 = 460.733$ nm and $\lambda_2^0 = 554.336$ nm in these experiments. In the following subsections the behavior of LEI is studied when one of the two lasers is scanned under a variety of conditions.

1.9.3.1. Scan of the Second-Step Laser for a Fixed First-Step Laser

The first four experimental curves are displayed in Fig. 1.15 and were obtained when the wavelength of the first step was detuned from the first-step resonance wavelength (0, -50, -100, and -200 pm, corresponding to curves a, b, c, and d, respectively) and the wavelength of the second step was scanned around the second-step resonance wavelength for each of the detunings. As can be seen from the scans, two peaks appear when the first-step laser is detuned from its resonance wavelength and the second-step laser is scanned. These two peaks correspond to a two-step and a two-photon transition, respectively.

The leftmost peak corresponds to a two-step excitation that populates the intermediate level (on the wing of the first transition for a detuned first-step excitation) before the absorption of a second photon. This peak always appears at the resonance wavelength, λ_2^0 (554.336 nm). The other peaks represent two-photon excitations in which the atom simultaneously absorbs two photons. Consequently, these peaks appear at wavelengths at which the sum of the energy of the two photons corresponds exactly to the total energy difference between the initial and final state. Hence, in the latter case, the atoms are excited directly from the ground state to the upper level (69, 71). The strengths of both the two-step and two-photon peaks decrease as a function of detuning (approximately inversely proportional to the square of the detuning from resonance) (59, 76), as can be seen by a comparison of curves b, c, and d in Fig. 1.15. This implies, furthermore, that the measured signal when both lasers

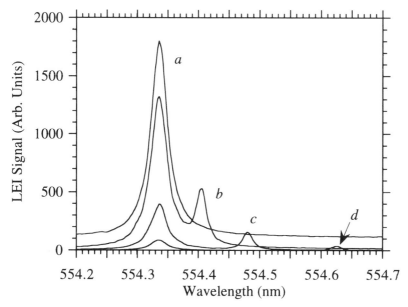

Figure 1.15. Experimental stepwise LEI signals obtained from scans of the second laser for various detunings of the first laser. The four curves represent cases with detunings of (a) 0 pm, (b) −50 pm, (c) −100 pm, and (d) −200 pm. The pulse energies were 50 µJ for both laser beams (20 kW/cm²).

are at resonance (curve a) has contributions from both a two-step and a two-photon signal. Although both the two-step and two-photon signals are of considerable size when óne laser is slightly detuned, the two-photon contribution to the total signal at resonance can generally be both positive and negative (77).

1.9.3.2. Scan of the First-Step Laser for a Fixed Second-Step Laser

When a stepwise LEI experiment is performed in which a strong first-step laser is scanned across the resonance wavelength, λ_1^0, and the second-step laser is held fixed at a certain detuning, a multiple-peak structure can appear. This is illustrated by curves b, c, and d in Fig. 1.16a,b (curves a, b, c, and d correspond to detunings of the second-step laser of 0, −72, −145, and −290 pm, respectively; Fig. 1.16b was taken at significantly higher intensities than Fig. 1.16a).

This time the curves seem to be composed of *three* different peaks. In curve c in Fig. 1.16a these three peaks have been labeled α, β, and γ. The peak denoted γ, whose position is shifted from the nominal transitions wavelength,

λ_1^0 (in this case 460.733 nm), represents the two-photon transition. The peak at the nominal transition wavelength seems, however, to be composed of two separate peaks. One then finds, in accordance with the discussion in Section 1.8 above, that the broad peak β originates from two-step excitation of saturated atoms in the true interaction region whereas the narrow peak α, on top of the broad peak β, originates from two-step excitation of atoms in other parts of the flame, exposed to weak, nonsaturating scattered resonance light.

When a stepwise LEI experiment is performed in which the second-step laser is held fixed at resonance, λ_2^0, while the first-step laser is scanned across its resonance wavelength, λ_1^0, a wide, double-peak structure might appear, as demonstrated by the two curves a. The reason for the appearance of this double-peak structure when the second step is at resonance is the following.

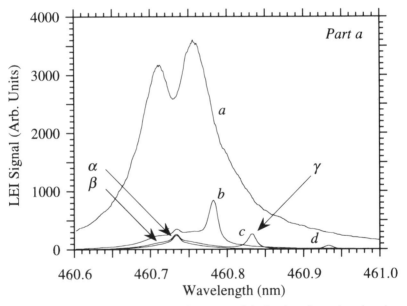

Figure 1.16. Stepwise LEI signals obtained from scans of the first laser for various detunings of the second laser. The four curves represent cases with detunings of (a) 0 pm, (b) −72 pm, (c) −145 pm, and (d) −290 pm. The pulse energies were 60 μJ for the first-step laser and 40 μJ for the second-step laser (24 and 16 kW/cm², respectively) in part a and 3 mJ for the first-step laser and 400 μJ for the second-step laser (1200 and 160 kW/cm², respectively) in part b. Curves b, c, and d are all composed of three different peaks. In curve c of part a the three peaks are labeled α, β, and γ for clarity. Peak γ represents the two-photon transition. The broad peak β originates from saturated atoms in the interaction region, while the narrow peak α, on top of the broad peak β, originates from atoms in other parts of the flame that are exposed to the weak, nonsaturating light.

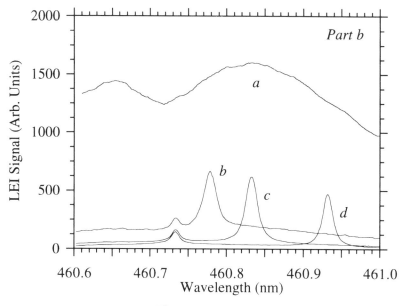

Figure 1.16. (*Contd.*)

As the first-step laser is tuned closer and closer to the resonance wavelength, the populations of both the intermediate and the upper level increase (and hence so does the two-step LEI signal). This will continue until a situation occurs when the power broadening starts to broaden the first transition so that the transition probability (and atomic population) per unit frequency interval for the intermediate level is diminished. If the first-step laser is narrower than the induced width of the transition and strong enough, the Rabi flop frequency might become so large that a certain dynamic Stark splitting of the levels occurs. In this case the intermediate level might be (partly) split into two states, more and more separated the closer the first step laser is scanned toward the resonance wavelength. If this is not the case, a broadening of the state will result (which is larger the closer the first-step laser is scanned toward the resonance wavelength).

When the intermediate level is being significantly broadened (or split) the second step laser, positioned at its resonance wavelength λ_2^0, will act as a narrowband probe of the population only at the center of the intermediate level (frequency-wise). A broadening or a splitting of the intermediate level *diminishes the population of the level in the energy interval where it is probed by the second laser* (at the unperturbed energy of the state). Thus, although the total number of excited atoms in the intermediate state is increasing monotonically as the first-step laser is tuned toward resonance, the upper-state population is not since the atoms at the intermediate state are distributed across a wider frequency range. Consequently, fewer excited atoms are accessible for further excitation by the second-step laser when both lasers are on or very close to "resonance" than when the strong first-step laser is slightly detuned. A dip in the LEI signal will therefore appear when the first-step laser is scanned across the resonance. This behavior is demonstrated experimentally by the two curves *a* in Fig. 1.16a,b and is in agreement with theoretical predictions (see Section 1.10, below). It is important to point out that this dip can occúr as a consequence of a laser-induced broadening as well as a split of the intermediate level. Hence, it is not necessary to induce a split in the intermediate level in order to get a dip in the LEI signal—a sufficiently large laser-induced broadening suffices.

The width of this dip is, naturally, dependent on the laser light intensities (i.e., the Rabi flop frequencies). This can be seen from a comparison between the two curves *a* in Fig. 1.16a,b and is more clearly visualized in Fig. 1.17. The four curves in Fig. 1.17 correspond to different first-step pulse energies, with the second-step laser held at resonance with constant pulse energy. The four

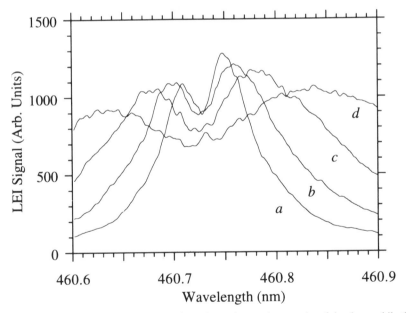

Figure 1.17. A set of scans of the first-step laser for various pulse energies of that laser while the second laser was tuned to resonance. The four curves correspond to pulse energies of the first laser of (*a*) 50 μJ, (*b*) 200 μJ, (*c*) 750 μJ, and (*d*) 2.7 mJ (corresponding to 20, 80, 300, and 1100 kW/cm^2, respectively). The pulse energy of the second laser was 50 μJ (20 kW/cm^2).

scans show that the width of the double-peak resonance increases significantly as a function of laser pulse energy of the first step. The laser pulse energy was increased approximately 50 times (from 50 μJ to 2.7 mJ) between curves a and d. This is in agreement with the assumption that the saturation-broadening phenomenon is responsible for this structure.

The width of this dip is, in fact, given roughly by the detuning needed to diminish the excitation and deexcitation rates to the extent that the broadening/split of the intermediate level decreases [and not the Rabi flop frequencies themselves; see Eq. (121) below]. This width can therefore be considerably larger than the actual Rabi-flop-induced Stark splitting of one particular level in the system. The width of these double-peak structures as well as the distance between two peaks (approximately 180 pm for the case corresponding to curve a in Fig. 1.16b) may clearly be much larger than the Rabi flop frequencies of the transitions.

Another illustration of the rather complicated width dependence of the system is given by a comparison between the nondetuned situations in Figs 1.15 (curve a) and 1.17 (curve a) (these two curves are measured under similar conditions with the only exception being that different lasers are scanned). It can be seen that both a dip and a large width can be seen when the first-step laser is scanned (curve a in Fig. 1.17) whereas no dip at all can be discerned when the second-step laser is scanned (curve a in Fig. 1.15); the Stark broadening/split is so small that it is obscured by other broadening effects.

1.9.3.3. *Saturation Curves That Bend Downward*

It is also worth noticing that the signal at resonance (no detuning) in fact decreases as a function of laser light intensity of the first-step laser due to this effect. This can be seen by a comparison of the signals at the resonance wavelengths in the four curves in Fig. 1.17. This exemplifies the situation in which the dynamic Stark effects give rise to saturation curves that, in fact, bend downward, i.e., have a negative derivative.

1.10. LASER-ENHANCED IONIZATION IN THE DENSITY-MATRIX FORMALISM

1.10.1. Introduction

In general, the rate-equation formalism is adequate for a satisfactory description of most of the phenomena associate with interactions between atoms and light in flames. The reason for this is that the high collision rates in flames destroy much of the coherence of the system. It has, for example, been previously shown that in media with high collision rates the density-matrix formalism approaches the rate-equation formalism for one-step excitation. Concerning excitations of atomic and molecular species in flames by laser light, Daily (12) concluded: "The range of application of the conventional rate-equations is examined and, for flames, shown to be valid for sufficiently slow laser pulse rise times under single-mode excitation and for certain special cases of multimode excitations." However, as indicated above (though not generally known), the foregoing statement is primarily valid only for one-step excitation and consequently not for step-wise excitation LEI.

The simplification of the density-matrix formalism to the the rate-equation formasism is a particularly good approximation for one-step excitation (12), for excitation in general with light significantly broader than the atomic transitions (78), or when the light-induced rates are smaller than the collisional rates (78). In the case of two-step excitation LEI when the laser light is not necessarily significantly broader than the transitions or when the laser-induced rates are comparable or larger than the collisional rates, however, not all physical properties of the combined atomic and light system will be described correctly by the rate-equation formalism (15). Effects that are not properly taken care of by the rate-equation formalism are those primarily caused by intense light fields, such as two-photon excitation and dynamic Stark broadening, splitting, and shifting. However, due to the simplicity of the rate-equation formalism, extensive derivations and calculations of the LEI signal strengths using the rate-equation formalism have frequently occurred in the literature (4, 9, 18–20).

In order to include two-photon excitations and dynamic Stark effects in the description of LEI, a theory based on the density-matrix formalism will be briefly presented. The intention of the density-matrix description of LEI flames is to model the processes involved (processes not modeled by the rate-equation formalism) in the simplest possible way. Therefore, certain simplifications of the description of the light–matter interactions, the nature of collisions in the flame, and the characteristics of laser light will be made. In a semiclassical approach, we will treat this as a simple three-level system (in which the levels all have a degeneracy factor of 1), in the steady state limit, with phenomenologically introduced collision rates (elastic and inelastic), and finite (but not necessarily a very broad or a very narrow) laser bandwidths (included as phase fluctuations).

The latter assumption implies that the laser has a Lorentzian shape. Non-Lorentzian lineshapes have been studied by, for example, Zoller and Lambropoulos (79) and Dixit et al. (80), suggesting a more realistic description of the laser light but at the expense of a higher mathematical complexity.

The assumption that the impact theory of line broadening is valid (i.e., that all collisions occur instantaneously) also leads to a Lorentzian shape of the atomic absorption lines. This is done, although it has been found that atomic lineshapes more correctly should be represented as non-Lorentzian (6, 81).

In addition, we have assumed that the atoms can be described by a simple three-level system without any degeneracy of the levels. Normally, in the full density-matrix approach, the number of levels that has to be considered is equal to the total number of sublevels included in the transitions, n (counting all sublevels with different m_F values in each electronic state). This gives rise to a system of density-matrix equations consisting of n^2 equations (82). Since the actual number of sublevels included often is large [e.g. $n = 24$ for the 589.0 nm one-step transition in Na, $3p\ ^2P_{3/2}\ (F = 3, 2, 1, 0)$–$3s\ ^2S_{1/2}\ (F = 2, 1)$; and $n = 9$ for the two-step transition $5s^2\ ^1S_0\ (F = 0) - 5s5p\ ^1P_1\ (F = 1)$–$5s6d\ ^1D_2\ (F = 2)$ in Sr], the number of equations needed for a complete treatment of such a system can be enormous (576 equations for the Na case, and 81 equations for the Sr case).

For an analytical treatment of the two-step excitation system, however, the n-level atomic system can be simplified to a three-level system without the loss of significant amounts of information under certain conditions. This can be done in basically two different ways—either by treating the atomic system as having three states with no degeneracy at all, as in the treatment given below, or by assuming that all substates of each of the three electronic states will be more or less equally populated and thus treat the system as if each electronic state is composed of a number of identical substates (and thus with the same transition probability between all sublevels of the electronic states), as was done by Axner and Ljungberg (60). In addition, further refinements of this concept (in which the different substates are treated differently depending on their possible involvement in various types of excitation processes) have recently been published (60, 61).

The simplification of a real atomic system to a nondegenerate three-level system is a very common one and therefore need not be discussed in more detail here. The simplification to a system in which each electronic state is composed of a number of identical substates (all equally populated) in conjunction with the use of the density-matrix formalism is not as common, although there are several justifications for its use in the description of LEI in flames (e.g., both the frequency width of the laser light and the collisions are significantly broader than any hyperfine splitting and the collision rates are larger than the inverse of the pulse durations). Such a simplification implies also that the Einstein A factors for spontaneous emission can be used for calculating excitation rates (rather than the dipole matrix elements for specific transitions that otherwise have to be used and are not always readily available). This simplification, whose limitations and implications are further discussed by Axner and Ljungberg (60) and Ljungberg et al. (61), implies, among other things, that the simplified treatment with fewer than n^2 equations cannot describe the coherence between states of the same m_F value but different F value that under single-mode excitation gives rise to quantum beats (82). The neglect of quantum beats is, however, not such a servere sacrifice since the use of a broadband laser, as is generally the case for pulsed excitations, often washes out much of that coherence. Another aspect of this simplification is that dynamic Stark effects (broadening, splitting, and shifting) might not be described completely correctly (quantitatively) since the total effect on the system, for a full description with n^2 equations, is given as a combination (a weighted average) of a number of individual Stark effects.

In addition, an aspect not considered here (and normally neither in the rate-equation formalism) is the effect of Doppler broadening.

Moreover, it is important to point out that the coherent effects that appear in multistep excitation (such as two-photon excitation) do not require that the laser light fields be single mode or coherent over long periods of time. On the contrary, the more amplitude fluctuations there are in the fields, the stronger the two-photon signal will be (78). This is in contrast to the restrictions which must be assessed to the laser light field in order to detect transient effects such as photon echoes. However, due to present meager knowledge of the exact form of laser pulses from dye lasers, the laser fields will be treated as having a fluctuating phase but a constant amplitude.

Although the theoretical description of each of the aforementioned fields can be improved, the full description of the whole system of laser light, atoms, and collisions tends, in such a case, to be very complicated and the formalism and calculations quite cumbersome. To avoid this, we have chosen to present here the use of the density-matrix formalism for a description of LEI in flames in the most simplified way as described above, i.e., to describe the atoms as a nondegenerate three-level system exposed to instantaneously occurring collisions, exposed to single-mode lasers whose frequency widths are given solely by phase fluctuations. A more thorough description of the use of the density-matrix formalism for excitations of three-level atoms with degenerate states is given in Axner and Ljungberg (60) and Ljungberg et al. (61).[5]

1.10.2. Formulation of the Density-Matrix System of Equations

The theoretical assumptions that lead to the density-matrix formulation of light and matter interactions have been dealt with previously in the literature,

---

[5] Note that the density-matrix description of excitations of three-level atoms given in Axner and Sjöström (15) do not correctly include the aspect of degeneracy of the levels. The equations derived in that work are only valid for the case with all degeneracy factors equal to unity. The reader should therefore consult the more recent works of Axner and Ljungberg (60), Ljungberg et al. (61), and Boudreau et al. (63) for the treatment of degenerate levels.

both in textbooks (13, 14, 83) and in a multitude of other publications (12, 15, 59, 67, 68, 71, 72, 84–86). Therefore, only the major steps leading to the derivation of the density-matrix equations will be given here. The presentation follows, in brief, that of Sections 6.1 and 6.2 in Axner and Ljungberg (59).

In short, the density matrix-equations arise from the evolution of the wave function (describing the state of the atom as a linear combination of the probability amplitudes of the wave function to be in one of several atomic states) according to the Schrödinger equation in which the interaction Hamiltonian describes ordinary dipole-allowed transitions induced by the laser field. The density-matrix elements themselves come from various (multiplicative) combinations of the probability amplitudes of the wave function for the system to be in various atomic states. Hence, the diagonal terms (ρ_{ii}) of the density matrix represent the population density of the various levels, while the off-diagonal elements $(\rho_{ij}, i \neq j)$ are the induced dipole moments between the two states.

The density-matrix system of equations for an atomic system consisting of three nondegenerate levels (where level 1 is the ground state and level 3 the highest lying excited state) can generally be written as (15, 59)

$$\frac{d\rho_{11}}{dt} = \omega_R^{12} \operatorname{Im}(\rho_{21}) + k_{21}\rho_{22} + k_{31}\rho_{33} \tag{114}$$

$$\frac{d\rho_{22}}{dt} = -\omega_R^{12} \operatorname{Im}(\rho_{21}) + \omega_R^{23} \operatorname{Im}(\rho_{32}) - k_{21}\rho_{22} + k_{32}\rho_{33} \tag{115}$$

$$\frac{d\rho_{33}}{dt} = -\omega_R^{23} \operatorname{Im}(\rho_{32}) - (k_{31} + k_{32})\rho_{33} \tag{116}$$

$$\frac{d\rho_{21}}{dt} = -i\Delta_{12}\rho_{21} + i\frac{\omega_R^{12}}{2}(\rho_{22} - \rho_{11}) - i\frac{\omega_R^{23}}{2}\rho_{31} \tag{117}$$

$$\frac{d\rho_{32}}{dt} = -i\Delta_{23}\rho_{32} + i\frac{\omega_R^{23}}{2}(\rho_{33} - \rho_{22}) + i\frac{\omega_R^{12}}{2}\rho_{31} \tag{118}$$

$$\frac{d\rho_{31}}{dt} = -i\Delta_{13}\rho_{31} + i\frac{\omega_R^{12}}{2}\rho_{32} - i\frac{\omega_R^{23}}{2}\rho_{21} \tag{119}$$

and

$$\rho_{\text{atom}} = \rho_{11} + \rho_{22} + \rho_{33} \tag{120}$$

where the remaining equations are given by the relation $\rho_{ij} = \rho_{ji}^*$.

Here, the expressions $-\omega_R^{12} \operatorname{Im}(\rho_{21})$ and $-\omega_R^{23} \operatorname{Im}(\rho_{32})$ are the number rates of absorbed photons for the first and second transition, respectively, while ρ_{atom} represents the total number density of neutral atoms in the interaction region.

Furthermore, the ω_R^{12} and ω_R^{23} are the Rabi flop frequencies, i.e., the rates at which the atoms are pumped between the excited levels (between the levels 1 and 2 and between the levels 2 and 3, respectively), given in angular units [rad/s],

$$\omega_R^{12} = \frac{1}{\hbar} E_{12} X_{12} \tag{121}$$

for linearly polarized light (and similarly for ω_R^{23}), where X_{12} is the dipole matrix element for the transition [C m] and E_{12} is the electric field amplitude of the laser light [V/m].

The dipole matrix element, X_{12}, is (for linearly polarized light between two nondegenerate levels) related to the Einstein A factor according to

$$A_{21} = \frac{1}{4\pi\varepsilon_0} \frac{8\pi}{h} \left(\frac{2\pi}{\lambda_{12}}\right)^3 |X_{12}|^2 \tag{122}$$

where ε_0 is the permittivity for free space [8.85×10^{-12} C/Vm]. By relating the electric field amplitude E_{12} to the laser light intensity I_{12}, using the relation

$$I_{12} = \frac{1}{2} c\varepsilon_0 E_{12}^2 \tag{123}$$

we can express the Rabi flop frequencies of the probability amplitude of the wave function for linearly polarized light for the two transitions (between levels 1 and 2 and between levels 2 and 3, respectively) as

$$\omega_R^{12} = \left(A_{21} \frac{\lambda_{12}^3}{2\pi hc} I_{12}\right)^{1/2} \tag{124}$$

$$\omega_R^{23} = \left(A_{32} \frac{\lambda_{23}^3}{2\pi hc} I_{23}\right)^{1/2} \tag{125}$$

The Δ_{12}, Δ_{23}, and Δ_{13} are the complex detunings:

$$\Delta_{12} = \omega_{12} - \Omega_{12} - i\gamma_{21} \tag{126}$$

$$\Delta_{23} = \omega_{23} - \Omega_{23} - i\gamma_{32} \tag{127}$$

$$\Delta_{13} = (\omega_{12} + \omega_{23}) - (\Omega_{12} + \Omega_{23}) - i\gamma_{31} \tag{128}$$

where the ω_{12} (ω_{23}) is the energy between the levels 1 and 2 (2 and 3) in angular frequency units [rad/s] and Ω_{12} (Ω_{23}) the angular frequency of the laser light, corresponding to the first (second) excitation step [rad/s]. Hence, Ω_{12} (Ω_{23}) is related to the wavelength of the light by the relation $\Omega_{12} = 2\pi c/\lambda_{12}$ (and similarly for Ω_{23}). The γ_{12}, γ_{13}, and γ_{23} are the "off-diagonal" decay rates between levels 1 and 2, 1 and 3, and 2 and 3, respectively, which in turn are given by

$$\gamma_{21} = \frac{k_{21}}{2} + \gamma_C + \gamma_L \tag{129}$$

$$\gamma_{31} = \frac{k_{31} + k_{32}}{2} + \gamma_C + 2\gamma_L \tag{130}$$

$$\gamma_{32} = \frac{k_{21} + k_{31} + k_{32}}{2} + \gamma_C + \gamma_L \tag{131}$$

where the k_{21}, k_{31}, and k_{32} are the inelastic collision rates between the levels 2 and 1, 3 and 1, etc., as before (i.e., ordinary deexciting and exciting collisions) given in hertz, γ_C is the elastic collision rate [Hz] (in this case assumed to be the same for all states); and γ_L the hwhm (half-width at half-maximum) laser bandwidths in angular frequency units [rad/s] (assumed to be of Lorentzian shape, having no mode structure and being equal for the two wavelengths, for simplicity).

Although the assumption that the laser frequency profile is Lorentzian tends to overestimate the light intensity in the far wings (87, 88), we have chosen to use this description of the laser light in the simulations below due to its simplicity. Some other descriptions of the laser light and considerations of their impact upon the present system are found in the literature (15, 67–74, 87, 88).

In the same way, the introduction of relaxation rates in the equations of motion to describe collisional effects upon the atomic system (the impact theory of line broadening), also has its physical limitations. This time, however, the far wings of the absorption lines will be underestimated since they tend to decrease less than a Lorentzian [i.e., as $(\omega_1 - \Omega_1)^{-2}$]. In general, the far wings of a collisionally broadened absorption line of an atom in a flame are asymmetric and decrease approximately as the detuning raised to a power in the range of $-3/2$ to $-5/4$ (6). However, in order to also treat the effects of collisions in the simplest way, we have assumed that the impact theory of line broadening is a sufficiently good approximation for the present description of atoms in the flame.

1.10.3. Solution of the Density-Matrix System of Equations

When solving the system of equations for excitations of atoms in flames by light with pulse durations longer than a few nanoseconds, we assume that the equations for the off-diagonal matrix elements (ρ_{ij}, $i \neq j$) can be solved in a steady-state approximation (since the off-diagonal decay rates are much larger than the inverse of the laser pulse duration). Equations (117)–(119) can then be solved for the imaginary part of the off-diagonal matrix elements (ρ_{21} and ρ_{32}):

$$\omega_R^{12} \operatorname{Im}(\rho_{21}) = R_{12}(\rho_{22} - \rho_{11}) - R_{13}(\rho_{11} - \rho_{33}) \tag{132}$$

$$\omega_R^{23} \operatorname{Im}(\rho_{32}) = R_{23}(\rho_{33} - \rho_{22}) - R_{13}(\rho_{11} - \rho_{33}) \tag{133}$$

where R_{12}, R_{23}, and R_{13} are the symmetrized transition rates for the atoms,

$$R_{12} = \frac{(\omega_R^{12})^2}{2} \operatorname{Im}\left(\frac{4\Delta_{23}\Delta_{13} - (\omega_R^{12})^2 + (\omega_R^{23})^2}{D} \right) \tag{134}$$

$$R_{23} = \frac{(\omega_R^{23})^2}{2} \operatorname{Im}\left(\frac{4\Delta_{12}\Delta_{13} - (\omega_R^{23})^2 + (\omega_R^{12})^2}{D} \right) \tag{135}$$

$$R_{13} = -\frac{(\omega_R^{12})^2 (\omega_R^{23})^2}{2} \operatorname{Im}\left(\frac{1}{D} \right) \tag{136}$$

and

$$D = 4\Delta_{12}\Delta_{23}\Delta_{13} - (\omega_R^{12})^2 \Delta_{12} - (\omega_R^{23})^2 \Delta_{23} \tag{137}$$

Since the diagonal terms (ρ_{ii}) of the density-matrix represent the population density of the various levels, they can, in accordance with the commonly adopted notation for rate-equation formulations, be denoted n_i. In the same way the total number of neutral atoms in the system, ρ_{atom}, can conveniently be denoted n_{atom}. By inserting the results from Eqs. (132) and (133) for the number rate of absorbed photons in Eqs. (114)–(116), we now have a set of equations for the level populations only, with all the off-diagonal equations eliminated:

$$\frac{dn_1}{dt} = -R_{12}(n_1 - n_2) - R_{13}(n_1 - n_3) + k_{21}n_2 + k_{31}n_3 \tag{138}$$

$$\frac{dn_2}{dt} = R_{12}(n_1 - n_2) - R_{23}(n_2 - n_3) - k_{21}n_2 + k_{32}n_3 \tag{139}$$

$$\frac{dn_3}{dt} = R_{23}(n_2 - n_3) + R_{13}(n_1 - n_3) - (k_{31} + k_{32})n_3 \tag{140}$$

and

$$n_{atom} = n_1 + n_2 + n_3 \tag{141}$$

These equations predict contributions to the excitation rates to the third level that are directly due to the population of level 1 (not dependent on the population of the second level) through the R_{13} term.

These equations converge to the conventional rate equations if $|\Delta_{13}|$ (the complex detuning between the energy difference between the upper and the lower states and the sum of the energies of the two photons) goes to infinity, whereby R_{13} (the coherent excitation rate from level 1 to level 3) goes to zero and the expressions for R_{12} and R_{23} become simpler. The R_{12} and R_{23} terms (for zero detuning) correspond to $B_{12} I_v^{12}$ ($= B_{21} I_v^{12}$) and $B_{23} I_v^{23}$ ($= B_{32} I_v^{23}$), respectively, in the rate-equation approximation and hence reduce to the ordinary excitation rates.

The above equations can be solved under steady-state conditions when the time dependencies of the level populations are neglected. This simplification is valid if the pulse duration is substantially longer than the inverse of the deexcitation collision rate. This gives us the following expressions for the fractions of atoms excited for the density-matrix formalism, $C_2^{D.M.}$ and $C_3^{D.M.}$:

$$C_2^{D.M.} = \frac{n_2}{n_{atom}} = \frac{1}{3} \frac{R_{12}(R_{23} + k_{31} + k_{32}) + R_{13}(R_{12} + R_{23} + k_{32})}{\bar{D}} \tag{142}$$

$$C_3^{D.M.} = \frac{n_3}{n_{atom}} = \frac{1}{3} \frac{R_{12} R_{23} + R_{13}(R_{12} + R_{23} + k_{21})}{\bar{D}} \tag{143}$$

where the denominator $\bar{D}$ is given by

$$\bar{D} = R_{12} R_{23} + R_{13}(R_{12} + R_{23}) + 2R_{12}(k_{31} + k_{32})/3 + R_{23}(k_{21} + k_{31})/3$$
$$+ R_{13}(2k_{21} + k_{32})/3 + k_{21}(k_{31} + k_{32})/3 \tag{144}$$

These equations, which are analyzed in more detail by Axner and Ljungberg (59), thus describe features such as two-photon excitations, Stark broadening, splits, and shifts. Some of the main properties of these equations are briefly examined in the following section.

1.11. COMPARISON BETWEEN THE DENSITY-MATRIX FORMALISM AND EXPERIMENTAL RESULTS

1.11.1. Density-Matrix Equation Simulated LEI Scans

As will be demonstrated below, the main features of the experimental curves from measurements on Sr atoms presented in Section 1.9 are fairly well reproduced by the computer simulations based on the theory in Section 1.10, although the exact lineshapes and peak heights are not always satisfactorily reproduced.[6]

A suitable starting point for a comparison between experiment and theory is to calculate the Rabi flop frequencies for the two transitions. If the constants and parameters used in the Sr experiments are inserted directly into the formulas for the Rabi flop frequency for non degenerate levels [Eqs. (124) and (125)] the Rabi flop frequencies will be given by

$$\omega_R^{12} = 8.2\sqrt{I_{12}} \cdot 10^9 \text{ rad/s} \tag{145}$$

$$\omega_R^{23} = 9.2\sqrt{I_{23}} \cdot 10^9 \text{ rad/s} \tag{146}$$

where the intensities are given in kilowatts per square centimeter. This implies that for an intensity of $100 \, \text{kW/cm}^2$ the corresponding Rabi flop frequencies are

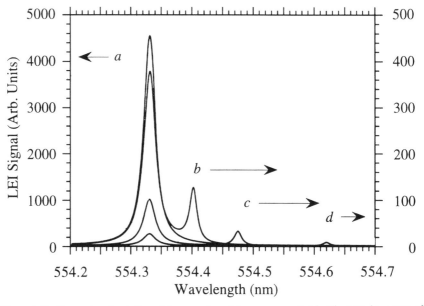

Figure 1.18. Computer simulation of the experimental scans presented in Fig. 1.11, i.e., scans of the second laser for various detunings of the first laser. The four curves represent cases with detunings of (a) 0 pm, (b) − 50 pm, (c) − 100 pm, and (d) − 200 pm. The pulse energies were 50 μJ for both beams (corresponding to $20 \, \text{kW/cm}^2$).

---

[6] The present section (1.11) is based mainly upon an article by Axner and Sjöström (15).

Table 1.3. Parameter Values Used for Calculations of the Simulated Scans[a]

Parameter	Value
k_{21}	1 GHz
k_{32}	1 GHz
k_{32}	1.5 GHz
$k_{2,\text{ion}}$	3 kHz[b]
$k_{3,\text{ion}}$	0.1 GHz[b]
γL	16×10^9 rad/s
γC	5 GHz
$g_2 A_{21}$	0.85×10^8 s^{-1}
$g_3 A_{32}$	0.63×10^8 s^{-1}

[a]The above parameters are identical for all simulations. Other parameter values are given in the figure legends of the simulations.
[b]Since depletion is not included, only the ratio of $k_{2,\text{ion}}$ to $k_{3,\text{ion}}$ is of importance.

both roughly around 10^{11} rad/s (corresponding to approximately 15 GHz) (15).

Let us first compare the experiment in which the second-step laser was scanned for various detunings of the first-step laser (shown in Fig. 1.15) with that of a computer simulation, shown in Fig. 1.18. The values of the various rates and widths used throughout all simulations are collected in Table 1.3 for clarity. The laser pulse energies in the simulations are identical to those in the experiment, i.e., 50 μJ for both the laser beams (corresponding to 20 kW/cm^2). A comparison of the experimental and simulated curves (Figs. 1.15 and 1.18) reveals a general agreement of the various signals: both a two-step and a two-photon signal are obtained whenever the first-step laser is detuned, as can be seen in curves b–d in Fig. 1.18. The positions of the two-photon signals in the simulations agree with those of the experiments. The relative magnitudes between the two-step and two-photon peaks in the simulation are in good agreement for the three detuned curves. However, the theory gives far too large a resonance signal in the nondetuned case (curve a) as compared to the detuned cases (curves b–d). In addition, the signal enhancement [the ratio of the LEI signals with and without the second-step laser, or (wihch is equivalent) the ratio of the LEI signals when the second step is at resonance to that when it is considerably detuned] is also exaggerated by the theory.

There are at least three possible reasons for these discrepancies. First, the simplified description of the laser light as consisting of only a single mode with phase fluctuations that accunte for its full spectral bandwidth can be partly responsible for the discrepancy between the two-step and two-photon peaks.

Second, the simplification of the true atomic system to a nondegenerate
three-level system will affect the simulation quantitatively (the relation between the two-photon and two-step signals will be affected). Third, it is
possible that the existence of Sr molecules in the flame [SrO and Sr(OH)$_x$]
might significantly affect the ionization mechanisms through associative
ionization of excited atoms at the intermediate level. Hence, the second
excitation step will not increase the ionization rate as much as predicted by the
theory (by the ratio of $k_{3,ion}$ to $k_{2,ion}$). In addition, a forth possible explanation
for the discrepancies found is that depletion of neutral atoms in the interaction
region mighy play a certain role. However, the sharply peaked signals indicate
that depletion cannot be the dominating reason for the observed discrepancies
in this particular case.

The experimental findings that there is a split double-peak when scanning
the first-step laser and holding the second-step laser fixed and that the
broadening of the split double-peak increases with increased first-step laser
power (Fig. 1.17) are in fairly good agreement with theory (the general trends
are reproduced by the simulations), as can be seen from Fig. 1.19.

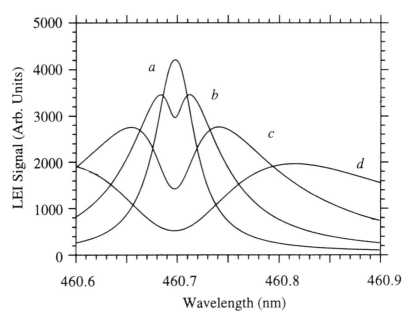

Figure 1.19. Computer simulation of the experimental scans presented in Fig. 1.17, i.e., scans of
the first-step laser for various pulse energies. The second laser was tuned to resonance. The four
curves correspond to pulse energies of the first laser of (a) 200 μJ, (b) 800 μJ, (c) 3 mJ, and (d)
−10.8 mJ (corresponding to 80, 320, 1200, and 4300, and 4300 kW/cm², respectively). The pulse
energy of the second laser was 50 μJ (20 kW/cm²).

Figure 1.19 shows the splitting of the curve in the nondetuned case for four different first-step laser intensities. In this case, the first-step energy has been taken as four times the experimentally used values in order to improve the qualitative agreement between experiment and theory. As can be seen from the figure, splitting occurs in three of the four curves. This splitting increases as a function of first-step laser intensity, in general agreement with experiments (Fig. 1.15). However, when comparing the simulation with the experiment we find quantitatively that the theory predicts a much larger increase in splitting as a function of laser light intensity as compared to the experiment. This discrepancy is not yet understood but might originate from the oversimplified theoretical description of the laser light and atoms.

In short, it has been found that there is a qualitative agreement between experiment and theory in predicting the LEI signal lineshapes. All the main features of the experiments, such as two-photon peaks, power and saturation broadening, and splitting of peaks, are present in the theoretical simulations. The theory, in fact, also predicts that saturation curves that bend downward (i.e., signals that decrease with increasing laser intensity) can occur in certain cases. This has been verified by experiment recently, and its implications are considered in Section 1.11.3, below.

The quantitative agreement is, however, still unsatisfactory, and a refinement of the theory, especially in the description of the properties of the laser light and the collisional broadening and ionization mechanisms, is required. Several studies have been published on these subjects (60, 61, 63, 70, 89, 90), but space constraints prevent our treatment of them in the present chapter.

1.11.2. Approximations in the Theory

In Section 1.10, where the theoretical treatment of LEI based on the density-matrix formalism was presented, a number of approximations were introduced. Many of these approximations will indeed affect the final shapes and magnitudes of the predicted curves in one way or another. It has been shown, however, that despite some quantitative disagreements between theory and experiments, the use of his simple density-matrix-based model for qualitative modeling of stepwise excitations of atoms in flames is justified, since the outcome of this simplified theoretical description has all its main features in common with experimental findings.

The most significant simplifications made in the theory presented in Section 1.10 were the following: treatment of the laser light as consisting of a single mode (i.e., neglect of the mode structure of the laser light) with phase fluctuations that give the laser its experimentally measured bandwidth (Lorentzian-shaped wings); reduction of the n-level system of equations (n being the total number of sublevels included in the excitations) to a three-level nondegenerate system; and the neglect of Doppler broadening, depletion, and the influence of other non-laser-connected levels. Other significant simplifications are the neglect of collision mechanisms beyond the impact theory (which only gives rise to Lorentzian shaped wings); the neglect of associative ionization routes for alkaline earth elements; spatial variations and fluctuations of the laser light due to inhomogeneous irradiation; amplitude fluctuations of the laser light (e.g., temporal fluctuations due to shot-to-shot fluctuations, temporal fluctuations due to intensity variations within the duration of the pulse, and temporal fluctuations due to mode jumping within the laser bandwidth); and the influence of amplified spontaneous emission from the lasers.

By refining the theoretical description of the experimental situation along the foregoing lines, a better agreement between theory and experiment is to be expected. Such an improvement will be achieved, however, at the cost of more complexity in the treatment. Some studies have been published suggesting improvements in the description of particular subjects, such as the properties of the laser light and collisions (70, 89, 90), but it would take us too far afield to comment upon them here.

1.11.3. Implications for LEI Spectrometry

There are some important implications of the existence of both two-photon and two-step signals in LEI spectrometry (59).

Let us assume, for example, that the wavelength setting for the first-step laser is slightly erroneous (e.g., due to a very weak one-step signal or a large background signal from flame molecules) and one tries to find the proper second-step transition wavelength by optimizing the signal. Then it is possible that the wavelength for which the signal has its maximum (when the second-step wavelength is scanned) is the two-photon peak and not the two-step peak. This two-photon peak, however, might be only a local maximum in the two-dimensional space spanned by the two laser wavelengths. Therefore, it is necessary to iteratively adjust the first step and the second step laser wavelengths for maximum signal.

Furtermore, the existence of a splitting of the resonances may cause unexpected features. It is possible that, when high intensities in the first step are used, the signal at resonance might decrease with increasing first-step intensity due to an increased dynamic Stark splitting, as has been demonstrated both experimentally (see Fig. 1.17) and theoretically (see Fig. 1.19).

Moreover, due to the existence of double peaks in the scans (even when one laser is held at its resonance wavelength) the strongest LEI signals will not always be found at the resonance wavelengths; only a scan across the resonances at the actual intensities to be used in the experiment truly reveals the laser wavelengths that give the strongest signals.

For many practical uses of LEI, e.g., ultrasensitive trace element analysis using LEI, most of the nonlinear effects associated with two-photon excitations or dynamic Stark effects are only of minor importance. The user should, however, know about their existence so that necessary precautions can be taken. On the other hand, anomalous contributions from atoms outside the "true" interaction region, as was discussed in Section 1.8, might have more detrimental effects upon the accuracy of measurements using the LEI technique in many applications and therefore requires more attention.

APPENDIX: THE DEGREE OF IONIZATION FOR HIGH-CHOPPING-FREQUENCY CONTINUOUS-WAVE EXCITATION

Statement. For high-chopping-frequency cw excitations, the total degree of ionization of illuminated atoms, $\Phi(M, \tau_{\text{laser}}^{\text{cw}})$, exposed to a series of M consecutive laser pulses, each with a duration of $\tau_{\text{laser}}^{\text{cw}}$, is given by $\Phi(\tau_{\text{atom}}^{\text{cw,high}})$, where $\tau_{\text{atom}}^{\text{cw,high}}$ is the total illumination time, which is equal to $M\tau_{\text{laser}}^{\text{cw}}$.

Proof. An expression for the total degree of ionization of the illuminated atoms, $n_{\text{ion}}(M, \tau_{\text{laser}}^{\text{cw}})$, exposed to M consecutive laser pulses, each with a duration of $\tau_{\text{laser}}^{\text{cw}}$, can be written [considering the definition of the degree of ionization, i.e., Eqs. (51) and (52)] as

$$n_{\text{ion}}(M, \tau_{\text{laser}}^{\text{cw}}) = \Phi(M, \tau_{\text{laser}}^{\text{cw}})n_{\text{tot}} \tag{A1}$$

Here is M is the number of illuminations experienced by the atoms during their journey across the laser beam. Now $\Phi(M, \tau_{\text{laser}}^{\text{cw}})$ can be calculated carefully quantifying the number density of ions produced in an ensemble of atoms passing through the interaction region, exposed to a series of M laser pulses, one by one.

The number density of ions produced in the atomic system that entered the interaction region between two laser illumination periods (which thus is exposed to laser light for the first time) can be written

$$n_{\text{ion}}^{(1)}(\tau_{\text{laser}}^{\text{cw}}) = \Phi(\tau_{\text{laser}}^{\text{cw}})n_{\text{tot}}^{(0)} \tag{A2}$$

Hence, $n_{\text{ion}}^{(1)}(\tau_{\text{laser}}^{\text{cw}})$ is the density of ions produced from the first laser pulse of duration $\tau_{\text{laser}}^{\text{cw}}$ (i.e., pulse number 1) by illumination of a system of atoms previously not exposed to any light, i.e., exposed to zero (0) light interactions. Thus $n_{\text{tot}}^{(0)}$ is the original density of atoms in this ensemble. The remaining density of (neutral) atoms after one interaction, $n_{\text{tot}}^{(1)}$, is equal to $n_{\text{tot}}^{(0)} - n_{\text{ion}}^{(1)}$. The atoms will then be exposed to a second laser pulse. This laser pulse will produce a density of ions of $n_{ion}^{(2)}(\tau_{laser}^{cw})$ according to

$$n_{ion}^{(2)}(\tau_{laser}^{cw}) = \Phi(\tau_{laser}^{cw})n_{tot}^{(1)} \tag{A3}$$

This continues until the atoms have passed the interaction region (i.e., been exposed to M pulses, each of duration τ_{laser}^{cw}). Hence, the total density of ions produced in this ensemble of atoms during the passage of the interaction regions, $n_{ion}(M, \tau_{laser}^{cw})$, can be written as a sum of all the ions produced, $n_{ion}^{(i)}(\tau_{laser}^{cw})$, after M such interactions:

$$n_{ion}(M, \tau_{laser}^{cw}) = \sum_{i=1}^{M} n_{ion}^{(1)}(\tau_{laser}^{cw}) = \sum_{i=1}^{M} \Phi(\tau_{laser}^{cw})n_{tot}^{(i-1)} \tag{A4}$$

This expression can be further simplified by noting that all the degree of ionization factors in the summation, $\Phi(\tau_{laser}^{cw})$, are the same. One further simplification is to change the index in the summation, letting $j = i - 1$. We can then write

$$n_{ion}(M, \tau_{laser}^{cw}) = \Phi(\tau_{laser}^{cw}) \sum_{j=0}^{M-1} n_{tot}^{(j)} \tag{A5}$$

We can then easily conclude that the density of (neutral) atoms after j pulses, $n_{tot}^{(j)}$, can be related to the original density of (neutral) atoms, $n_{tot}^{(0)}$, by the relation

$$n_{tot}^{(j)} = (1 - \Phi(\tau_{laser}^{cw}))^j n_{tot}^{(0)} \tag{A6}$$

Hence, Eq. (A5) can be written

$$n_{ion}(M, \tau_{laser}^{cw}) = \left[\Phi(\tau_{laser}^{cw}) \sum_{j=0}^{M-1} (1 - \Phi(\tau_{laser}^{cw}))^j \right] n_{tot}^{(0)} \tag{A7}$$

If we now make use of the expressions for the degree of ionization of a system of atoms, Eqs. (53) and (54), i.e. (for a one-step excitation case and correspondingly for a two-step excitation case),

$$\Phi(\tau_{laser}^{cw}) = 1 - \exp\left(- C_2^{off} k_{2,ion}^{eff} \tau_{laser}^{cw}\right) \tag{A8}$$

we find that Eq. (A7) can be simplified further:

$$n_{ion}(M, \tau_{laser}^{cw}) = \Big[\left(1 - \exp\left(- C_2^{off} k_{2,ion}^{eff} \tau_{laser}^{cw}\right)\right)$$
$$\times \sum_{j=0}^{M-1} \left(\exp\left(- C_2^{off} k_{2,ion}^{eff} \tau_{laser}^{cw}\right)\right)^j \Big] n_{tot}^{(0)} \tag{A9}$$

By inspection of this series, we can easily conclude that all terms cancel except the first and last. Hence, the final expression for the number density of ions is

$$n_{\text{ion}}(M, \tau_{\text{laser}}^{\text{cw}}) = (1 - \exp(-C_2^{\text{off}} k_{2,\text{ion}}^{\text{eff}} M \tau_{\text{laser}}^{\text{cw}})) n_{\text{tot}}^{(0)} \tag{A10}$$

Using the fact that the total illumination time, $t_{\text{atom}}^{\text{cw,high}}$, is given as the product between the number of laser pulses, M, and the laser pulse duration, $\tau_{\text{laser}}^{\text{cw}}$, we can write this expression as follows:

$$n_{\text{ion}}(M, \tau_{\text{laser}}^{\text{cw}}) = (1 - \exp(-C_2^{\text{off}} k_{2,\text{ion}}^{\text{eff}} \tau_{\text{laser}}^{\text{cw,high}})) n_{\text{tot}}^{(0)} \tag{A11}$$

As can seen from an inspection of the right-hand side of this equation, the expression within bold parentheses is simply the degree of ionization of an atomic system illuminated by laser light for a period of $\tau_{\text{atom}}^{\text{cw,high}}$. Hence, expression (A11) can be written

$$n_{\text{ion}}(M, \tau_{\text{laser}}^{\text{cw}}) = \Phi(\tau_{\text{laser}}^{\text{cw,high}}) n_{\text{tot}}^{(0)} \tag{A12}$$

From a comparison with Eq. (A1) we can conclude that the total degree of ionization of atoms exposed to a series of M consecutive laser pulses, each with a duration of $\tau_{\text{laser}}^{\text{cw}}$ [i.e., $\Phi(M, \tau_{\text{laser}}^{\text{cw}})$], is given by $\Phi(\tau_{\text{atom}}^{\text{cw,high}})$, where $\tau_{\text{atom}}^{\text{cw,high}}$ is the total illumination time given by the product of M and $\tau_{\text{laser}}^{\text{cw}}$:

$$\Phi(M, \tau_{\text{laser}}^{\text{cw}}) = \Phi(\tau_{\text{laser}}^{\text{cw,high}}) \tag{A13}$$

Hence, the atomic system behaves basically as if it were exposed to only one laser pulse with a duration equal to the total illumination time of the system.

REFERENCES

1. J. C. Travis and J. R. DeVoe, in *Laser in Chemical Analysis* (G. M. Hieftje, J. C. Travis, and F. E. Lytle, Eds), pp. 93–124. Humana Press, Inc., Clifton, NJ, 1981.

2. J. C. Travis, G. C. Turk, and R. Green, *Anal. Chem.* **9**, 1006A (1982).

3. J. C. Travis, G. C. Turk, J. R. DeVoe, P. K. Schenck, and C. A. van Dijk, *Prog. Anal. At. Spectrosc.* **7**, 199 (1984).

4. O. Axner and H. Rubinsztein-Dunlop, *Spectrochim. Acta* **44B**, 835 (1989).

5. K. Niemax, *Fresenius' Z. Anal. Chem.* **337**, 551 (1990).

6. C. Th. J. Alkemade, Tj. Hollander, W. Snelleman, and P. J. Th. Zeegers, *Metal Vapours in Flames.* Pergamon, Oxford, 1982.

7. Tj. Hollander, P. J. Kalff, and C. Th. J. Alkemade, *J. Chem. Phys.* **39**, 2558 (1963).

8. I. Magnusson, *Spectrochim. Acta* **43B**, 113 (1987).

9. O. Axner, T. Berglind, J. L. Heully, I. Lindgren, and H. Rubinsztein-Dunlop, *J. Appl. Phys.* **55**, 3215 (1984).

10. G. D. Boutilier, M. B. Blackburn, J. M. Mermet, S. J. Weeks, H. Haraguchi, J. D. Winefordner, and N. Omenetto, *Appl. Opt.* **17**, 2291 (1978).

11. G. D. Boutilier, N. Omenetto, and J. D. Winefordner, *Appl. Opt.* **19**, 1838 (1980).

12. J. W. Daily, *Appl. Opt.* **16**, 2322 (1977).

13. M. Sargent, III, M. O. Scully, and W. E. Lamb, Jr., *Laser Physics.* Addison-Wesley, Reading, MA, 1974.

14. R. Balian, S. Haroche, and S. Liberman, Eds., *Aux Frontiers de la Spectroscopie Laser.* North-Holland Publ., Amsterdam, 1977.

15. O. Axner and S. Sjöström, *Spectrochim Acta* **47B**, 245 (1992).

16. G. Zizak, N. Omenetto, and J. D. Winefordner, *Opt. Eng.* **23**, 749 (1984).

17. J. D. Bradshaw, N. Omenetto, G. Zizak, J. N. Bower, and J. D. Winefordner, *Appl. Opt.* **19**, 2709 (1980).

18. J. C. Travis, P. K. Schenck, and G. C. Turk, *Anal. Chem.* **51**, 1516 (1979).

19. N. Omenetto, B. W. Smith, and L. P. Hart, *Fresenius' Z. Anal. Chem.* **324**, 683 (1986).

20. O. Axner, M. Norberg, and H. Rubinsztein-Dunlop, *Spectrochim. Acta* **44B**, 693 (1989).

21. J. W. Daily, *ACS Symp. Ser.* **134**, 61–83 (1980).

22. C. H. Muller, III, K. Schofield, and M. Steinberg, *J. Chem. Phys.* **72**, 6620 (1980).

23. C. H. Muller, III, K. Schofield, and M. Steinberg, *ACS Symp. Ser.* **134**, 103–130 (1980).

24. M. Iino, H. Yano, Y. Takubo, and M. Shimazu, *J. Appl. Phys.* **52**, 6025 (1981).

25. G. Zizak, J. D. Bradshaw, and J. D. Winefordner, *Appl. Opt.* **19**, 3631 (1980).

26. M. B. Blackburn, J.-M. Mermet, G. D. Boutilier, and J. D. Winefordner, *Appl. Opt.* **18**, 1804 (1979).

27. C. A. van Dijk, N. Omenetto, and J. D. Winefordner, *Appl. Spectrosc.* **35**, 389 (1981).

28. N. Omenetto, C. A. van Dijk, and J. D. Winefordner, *Spectrochim. Acta* **37B**, 703 (1982).

29. C. Th. J. Alkemade, *Spectrochim. Acta* **40B**, 1331 (1985).

30. J. W. Daily, *Appl. Opt.* **17**, 225 (1978).

31. N. Omenetto, T. Berthoud, P. Cavalli, and G. Rossi, *Appl. Spectrosc.* **39**, 500 (1985).

32. O. Axner, P. Ljungberg, and Y. Malmsten, *Appl. Phys. B* **54**, 144 (1992).

33. O. Axner, P. Ljungberg, and Y. Malmsten, *Appl. Opt.* **6**, 899 (1993).

34. O. Axner, P. Ljungberg, and Y. Malmsten, *Appl. Phys. B* **56**, 355 (1993).

35. G. C. Turk, W. G. Mallard, P. K. Schenck, and K. C. Smyth, *Anal. Chem.* **51**, 2408 (1979).

36. O. Axner and T. Berglind, *Appl. Spectrosc.* **43**, 940 (1989).

37. G. J. Havrilla, S. J. Weeks, and J. C. Travis, *Anal. Chem.* **54**, 2566 (1982).

38. G. S. Hurst, M. G. Payne, S. D. Kramer, and J. P. Young, *Rev. Mod. Phys.* **51**, 767 (1979).

39. V. S. Letokhov, *Laser Photoionization Spectroscopy.* Academic Press, Orlano, FL, 1987.

40. G. S. Hurst and M. G. Payne, *Principles and Applications of Resonance Ionisation Spectroscopy.* Adam Hilger, Bristol, 1988.

41. C. M. Miller and J. E. Parks, Eds., *Resonance Ionization Spectroscopy 1992.* (*Proceeding of the 10th International Symposium on Resonance Ionization Spectroscopy and Its Applications*). IOP Publishing, Bristol, 1992.

42. F. M. Curran, K. C. Lin, G. E. Leroi, P. M. Hunt, and S. R. Crouch, *Anal. Chem.* **55**, 2382 (1983).

43. O. Axner and S. Sjöström, *Appl. Spectrosc.* **44**, 144 (1990).

44. N. Omenetto, B. W. Smith, B. T. Jones, and J. D. Winefordner, *Appl. Spectrosc.* **43**, 595 (1989).

45. B. W. Smith, L. P. Hart, and N. Omenetto, *Anal. Chem.* **58**, 2147 (1986).

46. W. L. Wiese, M. W. Smith, and B. M. Glennon, *Atomic Transition Probabilities,* Vol. I: *Hydrogen Through Neon–A Critical Data Compilation.* Natl. Stand. Ref. Data Ser., Natl. Bur. Stand. (U.S.), No. 4. U.S. Gov. Printing Office, Washington, DC, 1966.

47. W. L. Wiese, M. W. Smith, and B. M. Miles, *Atomic Transition Probabilites,* Vol. II: *Sodium Through Calcium—A Critical Data Compilation.* Natl. Stand. Ref. Data Ser, Natl. Bur. Stand. (U.S.), No. 22. U.S. Gov. Printing Office, Washington, DC, 1969.

48. A. Lindgård and S. E. Nielsen, *Atomic Data and Nuclear Data Tables. Transition Probabilities for the Alkali Isoelectronic Sequences Li I, Na I, K I, Rb I, Cs I, Fr I,* Vol. 19, No. 6. Academic Press, New York and London, 1977.

49. O. Axner, *Spectrochim. Acta* **45B**, 561 (1990).

50. K. C. Smyth, P. K. Schenck, and W. G. Mallard, *ACS Symp. Ser.* **134**, 175–181 (1980).

51. O. Axner, unpublished results (1990).

52. M. D. Seltzer and R. B. Green, *Appl. Spectrosc.* **43**, 257 (1989).

53. O. Axner and S. Sjöström, *Appl. Spectrosc.* **44**, 864 (1990).

54. O. Axner and S. Sjöström, *Conf. Ser. Inst. Phys.* **114**, 31 (1990).

55. G. Turk, *Anal. Chem.* **64**, 1836 (1992).

56. L. P. Hart, B. W. Smith, and N. Omenetto, *Spectrochim. Acta* **40B**, 1637 (1985).

57. N. Omenetto, L. P. Hart, B. W. Smith, and G. C. Turk, *Opt. Commun.* **62**, 86 (1987).

58. G. C. Turk, P. Cavalli, and N. Omenetto, *Appl. Spectrosc.* **41**, 698 (1987).

59. O. Axner and P. Ljungberg, *Spectrochim. Acta Rev.* **15**, 181 (1993).

60. O. Axner and P. Ljungberg, *J. Quant. Spectrosc. Radiat. Transfer* **50**, 277 (1993).

61. P. Ljungberg, D. Boudreau, and O. Axner, *Spectrochim. Acta* **49B**, 1491 (1994).

62. B. W. Shore, *The Theory of Coherent Atomic Excitations*, Vols. 1 and 2. Wiley, New York, 1990.

63. D. Boudreau, P. Ljungberg, and O. Axner, *Spectrochim. Acta Electron.* (in press) (1995).

64. J. E. M. Goldsmith, *Opt. Lett.* **10**, 116 (1985).

65. G. C. Turk, F. C. Reugg, J. C. Travis, and J. R. DeVoe, *Appl. Spectrosc.* **40**, 1146 (1986).

66. G. J. Havrilla and C. C. Carter, *Appl. Opt.* **26**, 3510 (1987).

67. R. Salomaa and S. Stenholm, *J. Phys. B* **8**, 1795 (1975).

68. R. Salomaa and S. Stenholm, *J. Phys. B* **9**, 1221 (1976).

69. J. E. Bjorkholm and P. E. Liao, *Phys. Rev. A* **14**, 751 (1976).

70. P. R. Berman, *Adv. At. Mol. Phys.* **13**, 57 (1977).

71. R. Salomaa, *J. Phys. B* **10**, 3005 (1977).

72. R. Salomaa, *Phys. Scr.* **15**, 251 (1977).

73. A. Gallagher, in *Spectral Line Shapes* (R. J. Exton, Ed.), Vol. 4, pp. 215–234. A. Deepak Publ. Hampton, VA, 1987.

74. K. C. Lin, S. H. Lin, P. M. Hunt, G. E. Leroi, and S. R. Crouch, *Appl. Spectroc.* **43**, 66 (1989).

75. A. M. Lau, *Phys. Rev. A* **33**, 3602 (1986).

76. J. H. Eberly and S. V. O'Neil, *Phys. Rev. A* **19**, 1161 (1979).

77. P. Ljungberg, *The Density-Matrix Model for Two-Step Laser-Enhanced Ionization (LEI) and Laser-Induced Fluorescence (LIF) Spectrometry.* Internal report, Department of Physics, Chalmers University of Technology, Göteborg, Sweden, 1990.

78. R. Loudon, *The Quantum Theory of Light.* Oxford Univ. Press, Oxford, 1985.

79. P. Zoller and P. Lambropoulos, *J. Phys. B* **12**, L547 (1979).

80. S. N. Dixit, P. Zoller, and P. Lambropoulos, *Phys. Rev. A* **21**, 1289 (1980).

81. R. Walkup, A Migdall, and D. E. Pritchard, *Phys. Rev. A* **29**, 2651 (1984).

82. P. M. Farrell, W. R. MacGillivray, and M. C. Standage, *Phys. Rev. A* **37**, 4240 (1988).

83. A. Yariv, *Introduction to Optical Electronics* (HRW Series in Electrical Engineering, Electronics, and Systems). Holt, Rinehart & Winston, New York, 1971.

84. M. S. Feld and A. Javan, *Phys. Rev* **177**, 540 (1969).

85. Th. Hänsch and P. Toschek, *Z. Phys.* **236**, 213 (1970).

86. R. G. Brewer and E. L. Hahn, *Phys. Rev. A* **11**, 1641 (1975).

87. M. Przbylski, B. Otto, and H. Gerhardt, *Lambda Highlights* **11**, 3 (1988).

88. A. T. Georges and P. Lambropoulos, *Phys. Rev. A* **18**, 587 (1978).

89. J. H. Eberly, H. Walter, and K. W. Rothe, Eds. *Laser Spectroscopy IV*, pp. 80–87. Springer-Verlag, Berlin, 1979.

90. T. J. McIlrath and J. L. Carsten, *Phys. Rev. A* **6**, 1091 (1972).

CHAPTER

2

FUNDAMENTAL MECHANISMS OF LASER-ENHANCED IONIZATION: SIGNAL DETECTION

JOHN C. TRAVIS and GREGORY C. TURK

Analytical Chemistry Division, Chemical Science and Technology Laboratory, National Institute of Standards and Technology, Gaithersburg, Maryland 20899

2.1. OVERVIEW

The production of ions and electrons in a flame by laser-enhanced ionization (LEI) would be useless without a means of detecting them. Although detection by optical means has been reported (1), the emphasis of this chapter will be on the direct electrical detection of laser-produced charges by means of an electric field applied to the flame. Indeed, a major advantage of LEI is the direct generation of an electrical signal. Not only is a (potentially noisy) signal transducer eliminated, but the ability to control charges with fields leads to an inherently higher detection probability than that of typical fractional solid angles for the optical detection of photons.

2.1.1. Approaches to Deriving the LEI Signal Response in One Dimension

The goal of Sections 2.2–2.7, below, is to establish a model for the behavior of natural and laser-produced ions and electrons in an applied field and for the current induced in the external circuit by this behavior. Most of the derivation superimposes the transient LEI event (for pulsed lasers) on the steady-state perturbed flame treatment of Lawton and Weinberg (2), which makes generous use of approximations to yield useful solutions in analytical closed form. The functional form of such solutions permits very useful generalizations about the nature of the LEI detection process and its optimization. However, the highly stylized predicted pulse shapes provide a poor fit to actual experimental data.

Laser-Enhanced Ionization Spectrometry, edited by John C. Travis and Gregory C. Turk.
Chemical Analysis Series, Vol. 136.
ISBN 0-471-57684-0 © 1996 John Wiley & Sons, Inc.

For this reason, it is constructive to numerically solve the coupled differential equations for the charged particle and field distributions on a computer. Such numerical solutions may be used to produce "motion pictures" of the laser-induced charge distributions moving through the flame, as well as more realistic induced current waveforms.

2.1.2. The One-Dimensional Approximation

The mathematical description of the distributions and dynamics of ions, electrons, and fields in flames is greatly simplified by reducing the dimensionality of the problem to one. This approximation is not as great a compromise of reality as it may at first seem and does not preclude the effective use of two-dimensional visualizations or the use of such three-dimensional parameters as number density (cm^{-3}) or current density (A/cm^2). Mathematically, a one-dimensional problem is one for which all the motions, gradients, etc., of interest are parallel to a single coordinate axis and the appropriate three-dimensional solution can be generated by translations and/or rotations of the one-dimensional solution, depending on the choice of coordinate system.

Figure 2.1 represents a two-dimensional cross section of the three-dimensional geometry of a uniform flame between parallel plate electrodes. For simplicity, the free charges furnished by the flame are not shown in the figure. The charges shown are those added by the LEI event at a slightly earlier time.

If we assume that the figure in fact represents a small portion of plates that extend an infinite distance parallel to the long edge of the page and an infinite distance perpendicular to the plane of the page, then the only axis that must be accommodated mathematically is parallel to the short edge of the page, or perpendicular to the plane of the electrodes.

We denote this axis as x and will use the sign convention that the cathode plate is located $x = 0$ and the anode plate is located at $x = W$, where W is the spacing between the plates. All vector quantities, such as the electric field and the ion velocity, are now fully represented by their scalar amplitude, with the sign denoting the direction. Since only one dimension is considered, we omit the customary x subscript on the single vector component.

For simplicity, the flame is assumed to uniformly fill the space between the plates, though variation of flame properties with x would be permitted in the one-dimensional setting. Charges are deposited by the action of laser irradiation centered a distance X_L from the surface of the cathode. After the passage of time t, the ions and electrons deposited by the laser have been moved to positions denoted by x_+ and x_-, respectively.

In the one-dimensional approximation, both the flame and the laser light extend infinitely in the dimensions perpendicular to x, such that the laser excitation is approximated by a plane rather than a (more realistic) line. The

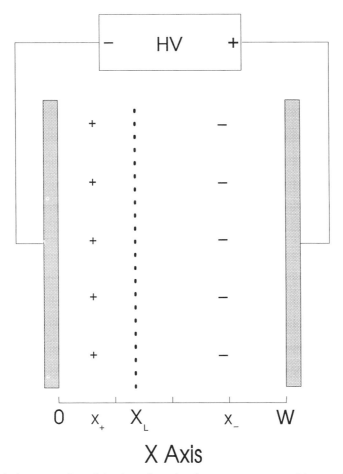

Figure 2.1. A cross section of the three-dimensional geometry represented by one dimension mathematically. The significant dimension is perpendicular to two semi-infinite parallel plate electrodes at $x = 0$ and $x = W$. A *planar* laser beam is taken to deposit equal areal densities of positive ions and electrons in the plane at $x = X_L$ at time $t = 0$. At a later time, t, the ions and electrons will have moved in opposite directions, under the influence of an applied field, to positions, x_+ and x_-, respectively. (HV: high-voltage source.)

planar beam shape is obviously a poorer approximation to reality than the electrode and flame geometry.

2.1.3. Experimental Results

Section 2.8 will examine the results of laboratory experiments in the light of detection theory, and Section 2.9 will consider some unfinshed business.

Topics covered include electrode design, pulse profiles, and ion/electron-induced charge ratios, as well as two different methods of ion imaging.

2.2. THE EFFECT OF AN EXTERNAL FIELD ON IONS AND ELECTRONS

2.2.1. Sign Conventions

The choice of the cathode for the origin ($x = 0$) serves to simplify LEI equations for common circumstances under which the position of the anode will be seen to be irrelevant. However, it produces some confusion with respect to conventional electrostatic notation. Specifically, the direction of the electric field vector is defined as the direction of motion of a positive charge in the field. Since positive ions move *away* from the anode and *toward* the cathode, our placement choice for the electrodes results in the electric field vector being directed in the $-x$ direction. In one-dimensional versions of familiar equations the electric field vector will be replaced with its scalar amplitude, E, without loss of generality. However, our choice of origin and sign convention causes the *value* of E to always be negative. This introduces a certain confusion into discussions of field "strength," which always refers to the absolute magnitude of E. Thus, for our convention, the "strongest" fields are the most negative ones. For this reason, it is often preferable to consider the field magnitude, $|E|$, and count on the reversed convention to keep track of the direction of motion of ions and electrons. Our convention is reversed with respect to that of Lawton and Weinberg (2), whose work forms the basis for this entire discussion, as well as the treatment of Magnusson (3). However, the conventions used here are consistent with our prior theoretical treatments (4–6).

2.2.2. Positive Ions in the Flame

In vacuo, an ion will experience constant acceleration in a uniform (constant magnitude) electric field. However, in the presence of a buffer gas at sufficiently high pressure (e.g., atmospheric pressure), the ion reaches a terminal velocity due to the frictional behavior of collision interaction. This ion velocity, v_+, scales linearly with the electric field through the ion mobility, μ_+, such that

$$v_+ = \mu_+ E \tag{1}$$

An ion number density of n_+ (cm^{-3}) moving collectively at the velocity v_+ constitutes an ion current density, j_+ (A/cm^2), in the flame, where

$$j_+ = e v_+ n_+ = e \mu_+ n_+ E \tag{2}$$

and e is the magnitude of the charge on an electron. The mobility is defined as a positive definite quantity, such that both the positive ion velocity and current density in Eqs. (1) and (2) are negative for our sign convention, which results in a negative value for the field amplitude.

The mobility of an ion in a particular medium can be derived from first principles and is shown to depend upon the masses of the ion and buffer gas molecules, collision cross sections, total pressure, and partial pressures of buffer species, etc. (7–9). Even the simplest of flames should require separate positive ion mobilities for a number of species present, especially when the flame is seeded with an analytical sample.

Thus it is purely for simplicity that a single positive ion mobility will be used throughout the following derivations. The fact that ion mobility *does not vary widely* is good for this approximation but bad for the prospects of utilizing ion mobility spectroscopy in LEI detection. (This concept will be discussed briefly at the end of the chapter.) During the derivations, however, one should be sensitive to derivations that could be extended to multiple mobilities by simple linear superposition, as well as to other derivations with more complicated implications of multiple mobilities. A simple rule of thumb is that ion mobilities are on the order of $20 \, \text{cm}^2 \, \text{s}^{-1} \, \text{V}^{-1}$, or the ion velocity increases about 20 cm/s for each volt per centimeter of applied field.

2.2.3. Electrons in the Flame

Negative charge in the flame is carried almost exclusively by electrons, rather than negative ions, because electron attachment potentials are relatively low compared to the thermal energy in the flame. The theory of electron drift in external fields is more complicated than for ions, and simple scaling with mass and/or cross section from ion mobilities is not appropriate (10,11). Indeed, the electron velocity does not increase linearly with applied field over all experimental conditions, so that the mobility is not even defined. As a practical matter, however, electron travel times for typical distances and fields in LEI are $< 1 \, \mu s$, and temporal resolution of electron events are often limited by $\sim$ MHz detection electronics. We have consistently employed an arbitrary value of $2000 \, \text{cm}^2 \, \text{s}^{-1} \, \text{V}^{-1}$ for the electron mobility and will use that value in the illustrations below. Thus, with some reservations, electron motion is treated by analogy with ion motion as

$$v_- = -\mu_- E \tag{3}$$

$$j_- = -ev_- n_- = e\mu_- n_- E \tag{4}$$

where the mobility is again a positive definite quantity and the explicit negative sign in Eq. (3) indicates that negative charge travels antiparallel to the field. The sign of the electron current density is the same as for the ion current density, even though the velocity is oppositely directed, because the charge on the electron is $-e$. With continual advances in high-speed electronics, future LEI pulse shape studies may well necessitate a more exacting look at electron drift theory.

2.3. THE EFFECT OF IONS AND ELECTRONS ON THE FIELD AND SUPPORTING CIRCUIT

2.3.1. The Induction of Surface Charges on Parallel Plates

In an insulating medium, the parallel plate construction of Fig. 2.1 can be treated as a simple capacitor of capacitance $C = Q/V$, where V is the potential applied across the plates and Q is the charge (of each sign) stored in the capacitor. The charge results from a transient current flow when the potential is applied, which exponentially decays to zero current due to the lack of a current conduction medium between the plates. The fact that even a transient current flows in such an "open" circuit is attributed to the mutual induction of surface charges on each plate by the surface charges on the other.

Charge induction may be understood as the natural perturbation of equilibrium charge densities by the coulombic interaction between charges. Thus, an evenly distributed surface density of positive charges on the anode trends to attract electrons in the cathode plate, forming an electron (negative charge) excess on the surface facing the anode. The positive surface charge that one would expect on the back side of the plate, resulting from the migration of electrons to the inner surface, is neutralized by electrons flowing through the connecting wire from the negative terminal of the power supply. A similar argument obviously applies to the response of the anode to charge on the cathode. Of course, the buildup occurs simultaneously rather than sequentially.

Considering plates of "infinite" extent leads to an infinite capacitance, but for the finite case in which each of the major dimensions of the electrodes is much greater than their separation, the charge is uniformly distributed over each electrode (neglecting edge effects).

2.3.2. Laplace's Equation and the "Lines of Force" Construct

If one were to "probe" the space between the electrodes of Fig. 2.1 with a charged particle, the surface charges on the plates would act to repel the probe from the plate containing like charge and attract the probe from the plate charged with unlike charge (to that of the probe). The collective force of all of the surface charges on the probe is characterized by the electric field, which is the force per unit charge on a positive point charge. Like force in mechanics, the electric field is a vector quantity, $\mathbf{E}$, with a direction as well as a magnitude. Also by analogy to mechanics, the potential, V, is the potential energy per unit of electric charge on the probe. The simple relationship between the two is that the vector gradient of the potential yields the field, or

$$\mathbf{E} = -\nabla V \tag{5}$$

The behavior of the electric field and potential in a charge-free region of space is described by the solution of the differential equation known as Laplace's equation,

$$\nabla \cdot \mathbf{E} = -\nabla^2 V = 0 \tag{6}$$

For our "one-dimensional" formulation of the problem, Laplace's equation simplifies to

$$\frac{dE}{dx} = -\frac{d^2 V}{dx^2} = 0 \tag{7}$$

The solutions are found by successive integration, and by evaluation of the constants of integration for the boundary conditions $V = 0$ at $x = 0$ and $V = V_{appl}$ at $x = W$:

$$E = -\frac{V_{appl}}{W} \tag{8}$$

$$V = \frac{x V_{appl}}{W} \tag{9}$$

The constant field between the plates, coupled with the mobility discussion in Section 2.2, above, implies that a probe ion in an atmospheric pressure buffer gas between the plates would traverse the distance between electrodes at constant velocity. The potential in Eq. (9) is referenced to a grounded cathode through the boundary conditions given, but the end results are the same for a grounded anode and a negative voltage applied to the cathode.

A convenient construct for describing the electric field is provided by *lines of force*. These imaginary lines follow the direction of the electric field, and their areal density in a plane perpendicular to their direction is proportional to the magnitude of the electric field. For the simple case of infinite parallel electrodes

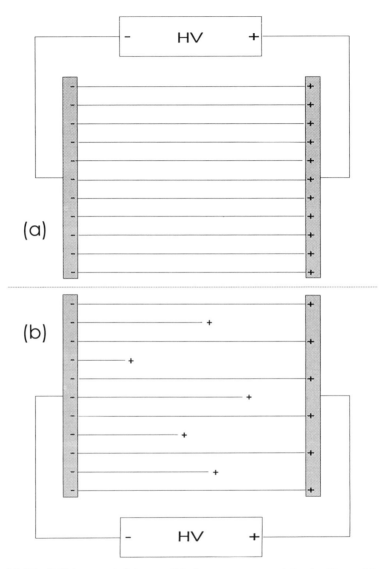

Figure 2.2. The field between infinite parallel plates as represented by the "lines of force" construct. (*a*) In the absence of free charges, the field is constant in *x*. (*b*) In the presence of free positive ions between the plates, the field magnitude decreases with distance from the cathode. In many discharges and flames, a positive ion excess exists as a dynamic equilibrium resulting from the constant production of ions and electrons in equal numbers, and the much more rapid removal of electrons by the field.

of the one-dimensional model, these lines are simply parallel to the x axis and equally spaced, as illustrated in Fig. 2.2a.

The lines of force terminate in the charges that give rise to the field. In our example, these charges are those induced on the electrodes and are represented by plus $(+)$ and minus $(-)$ signs along the electrode surfaces in the figure. Because of our "inverted" geometry, with the cathode at $x = 0$, the lines of force will terminate on charge pairs for which the negative charge is on the left in this discussion. The scaling between lines of force and charges is arbitrary, with one-to-one correspondence used for convenience here.

2.3.3. Poisson's Equation and the Interruption of Field Lines

Laplace's equation is only valid for regions of space containing no charge and is a special case of Poisson's equation:

$$\mathbf{V} \cdot \mathbf{E} = -\nabla^2 V = \frac{\rho}{\varepsilon_0} \tag{10}$$

where ρ is the charge density in coulombs (which may vary with position), and ε_0 is the permittivity of free space. In one dimension, Poisson's equation reduces to

$$\frac{dE}{dx} = -\frac{d^2 V}{dx^2} = (n_+ - n_-)\frac{e}{\varepsilon_0} \tag{11}$$

where we have used the number density of ions, n_+ (cm^{-3}), and electrons, n_- (cm^{-3}), along with the charge on a proton, e, in place of the charge density of Eq. (10).

The effect of positive charges between the electrodes in the simple geometry of our one-dimensional model is illustrated in Fig. 2.2b. Lines of force originating at negative charges in the cathode may be intercepted by positive charges in the intermediate volume. The density of lines of force obviously decreases from the cathode toward the anode, corresponding to a decrease in the electric field magnitude. [The positive slope of Eq. (11) for excess ion density is a decrease in field *magnitude* because the field *amplitude*, E, is negative.]

2.3.4. The Effect of Free Charges on the Surface Charge and External Circuit

The intermediate charges shield the anode from "seeing" some of the charges on the cathode, so that correspondingly fewer positive charges are induced on the anode. This effect is indicated in Fig. 2.2b by "missing" positive charges on the surface of the anode at the end of "intercepted" lines of force. Though the nonuniformity of charge distribution on the anode is unrealistic in this simple model, the reduction of charge on the anode is accurate.

The difference between Figs. 2.2a and 2.2b illustrates that free charge in the region between the plates influences the external circuit without requiring physical contact between the charges and either electrode. Specifically, the positive electrode of the power supply "sinks" fewer electrons from the anode in the latter case than in the former. This is an important concept in LEI that will be developed further below.

2.3.5. Image Charges and the Ionization Chamber Model

A nuclear ionization chamber furnishes a convenient analogy to LEI, without the complication of the flame between the electrodes. The following theory is taken directly from the nuclear ionization chamber literature of the 1950s (12) but applies reasonably well to LEI in a cool, non-hydrocarbon flame (e.g., H_2/air) with a reasonably high applied potential (> 1000 V). It will also adapt readily to other conditions, given an appropriate expression for the dependence of the electric field on x.

An isolated point charge, Q, in space gives rise to a potential given by

$$V = \frac{Q}{r} \tag{12}$$

and a field given by

$$\mathbf{E} = \frac{Q}{r^2}\left(\frac{\mathbf{r}}{r}\right) \tag{13}$$

where r is the radial distance from the charge, and $\mathbf{r}/r$ is a radial unit vector. Figure 2.3a illustrates such a charge with lines of force emanating radially. When a conducting plane is brought into proximity with the charge, the field must be altered to conform to the condition that the potential in a conductor is constant, or the vector component of the field is zero in the plane of the conductor.

The conditions required for the conducting plane are met at the plane bisecting the connecting axis between charges of equal magnitude and opposite sign, as illustrated in Fig. 2.3b. Thus, the "image charge" (13) construct shown in Fig. 2.3c is used to illustrate the solution of the "point-and-plane" problem. In reality, of course, the induced charges reside on the surface of the conducting plane, as illustrated by the minus signs at the intersections of the lines of force with the conducting surface. The surface density is not uniform,

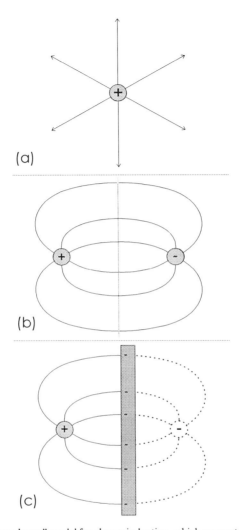

Figure 2.3. The "image charge" model for charge induction, which accounts for LEI signal being registered in an external circuit before charge reaches a sensing electrode. (*a*) The field of a free ion represented by radial lines of force. (*b*) The lines of force for a pair of opposite charges are normal to the plane bisecting the charges, indicating an equipotential surface. (*c*) An equipotential plane may also be imposed by a flat conductor held at a fixed potential. The equivalence between parts b and c is the image charge method for computing fields for certain simple geometries and for illustrating the concept of charge induction.

as it is for the parallel plates, and the sum of all of the negative charge on the surface is equal to $-Q$, the negative of the point charge. Motion of the external charge along x will not increase or decrease the total charge induced, although the distribution will change. Thus, if the plate were connected by

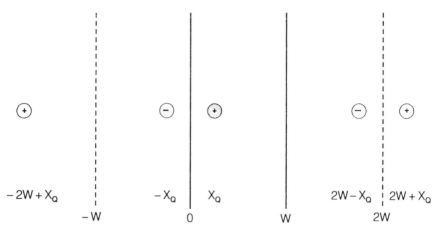

Figure 2.4. Two "orders" of the infinite progression of the image charge model for a real point charge (shaded) between real electrodes at $x = 0$ and $x = W$. The real positive charge shown results in the two negative images. The image at $-X_Q$ gives rise to the secondary image at $2W + X_Q$ through the $x = W$ electrode. The image at $2W - X_Q$ generates the image at $-2W + X_Q$ through the electrode at $x = 0$.

a wire to ground, no current would flow through the wire as the charge was moved.

However, a quite different circumstance exists if the charge is between *two* parallel plates. In this case, each plate "sees" not only the real point charge but also the image charges in the opposite plate. Figure 2.4. shows an example for a charge Q at position X_Q. (We will refer to the plate at $x = 0$ as the "cathode" and the plate at $x = W$ as the "anode" for consistency, even though the present proof requires no applied voltage.) The primary image charges (of value $-Q$), are located behind the plates at $x = -X_Q$ and $x = 2W - X_Q$. The second-order image charges are opposite in sign to the first order images and represent "reflections" of the first-order images in the opposite electrodes. This process continues and infinitum in the manner of an image in parallel mirrors.

As for the single conductor, charge conservation requires that the total charge actually induced on the two plates be $-Q$, but the distribution *between* the two plates is now

$$Q_c = -Q\left(\frac{W - X_Q}{W}\right)$$

$$Q_a = -Q\left(\frac{X_Q}{W}\right)$$

(14)

where Q_c is the charge induced on the cathode, and Q_a is the charge induced on the anode (12). This simple solution is not obvious from the recursive problem shown in Fig. 2.4, but the consistency of Eq. (14) with Fig. 2.4 may be readily confirmed for the limiting conditions of $X_Q \to 0$ and $X_Q \to W$.

Since the induced charge on each electrode changes with X_Q, it is clear that moving Q along the x direction would induce a current in a circuit connecting the two electrodes. Indeed, differentiating Eqs. (14) with respect to time gives cathode and anode currents of

$$i_c = \frac{dQ_c}{dt} = \frac{Q}{W}\left(\frac{dX_Q}{dt}\right)$$

$$i_a = \frac{dQ_a}{dt} = \frac{Q}{W}\left(\frac{dX_Q}{dt}\right)$$

(15)

In a simple ionization chamber, the electric field would be a constant, and hence the charge velocity in Eqs. 15 would be a constant. The current would be a "square" pulse beginning at the time the charge started to travel (e.g., when the voltage was turned on) and ending when the charge encountered an electrode.

Interestingly enough, the total charge transferred by integrating the current pulse is not Q, but rather is the fraction of Q corresponding to the fraction of the plate separation traversed by the moving charge. This would seem to violate some fundamental conservation law. However, an *ionization* event would produce equal and opposite charges, Q and $-Q$, at the laser position at time zero. These would move in opposite directions in response to the electric field and would produce current contributions of the same sign. The integral under the *total* current pulse resulting from the two charges is then equal to Q.

To proceed further with the derivation of the LEI pulse shape requires further development of the actual character of the field between the plates in the presence of an ion-producing flame.

2.4. THE PERTURBATION OF A FLAME BY AN EXTERNAL APPLIED FIELD

2.4.1. The Volume Ionization Rate and Three-Body Recombination Coefficient in an Ideal, Uniform Flame

Although there are many ionization and charge neutralization processes occurring simultaneously in the flame, we will consider primarily the thermal

ionization process discussed in detail in Chapter 1, Section 1.2. In Eqs.
(15)–(20) of that section, an ionization rate, k_{ion} (s^{-1}), is derived for a single
atom of species M immersed in an atmospheric pressure bath of species X at
a temperature T. Thus, if the number density of the species M in the flame is
given as n_M (cm^{-3}), then we can define a volume ionization rate, r_M (s^{-1} cm^{-3}),
for that species as

$$r_M = k_{ion} n_M \qquad (16)$$

Consider species M to be a metal (or matrix) aspirated into the flame with the
sample solution. It may be the analyte itself, but in general it suffices to
consider it to be a matrix element of low ionization potential and potentially
high concentration. Thus, r_M will scale linearly with the concentration of
species M in the solution being aspirated into the flame.

The unseeded flame is taken to contribute a constant volume ionization
rate of r_F (cm^{-3} s^{-1}) to the flame, yielding a total volume ionization rate,
r_c, of

$$r_c = r_F + r_M \qquad (17)$$

As discussed in Chapter 1, Section 1.2, a normal flame reaches ionization
equilibrium when the rate of ionization is equal to the rate of charge neutralization. As mentioned there, this process is of little relevance to the field-
perturbed flame of LEI, but, for completeness, we consider briefly the recombination rate, $k_{recomb.}$, of the three-body recombination process of Eq. (2)
of Chapter 1. This process, the inverse of collisional ionization [Eq. (1),
Chapter 1], requires the presence of a third body (X) to carry off excess energy
released in the recombination, but it may be treated kinetically as a second-
order process. We define a pseudo-second-order recombination coefficient,
α (s^{-1} cm^3), as

$$\alpha = k_{recomb.} n_X \qquad (18)$$

where n_X (cm^{-3}) is the number density of the bulk species X, which is
nominally constant for a given flame.

2.4.2. Ionization Equilibrium

At equilibrium, the volume ionization rate should just balance the volume
recombination rate, so that

$$r_c = \alpha n_+ n_- \qquad (19)$$

Without an applied field the plasma should not be charged, so that Eq. (19) yields

$$n_+ = n_- = \left(\frac{r_c}{\alpha}\right)^{1/2} \tag{20}$$

The treatment of Eqs. (19) and (20) simplifies the normal treatment of Saha equilibrium [see Boumans (14)] by treating r_c as a constant with respect to n_+ and n_- [whereas Eq. (16) shows that r_M depends on the neutral number density of species M, which would be reduced by increased ionization of M]. This simplification is consistent with our purpose in this chapter of establishing a reasonably simple and workable model for the detection mechanism in LEI, as opposed to ionization mechanisms, treated in Chapter 1.

2.4.3. Perturbation of Ionization Equilibrium by a Field

When an electric field exists throughout a flame, ions and electrons are transported through each volume element at the velocities and current densities described by Eqs. (1)–(4), above. If ions and electrons were extracted from each volume element at the same rate as they were introduced, then there would be no net change in changed particle density. Thus, at equilibrium in a field, the charged particle densities satisfy the solutions of the differential equations

$$\frac{dn_+}{dt} = r_c - \alpha n_+ n_- - \frac{1}{e}\frac{dj_+}{dx} = 0 \tag{21}$$

$$\frac{dn_-}{dt} = r_c - \alpha n_+ n_- + \frac{1}{e}\frac{dj_-}{dx} = 0 \tag{22}$$

For numerical modeling of field-perturbed flames, Eqs. (21) and (22) may be used as they stand. However, it is very helpful to have at hand the simple analytical solutions of Lawton and Weinburg (2), which follow from the assumption that

$$\frac{1}{e}\left|\frac{dj_+}{dx}\right|, \quad \frac{1}{e}\left|\frac{dj_-}{dx}\right| \gg \alpha n_+ n_- \tag{23}$$

or that field removal of ions and electrons totally dominates ion–electron recombination when a field of any magnitude exists. Using the assumptions of Eq. (23) in Eqs. (21) and (22), integrating the resulting equations, and applying appropriate boundary conditions yields

$$j_+ = -er_c(W - x) \tag{24}$$

$$j_- = -er_c x \tag{25}$$

$$j \doteq j_+ + j_- = -er_c W \tag{26}$$

Equations (24)–(26) show the reasonable result that the total current density, j (A/cm^2), is proportional to the volume ionization rate and the width of the flame, and that the proportion carried by ions and electrons varies linearly with position in the flame, assuming a nonzero field throughout the flame.

By equating Eqs. (24) and (25) with our earlier expressions for current density, defined in Eqs. (1)–(4), we can solve for the ion and electron number densities:

$$n_+ = -\frac{r_c(W - x)}{\mu_+ E} \tag{27}$$

$$n_- = -\frac{r_c x}{\mu_- E} \tag{28}$$

where the negative signs compensate for $E < 0$ in our sign convention. For a nearly constant electric field between parallel plates, segregation of *positive space charge* near the cathode and *negative space charge* near the anode is indicated. Also, since the electron mobility is 2–3 orders of magnitude greater that the ion mobility, it is clear that the positive space charge next to the cathode ($x = 0$) is much greater than the negative space charge near the anode ($x = W$). Since both charge densities vary linearly from zero at one electrode to their very different maximum value at the other, it is also clear that the flame exhibits a net positive charge, resulting from the higher electron mobility.

2.5. THE PERTURBATION OF THE FIELD BY THE FLAME

2.5.1. The Positive Ion Space Charge, or "Sheath"

The preceding discussion considered a "nearly constant" field, because Poisson's equation [Eq. (11)] states that the presence of a net charge introduces a field gradient. If we consider for the moment the behavior of the field near the cathode ($x = 0$), it is clear from Eqs. (27) and (28) that $n_+ \gg n_-$, and we solve

Poisson's equation for $n_- \approx 0$ and n_+ given by Eq. (27):

$$\frac{dE}{dx} = -\frac{e}{\varepsilon_0} \frac{r_c(W-x)}{\mu_+ E}$$

$$E\, dE = \frac{er_c}{\varepsilon_0\mu_+}(W-x)d(W-x) \tag{29}$$

$$E^2 = \frac{er_c}{\varepsilon_0\mu_+}(W-x)^2 + E_W^2$$

where E_W^2 is a constant of integration that is readily seen to be the square of the field at the anode ($x = W$). From Eq. (29), it may seen that the field magnitude decreases monitonically from the cathode ($x = 0$) to the anode. Indeed, it is possible to increase the plate separation (W) or the ionization rate (r_c) to the point where the field magnitude is reduced to zero at the anode, or even before the anode.

In this latter circumstance, the flame must be considered to consist of two distinct regions: (i) a positive ion space charge region, often referred to as the *cathode sheath* or simply the *sheath*, which extends from the cathode ($x = 0$) to some intermediate position ($x = X_s$), at which point the field reaches a value of zero; and (ii) an unperturbed region of the flame from $x = X_s$ to $x = W$.

Equations (24)–(29) were derived under the assumption that a field existed throughout the flame, i.e., for all values of $0 < x < W$. However, under the presently considered circumstance, the field only exists for $0 < x < X_s$, so that $W \to X_s$ in Eqs. (24)–(29). Equation (29) simplifies ($E_W \to 0$; $W \to X_s$) to give a field amplitude of

$$E = -\delta(X_s - x), \qquad x < X_s$$

$$\delta \doteq \left(\frac{r_c e}{\mu + \varepsilon_0}\right)^{1/2} \tag{30}$$

By substitution into Eq. (27), the positive ion number density in this region becomes

$$n_+ = \frac{r_c}{\mu_+\delta} \tag{31}$$

Recall that the electron density is taken to be approximately zero to obtain the analytical solutions given above. However, it is still necessary to consider an electron contribution to the current, which is given in the sheath by $W \to X_s$ in

Eqs. (24)–(26):

$$j_+ = -er_c(X_s - x) \qquad x \leqslant X_s \tag{32}$$

$$j_- = -er_c x, \qquad x \leqslant X_s \tag{33}$$

$$j = -er_c X_s \tag{34}$$

In region 2, $E = 0$ and the electron and ion number densities are given by Eq. (20). In fact, the density of free charges is higher than in region 1, and the region acts as a good conductor. Thus, the current of Eq. (34) is not restricted to region 1 but is constant across both regions. If the current flow in the absence of a field is troublesome to the reader, the region can be considered to be equivalent to a conducting wire, and one can use a "compressibility" argument that inserting an electron into one end of the wire results in an electron exiting the opposite end.

The foregoing treatment has been extended by Magnusson (3) to include "cold" regions between the flame edge and the electrode on both sides of the flame.

2.5.2. Extent of the Sheath

The position of the sheath edge may be found by requiring that the integral of the field from $x = 0$ to $x = X_s$ must be the applied potential, V_{appl}, which yields

$$X_s = \left(\frac{2V_{appl}}{\delta} \right)^{1/2} \tag{35}$$

Thus, the extent of the sheath is subject to some experimental control via the applied potential. Curve a of Fig. 2.5 shows the electric field magnitude determined from Eqs. (30) and (35) for an applied voltage of half the value required to extend the sheath to the anode. The sheath edge may be seen to occur for $x = X_s = W/\sqrt{2}$. On the other hand, the extent of the sheath may be seen from Eqs. (35) and (30) to depend inversely on the fourth root of r_c and thus on the presence of easily ionized elements in a sample introduced into the flame.

From Eq. (34) it may be seen that the current extracted from the flame by the applied field varies linearly with X_s and hence as the square root of V_{appl}.

2.5.3. Electric Field "Saturation"—and Beyond

For combinations of r_c, V_{appl}, and W that would yield a computed value of X_s [from Eq. (35)] greater than or equal to W, the electric field fills the total volume between the electrodes, and the flame is said to be *electrically saturated* (2). Trace b for Fig. 2.5 is an illustration of the exact saturation condition, corresponding to either Eq. (29) with $E_W = 0$ or to Eq. (30) with $X_s = W$. At

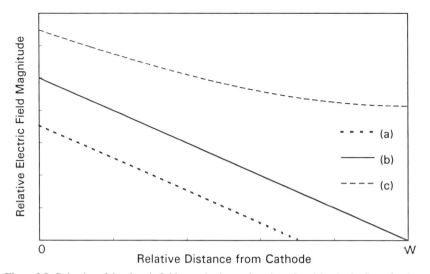

Figure 2.5. Behavior of the electric field magnitude as a function of position in the flame for three different applied voltages and identical flame conditions (i.e., volume ionization rate). Curves: (*a*) The applied voltage is half the value required for saturation; the flame experiences a region of zero electric field from the sheath edge at $W/\sqrt{2}$ to the anode. (*b*) The applied voltage is just that required for saturation, filling the flame, but with a slope equal to that of curve *a*, the subsaturation case. (*c*) The applied voltage is twice that required for saturation, and the field magnitude is no longer linear with distance from the cathode; in fact, the field magnitude eventually approaches the limit of being constant with position, as the applied voltage is increased.

saturation and beyond, Eqs. (24)–(29) are valid as they stand and the current that can be extracted from the flame is no longer a function of applied potential but has the constant value indicated by Eq. (26). The field, as given in Eq. (29), is no longer linear with position, and the constant of integration, E_W, is not trivial to evaluate analytically. Since the integral of the field from $x = 0$ to $x = W$ must be the applied voltage, it *is* easy to evaluate by numerical computation, however. Trace *c* of Fig. 2.5 illustrates the supersaturation condition, with E_W adjusted iteratively to result in an integral twice that of curve *b*. This is, the applied voltage for trace *c* is exactly twice that for the exact saturation case of trace *b*.

2.6. THE POINT CHARGE MODEL FOR LEI USING PULSED LASERS

2.6.1. The Electron Pulse

The *point charge model* of the LEI current pulse provides a reasonably simple analytical expression by which to understand the general features of the LEI signal (4–6). The model supposes that a pulsed laser deposits charges Q of ions and $-Q$ of electrons at a position X_L in the flame at time $t = 0$. (Of course, the ions and electrons result from the LEI production process as described in Chapter 1.) The model is derived for an electrically unsaturated flame, for which the field magnitude decreases linearly from the cathode to the sheath edge and the laser beam is within the sheath $(0 < X_L < X_s)$.

From Eq. (15) it may be seen that the electron point charge will induce an equivalent current in either electrode (and hence the external circuit) that is proportional to the charge and the velocity. The velocity of each electron is given by Eq. (3), with the electric field given by Eq. (30). Indeed, combining Eqs. (3), (15), and (30) yields an expression for the electron current, i, induced in the external circuit as a function of the *position* of the point change as it is moved by the field from X_L (where it was deposited) to X_s (where the field vanishes):

$$i_-(x_-) = - Qv_- = Q\mu_- E = - Q\mu_- \delta(X_s - x_-), \qquad 0 < x_- < X_s \quad (36)$$

where x_- is the instantaneous *position* of the negative point charge, $-Q$, at time t. To express the current in terms of its *time* dependence, which is what we register experimentally, we note that x_- is related to elapsed time through the velocity, or

$$\frac{dx_-}{dt} = - \mu_- E = \mu_- \delta(X_s - x_-)$$

$$\int_{X_L}^{x_-} \frac{d(X_s - x)}{X_s - x} = - \int_0^t \mu_- \delta \, dt \qquad (37)$$

$$\ln\left(\frac{X_s - x_-}{X_s - X_L}\right) = - \mu_- \delta t$$

$$(X_s - x_-) = (X_s - X_L)e^{-\mu_- \delta t}$$

Substituting the last line of Eq. (37) into equation 36 gives

$$i_- = - Q\mu_- \delta\left(\frac{X_s - X_L}{X_s}\right)e^{-\mu_- \delta t} \qquad (38)$$

The idealized electron pulse thus rises instantaneously to a maximum amplitude proportional to the deposited charge, the electron mobility, the field slope, and the fraction of the sheath width to be traversed by the electrons. The pulse decays exponentially with a rate constant given by $\mu_- \delta$, or a characteristic time constant, τ_-, defined by

$$\tau_- \doteq \frac{1}{\mu_-\delta} \tag{39}$$

The exponential decay of the pulse results from the direction of travel of electrons in the field, toward increasing x, which corresponds to decreasing field magnitude. Thus the electrons are traveling their highest velocity at $x = X_L$ and $t = 0$, and approach zero velocity asymptotically as they approach $x = X_s$. Curiously, the electrons are never required to "reach" the true anode. This does not imply that a negative charge builds up at X_s, since the unperturbed portion of the flame acts as a conductor. In fact, the plane $x = X_s$ may be thought of as a "virtual anode."

The electron pulse may be given a more experimentally realistic profile by convolving the raw current profile with an exponential circuit response function, with a characteristic electronic time constant of τ_e, to give an experimental electron "signal" of (15)

$$s_-(t) = \tau_e \int_0^{+\infty} \exp(-t'/\tau_e) i_-(t-t')\, dt'$$

$$= -Q\left(\frac{X_s - X_L}{X_s}\right)\left(\frac{\tau_- - \tau_e}{\tau_-}\right)[\exp(-t/\tau_-) - \exp(-t/\tau_e)] \tag{40}$$

2.6.2. The Ion Pulse

Ions travel in the direction of decreasing x, and hence increasing field. The ions are deposited at X_L at time $t = 0$, and travel with increasing velocity until they encounter the cathode at $x = 0$. The derivation parallels that for the electrons, with the result that

$$i_+ = -Q\mu_+\delta\left(\frac{X_s - X_L}{X_s}\right)e^{\mu_+\delta t}, \qquad 0 < t < t_{arr}$$

$$= 0, \qquad t > t_{arr}$$

$$t_{arr} \doteq \tau_+ \ln\left(\frac{X_s}{X_s - X_L}\right), \qquad X_L < X_s \tag{41}$$

$$\tau_+ \doteq \frac{1}{\mu_+\delta}$$

where t_{arr} is the arrival time for the ions at the cathode, or the time required for them to traverse the distance from X_L to the cathode. The ion pulse is seen to increase exponentially from the time the charges are deposited until they intercept the cathode, at which the current abruptly returns to zero. This is a very distinctive shape as compared to the electron pulse and is rarely observed in LEI because of the relative amplitudes and time scales of the electron and ion pulses. Note that the signal amplitude at time $t = 0$ differs in only one factor—the charged particle mobility—which results in a difference in signal of orders of magnitude. Similarly, the pulse time constants differ by the mobility ratio as well, so that the ion pulse is much lower in magnitude and very spread out in time relative to the electron pulse.

2.7. NUMERICAL MODELING

2.7.1. Diffusion and the Continuity Equation

The point charge model can be convolved with charge distribution functions to account for the finite extent of the laser beam depositing the charges and even to account for diffusional broadening of the charge cloud with time. However, it is equally feasible to consider a total numerical solution for the problem, including the effects of diffusion. The *continuity equations* regulating the time dependence of charge density at each point in the flame are now written to include the effect of diffusion, along with the mechanisms we have already considered for gain and loss of charge in an element of volume (4):

$$\frac{dn_+}{dt} = r_c - \alpha n_+ n_- - \frac{d}{dx}\left\{ + \mu_+ E n_+ - D_+ \frac{dn_+}{dx} \right\}$$

$$\frac{dn_-}{dt} = r_c - \alpha n_+ n_- - \frac{d}{dx}\left\{ - \mu_- E n_- - D_- \frac{dn_-}{dx} \right\}$$

(42)

where D_+ and D_- are diffusion coefficients for the ion and electron. The first two terms on the right of each equation represent the volume ionization and recombination rates, as discussed earlier. The terms in braces are fluxes (particles per square centimeter per second) across opposite faces of the volume element. The negative spatial differential of the flux, i.e., the flux *into* a volume element less the flux *out of* the same volume element, indicates the volumetric rate of number density increase due to nonuniform charge transport. The first term in braces is the field transport term and is related to the corresponding current densities (j_+ and j_-) through the charge on an ion (e) or an electron $(-e)$. The diffusional term, of course, represents the tendency of any particle—neutral as well as charged—to migrate from a region of high density to a region of low density. Thus, the diffusional flux term (the second in the braces) has a negative sign to indicate migration *in the opposite direction* to the density gradient. Mobility and diffusion are both limited by collisional ("frictional") retardation of motion, so it is not surprising that a relationship exists between the two coefficients (16):

$$D_{\pm} = \frac{kT}{e} \mu_{\pm} \qquad (43)$$

The calculations below will require some "typical" values for various coefficients (3). We will employ an ion mobility of $\mu_{+} = 20 \, \text{cm}^2 \, \text{V}^{-1} \, \text{s}^{-1}$, an electron mobility of $\mu_{-} = 2000 \, \text{cm}^2 \, \text{V}^{-1} \, \text{s}^{-1}$, a flame temperature of $T \approx 2500 \, \text{K}$ (which yields $kT/e \approx 0.2 \, \text{V}$), an ion diffusion coefficient of $D_{+} = 4 \, \text{cm}^2 \, \text{s}^{-1}$, and an electron diffusion coefficient of $D_{-} = 400 \, \text{cm}^2 \, \text{s}^{-1}$. For computing fields we will use the ratio $e/\varepsilon_0 = 1.807 \times 10^{-6} \, \text{V cm}$, and for recombination in the field free region of the flame we will use an estimate of $\alpha = 10^{-7} \, \text{cm}^3 \, \text{s}^{-1}$ (17).

2.7.2. Electron Charge Profiles for Laser-Induced Charges in a Fixed Field

A simple yet instructive numerical model treats only *excess* charges deposited by the LEI event and subject to a fixed field in the flame, as given by Eqs. (30) and (35). For this model, the first two terms in Eqs. (42) may be omitted as pertaining to the steady state background only. In fact, we will only be solving the equations in the sheath, for which the second term approaches zero anyway, due to the low electron density.

Equations (42) are further simplified by substituting Eq. (30) for the field, E, and substituting the unitless variables

$$\hat{n}_{\pm} \doteq \frac{n_{\pm} r_c}{\mu_{+} \delta}$$

$$\hat{x} \doteq \frac{x}{X_s} \qquad (44)$$

$$\hat{t}_{\pm} \doteq \frac{t}{\tau_{\pm}}$$

Here the number density has been ratioed to the steady-state background ion density in the sheath [Eq. (31)], the distance has been normalized to the sheath width [Eq. (35)], and the time has been ratioed to the characteristic time of the signal pulse [Eqs. (39) and (41)]. Equations (42) then simplify to

$$\frac{d\hat{n}_\pm}{d\hat{t}_\pm} = \frac{d}{d\hat{x}}\left[\pm(1-\hat{x})\hat{n}_\pm + \frac{kT}{2e|V_{appl}|}\frac{d\hat{n}_\pm}{d\hat{x}}\right] \tag{45}$$

where Eq. (43) has been used to substitute for a ratio of diffusion coefficient to mobility, and the absolute value of the applied voltage has been used to remove any potential sign ambiguity.

The appendix contains the source-code listing of a QuickBasic (Microsoft, Redmond, Washington)[1] computer program, LEIMODEL.BAS, based on Eq. (45). The program is designed to trace the time history of either ions or electrons deposited as a Gaussian distribution in an idealized sheath at reduced time zero ($t = 0$). The user is asked by the program to supply the initial position and distribution width of the laser-produced charges and the ratio of the maximum added charge density to the background ion density, as well as to specify whether ions or electrons are to be followed. [The program *could* be written to accommodate ions and electrons simultaneously, as in reality, but the formulation of Eq. (45) specifies two different time coordinates, which differ by more than 2 orders of magnitude, so it is convenient to examine the species independently.] The program writes output to "comma-separated variable" (filename.csv) files, which are imported into a spreadsheet (Excel 4.0, Microsoft) for graphic display. Figures 2.6 through 2.12 were generated using output from the program.

Figure 2.6 shows several stages, at intervals of 0.5 in reduced time units, in the time history of an electron distribution with initial position $X_L = 0.4$ and width $W_L = 0.2$ in the normalized, unitless spatial coordinate. It is apparent that the distribution *narrows* as it moves, consequently growing in amplitude (since the integral, or charge, is constant). This behavior is a consequence of the negative slope of the electric field magnitude, which causes the electrons at the trailing edge to advance on the electrons on the leading edge. It also points out that the effect of diffusion, the second term in Eq. (45), is small, since this effect would *broaden* the distribution. Indeed, for "typical" values of $kT/e \sim 0.2$ V and $|V_{appl}| \sim 1000$ V (for a C_2H_2/air flame) the leading coefficient requires steep spatial gradients in number density or positions near the anode to be comparable to the first term.

After several attempts to remove the diffusion term to speed up the calculation, it was discovered that the term stabilizes the computation for

---

[1] In order to adequately describe procedures, it is occasionally necessary to identify commercial products by the manufacturer's name or label. In no instance does such identification imply endorsement by the National Institute of Standards and Technology, nor does it imply that the particular products or equipment are necessarily the best available for that purpose.

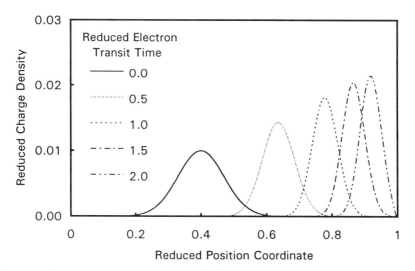

Figure 2.6. The time evolution of an electron distribution deposited at a reduced position of 0.4 with a Gaussian full-width of 0.2. The initial distribution and distributions at increments of 0.5 units in the reduced time parameter for electron transport are shown (as indicated in the legend on the art).

small values of W_L and large values of $\hat{x}$, where the density gradient becomes large. Thus, a slightly inflated diffusion term is accommodated in the program $(kT/eV = 10^{-3})$, which negates the narrowing effect of Fig. 2.6 for the minimum permitted value of $W_L = 0.1$ in the program.

2.7.3. Electron Current Induction for Laser-Induced Charges in a Fixed Field

Induced current is calculated as for the point charge model, with each element of charge giving an induced current proportional to the charge times the field [Eq. (36)]. Since the charge is now a distribution rather than a point charge, the total current is an integral over the current increments:

$$\hat{j}_{\pm}(\hat{t}) = - \int_{\tilde{x}=0}^{1} (1 - \tilde{x})\tilde{n}(\tilde{x}, \hat{t}) \, d\tilde{x} \tag{46}$$

where $\tilde{x}$ is a dummy integration variable and $\hat{j}$ is a unitless current density. In the program, of course, the integral becomes a sum and the unitless current density is arbitrarily normalized.

Figure 2.7 shows computed current pulses, compared to the point charge model [Eq. (38)], for several values of the starting position, X_L, and for an

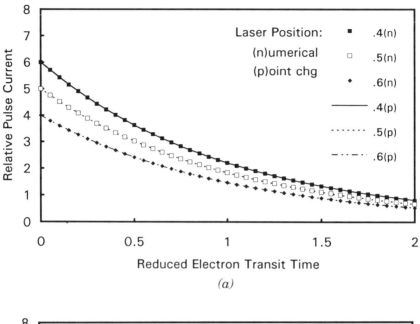

(a)

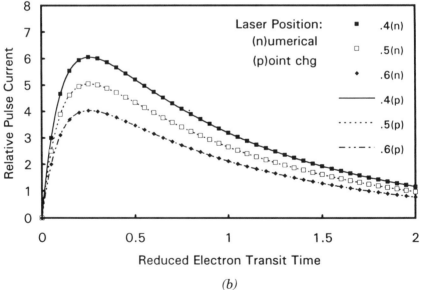

(b)

Figure 2.7. Relative electron pulse currents computed numerically (symbols) and analytically (lines) for three different laser positions, as indicated in the legend on the art: (*a*) raw current; (*b*) same currents as in part a, but subjected to an amplifier rise time of 0.1 reduced time units.

electronic time constant of $\tau_e = \tau_-/10$ for part b of the figure. The circuit response for the numerical model is a discrete, numerical version of the convolution given in Eq. (40) for the point charge model. It is apparent that the tendency of the field gradient to skew the initial Gaussian charge distribution has a negligible effect on the current computation, which is indistinguishable from the point charge model. Although only one distribution width is used for Fig. 2.7, the same result applies regardless of width (for electrons). The relative current scale was arbitrarily normalized to emphasize the 6:5:4 amplitude ratio corresponding to the starting positions of 0.4, 0.5, and 0.6, respectively, in accordance with Eqs. (38) and (40).

The "unitless" formulation reduces the number of figures required to represent electron distributions and currents, but it should be remembered that the abcissa scale of Figs. 2.6 and 2.7 is a function of the flame ionization rate (and hence the matrix aspirated) and the applied field. In real units the electron time constant, τ_-, may be on the order of 100 ns (3) and the sheath width, X_s, may be on the order of a centimeter (18). The variation in the decay time of the electron pulse with matrix and potential [Eqs. (30) and (39)] has rarely been observed, since "practical" LEI preamplifiers have often been too slow (~ 1 MHz bandwidth) to resolve the true shape. With the constant improvements in fast electronics, it is likely that electron pulses will be routinely resolved in the future.

2.7.4.   Positive Ion Charge Profiles for Laser-Induced Charges in a Fixed Field

Figure 2.8 shows several instantaneous ion profiles for an initial ion charge distribution centered at $X_L = 0.6$, with a width of $W_L = 0.2$. The expansion and amplitude reduction of the ion distribution with time is due not to diffusion but to the field gradient, as with the electrons, only now the field gradient is reversed with respect to the motion.

Unlike the electrons, whose velocity slows to zero at the anode, the ions reach the cathode at maximum velocity, and *boundary conditions* become an important consideration in the numerical computation. One may argue that "realistic" boundary conditions would require that the free charge density be zero in the electrodes, but as the charges approach the boundary this condition causes steep gradients and computational instabilities. Hence, the debatable boundary condition employed here is that the first derivative of the number density be constant across the boundary. This condition causes the ions to appear to traverse the cathode as though it were "transparent" to them in Fig. 2.8.

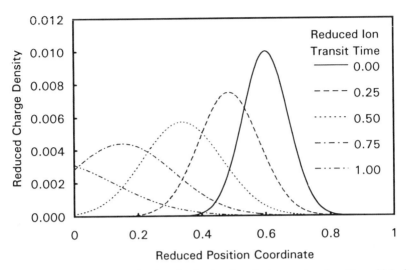

Figure 2.8. The time evolution of an ion distribution deposited at a reduced position of 0.6 with a Gaussian full-width of 0.2. The initial distribution and distributions at increments of 0.25 units in the reduced time parameter for ion transport are shown, as indicated in the legend on the art.

2.7.5. Positive Ion Current Densities for Laser-Induced Charges in a Fixed Field

Figure 2.9 shows the current density profile corresponding to the conditions of Fig. 2.8, as well as for two different initial positions, X_L. The initial value scales with X_L in the same manner as the electrons, but the amplitude is determined by both the initial position and the resulting width at the time of arrival. The effect of the distribution width clearly distinguishes the ion current pulse from the prediction of the point charge model [Eqs. (41)], which is shown in the figures for comparison.

The characteristic time for ions is on the order of tens of microseconds, and the ion pulse may be readily resolved by conventional electronics. Consequently, no time response filter is employed in Fig. 2.9.

2.7.6. Field-Distortion Effects for Large Laser-Induced Charge Densities

Since the ion density was normalized [Eq. (44)] to the ion density of the steady-state sheath, Eq. (45) is *actually* valid only for reduced number densities $\ll 1$. That is, the simple linear field assumed that the laser-produced charge densities are negligible with respect to the steady state background positive charge density. Indeed, an *added* (by the laser) number density of 1 would bring

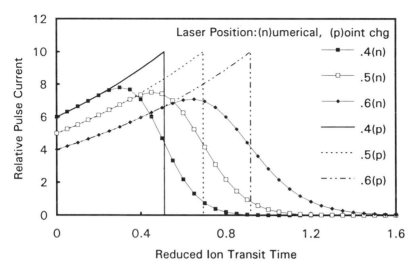

Figure 2.9. Relative ion pulse currents computed numerically (symbols) and analytically (lines) for laser positions of 0.4, 0.5, and 0.6 reduced position units, as indicated in the legend on the art.

the local number density to 2 reduced units, thereby doubling the local field gradient [see Eq. (11)]. Figure 2.10 illustrates the effect of large LEI charges on the field at $\hat{t}_- = 1.0$ for initial Gaussian amplitudes of 0.5, 1.0, and 2.0 reduced units.

Figure 2.10 was computed using the same program as that used for computing Figs. 2.6 and 2.8, but with larger values of the input variable representing the normalized charge added by the laser. The program uses the background electric field *magnitude* of $1 - \hat{x}$, with the slope perturbed by the local added charge density through Poisson's equation. From Fig. 2.10 it is evident that the perturbation tends to reduce the area under the field, which must remain constant, since the area under the field is the applied potential. Hence, the algorithm in the program includes a field offsert parameter adjusted to yield constant area, such that the perturbation actually changes the value and extent of the background field, in addition to the local perturbation at the position of the perturbing charges.

The "motion" of the y and x intercepts of the field occurs as a function of time as well as charge density, as shown in Fig. 2.11. The movement of the x-intercept, or sheath edge, during the electron transit causes appreciable shifts in the background charge density at the sheath edge, with consequences that are discussed later in this chapter. The unperturbed field at a reduced time of zero results from the fact that the initial electron and ion distributions perfectly cancel. Indeed, the program LEIMODEL retains a static "ion distribu-

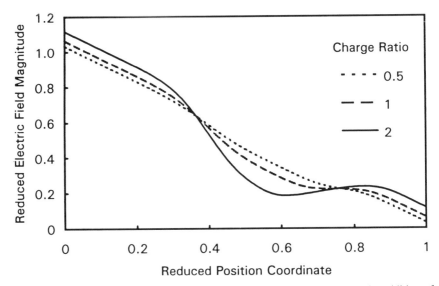

Figure 2.10. Profiles of the reduced electric field magnitude as perturbed by the addition of laser-produced charge distributions whose number densities are comparable to that of the background ions in the sheath. Ions and electrons were simultaneously deposited at a reduced position of 0.4 and a width of 0.2, with the ratio of maximum charge density to that of the background ions being 0.5, 1.0, and 2.0, as given in the legend on the art. The profiles are shown at a reduced time of 1.0 for electron transit.

tion" for calculations of electron transport, strictly for the computation of the electric field distortion. The effects of this static ion distribution may be observed about the reduced position of 0.4 in Figs. 2.10 and 2.11.

2.7.7.   Current Pulse Distortions

Obviously, distortion of the field in the manner illustrated in Figs. 2.10 and 2.11 will influence the transport of the charge distribution and hence the resulting current pulse. This can be accommodated in numerical computation by including an array for the reduced electric field and recomputing this field after every time iteration, in the manner of Figs. 2.10 and 2.11.

Figure 2.12a shows the field perturbation for a perturbing charge of two times the density of the background ions, with the corresponding electron and static ion density distributions superimposed, at a reduced time of 1.0. Also plotted is the "ideal" electron distribution that would have resulted without field distortion. Notice that the field slope at the highest electron density is actually reversed with respect to the original field. Thus, leading electrons are

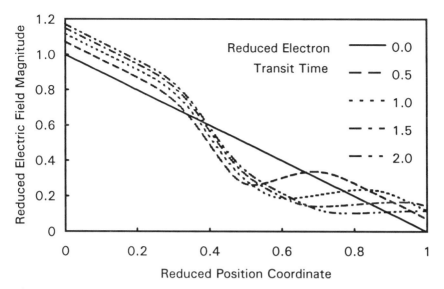

Figure 2.11. Time dependence of the perturbation of the reduced electric field magnitude by laser-induced ion and electron charge distributions whose maximum number density is twice the number density of the background ion distribution. The field profiles are shown at intervals of 0.5 in the reduced time parameter for electron transport. No perturbation is evident at a reduced time of zero because the effects due to the ion and electron distributions exactly cancel.

translated more rapidly than trailing electrons, unlike the situation shown in Fig. 2.4, where the electrons were compressed by the field. Similarly, the field slope magnitude near the ion maximum (at 0.4) is greater than the unperturbed case, eventually causing the ion broadening to exceed that of Fig. 2.7. Thus, both charge distributions experience *coulombic repulsion*—tending to explode outward—when their density becomes comparable to the steady-state ion density in the sheath (which establishes the "normal" field slope). In addition, much of the field magnitude between the ion and electron charge distributions is reduced with respect to the unperturbed field. The behavior may be interpreted as *coulombic **attraction*** between the adjacent edges of the opposite charge distributions. Thus, the combined effects of self-repulsion and mutual attraction contribute to the observed distortion of the charge distribution and current pulse.

Figure 2.12b shows the current pulse resulting from the entire time history of the computation from which Fig. 2.12a was taken at a particular time (1.0). Comparison with the point charge model shows a significant influence due to the field perturbation. Computations for lower charge ratios (included in the figure) approach agreement with the point charge model.

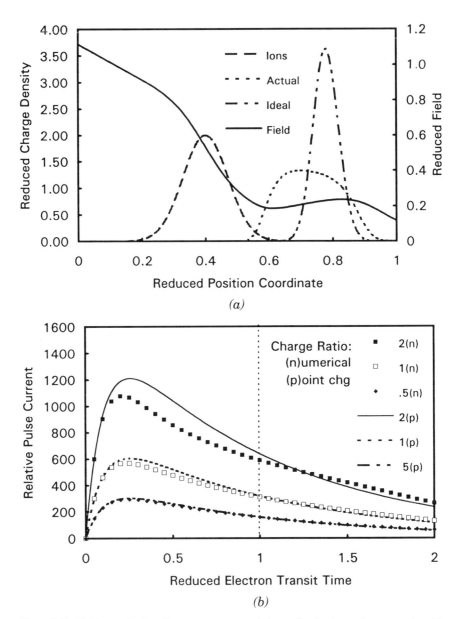

Figure 2.12. Field perturbation effects on a transported charge distribution and current pulses. (*a*) The electron and ion distribution for a charge ratio of 2, at a reduced time for electron transport of 1.0 following deposition at a position of 0.4 with a width of 0.2. The "ideal" electron distribution, as it would be in the absence of field perturbation, and the perturbed field are also shown, as indicated by the legend on the art. (*b*) Current pulse profiles arising from charge ratios of 0.5, 1.0, and 2.0, computed numerically (symbols) as compared with the point charge model (lines). The vertical dotted line indicates the "sampling" point for the distributions of part a. An amplifier time constant of 0.1 reduced time unit is assumed.

2.7.8. Modeling the Flame Background

It is entirely possible to numerically model the response of the flame background ions and electrons to an applied potential (19,20) using all of the terms of Eq. (45), instead of beginning with the analytical solutions of Lawton and Weinburg (2). Once the background computation has achieved a steady state, a packet of "laser-produced" ions and electrons may be "deposited" and the response of the entire system may be studied. As shown in the results of Schenck et al. (20), reproduced in Fig. 2.13, the LEI event creates interesting

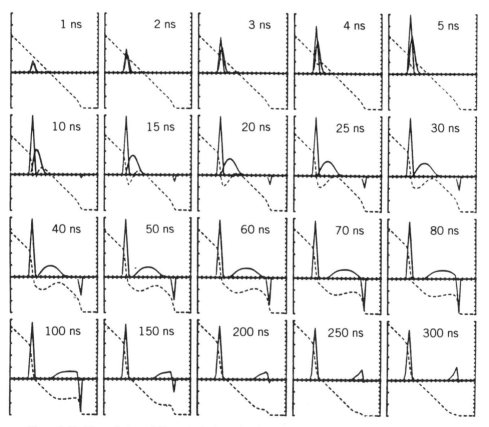

Figure 2.13. Numerical modeling calculations showing calculated changes in electron and ion number densities following incremental charge introduction (over the first 5 ns) by a laser. The calculation includes background ion and electron production and recombination, and the densities shown are *differences* with respect to the unperturbed background. Thus, the "negative" electron densities shown are electron density reductions with respect to the steady-state condition. Reprinted with permission from Schenck et al. (20).

perturbations in the electron population at the sheath edge. This effect in the background equilibrium near the sheath edge may be understood by further examination of Fig. 2.11, which implies that the sheath edge extends beyond the plotted range in response to the perturbation of added charges.

As the sheath edge moves, background electrons that had been in the field-free region suddenly experience a field and are removed by the field. Since the plot in Fig. 2.13 shows the *difference* in the electron density with that in equilibrium, the decrease in electron density appears as a negative excursion in the plot. Thus, as the electron distribution moves from the original spot of deposition toward the sheath edge, an electron deficiency opens up at the sheath edge, which is later filled by the arriving electron distribution.

The perturbation in background charge density at the sheath boundary can be viewed as a physical manifestation of charge induction when the anode is a plasma rather than a conducting solid. Thus, the "image charge" at the sheath boundary, which functions as a virtual anode, appears as soon as the charges are separated and long before the electrons reach the sheath boundary, at which point the real charges and "image charges" effectively neutralize each other.

Codes for fully modeling the flame background with the LEI event superimposed are not given here for several reason: they were written for earlier computer generations (19,20), have not been recoded for the current generation of computer, and were usually unstable. The principle reason for the instability is that the set of differential equations is considered "stiff," as defined by applied mathematicians. That is, because of the great difference in electron and ion mobilities, the appropriate incremental time intervals for electron motion were entirely different than those for ion motion; however, since the two species are always both present in the complete model, the electron-appropriate value must be used to prevent divergence of the calculation. Thus, the simple program given earlier considers electrons and ions separately, driven by a perturbed "point charge model" field that represents the effect of background charges indirectly.

Of course, there is nothing inherently wrong with coding the general problem, running the entire program at the electron-appropriate time interval, and simply letting it run overnight on a desktop computer. Also, a reasonably informative result would be obtained by assigning only a factor of 10 difference to the ion and electron mobilities, greatly shortening the computation. However, other instabilities and uncertainties in the early codes concerned the boundary conditions at the electrode surfaces and minor details of bin symmetry in the coding. For the calculations shown in Fig. 2.13, the boundary conditions call for all charge concentrations to be zero in the electrodes, leading to prominent diffusion layers, which are accentuated by an oversize effective diffusion coefficient. This "effective diffusion," which was discovered only in the course of writing this chapter, is a computational artifact resulting from the use of inappropriately large spatial bin widths with respect to the concentration gradients considered and the actual diffusion coefficient. Thus, it may be appropriate to revisit these calculations with this enlightenment and modern desktop computers.

Obvious benefits of total modeling of the flame background would include the capability of studying nonuniform flames, with temperature and concentration gradients in x. It would also permit the study of mixed, multiple-background ion sources.

2.8. RELATIONSHIP OF THEORY TO LABORATORY RESULTS

2.8.1. Current vs. Voltage Curves in Analytical Flames

Much information may be gleaned from the fairly simple expedient of plotting the current extracted from a flame as a function of the potential applied across the flame. Figure 2.14 shows the result of several experiments of this nature for parallel plate electrodes arranged very close to the opposite sides of a H_2/air laminar flow flame (1.25 cm plate separation) (21). The curves may be seen to rise with a square-root-type dependence on voltage, as predicted by Eqs. (34) and (35). The slope at a given low value of voltage may be seen to increase with the concentration of an easily ionizable element (EIE), sodium in this example. This effect may be understood through Eqs. (30), (34), and (35) and the linear dependence of the volume ionization rate, r_c, on the concentrations of ion-producing species.

Eventually, the curves of Fig. 2.14 saturate at a current density

$$j_s = - er_c W \tag{47}$$

derived from Eq. (34) by substituting the flame width for the sheath width. Equation (47) may be seen as a convenient way to estimate realistic volume ionization rates for measured saturation currents, flame–electrode overlap areas, and estimated flame widths.

Our one-dimensional model took the parameter W to describe both the width of the flame and the spacing of parallel plate electrodes. In reality, of course, plate electrodes may be spaced much further apart than the width of the flame (3), and such spacing permits cooler operation of the electrodes. The family of curves shown in Fig. 2.15 is again for the hydrogen/air flame with sodium aspirated, but this time several electrode spacings are shown, all including cold buffer zones between the flame and the electrodes. Since the flame conditions are identical for all cases shown, the saturation current is

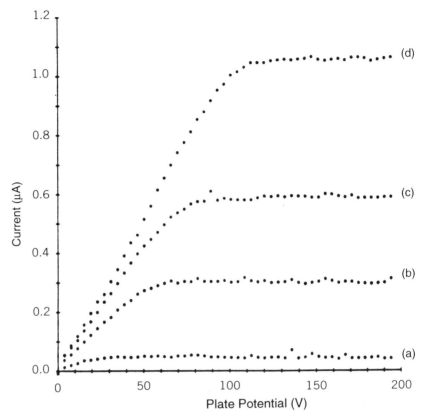

Figure 2.14. Background current as a function of applied voltage (V/I curves) for parallel plate electrodes arranged very close to the opposite sides of a H_2/air laminar flow flame (1.25 cm plate separation) and aspirated sodium solution concentrations of (a) 0, (b) 1, and (c) 2 ppm Na, without laser irratiation, and (d) 1 ppm Na with 200 mW laser irradiation at 589.0 nm. Reprinted with permission from Schenck et al. (21).

identical as well, indicating the collection of all charges created by thermal ionization, regardless of plate spacing, given adequate applied potential.

Below saturation, the functional dependence of current on applied voltage may be seen to depend on the plate spacing and differ significantly from the functional prediction for a uniform flame filling the space between the electrodes. This behavior is beyond the scope of our theoretical treatment, but the interested reader may refer to the treatment by Lawton and Weinberg (2) of the extreme of an infinitely thin ion source between parallel plates at atmospheric pressure and the treatment of Magnusson (3) for intermediate cases.

Current vs. voltage curves in the more common air/acetylene flame, and with the recommended immersed electrode configuration, rarely behave as

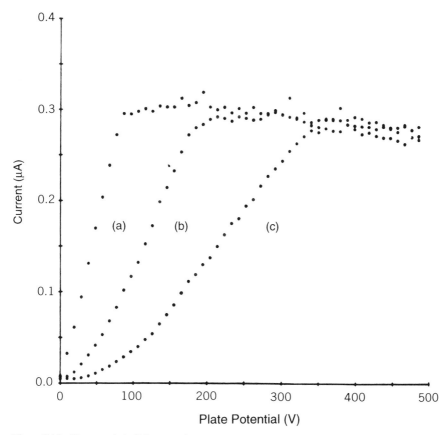

Figure 2.15. Characteristic V/I curves for a 1 ppm Na solution aspirated into a H_2/air laminar flow flame with parallel plate electrodes separated by (a) 1.5, (b) 2.0, and (c) 2.5 cm. Reprinted with permission from Schenck et al. (21).

much like the one-dimensional theory as the examples above for a number of reasons. (i) The volume ionization rate, r_c, arises from a number of species. In the seeded hydrogen/air flame, reasonable concentrations of a seeded EIE may dominate. (ii) The volume ionization rate is not constant over the volume of the flame, due to variations in temperature, species concentrations, and electric field across the probed region. The Saha equation explicitly predicts the variation with temperature and species concentration and implicitly predicts the variation with field resulting from the reduced concentrations of ions and electrons available for recombination. (iii) The combustion zone of the flame is an active participant in the immersed electrode configuration. The combustion zone is taken to be especially vulnerable to the field dependence of

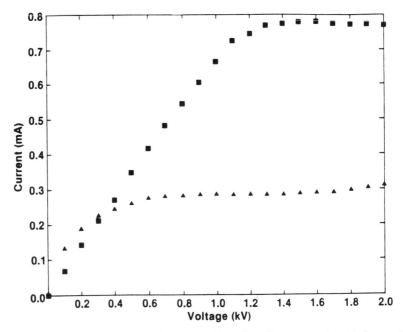

Figure 2.16. Characteristic V/I curves for 100 ppm Na aspirated into an acetylene/air flame, using external split cathode plate electrodes (■) and the water-cooled immersed cathode (▲). Reprinted with permission from Turk (22).

bulk ionization rate, as discussed in point ii, above. When the air/acetylene flame is unseeded, a flat saturation plateau is not observed. However, when the flame is highly seeded with an easily ionizable element, as shown in Fig. 2.16, a saturation plateau is more evident (22).

2.8.2.   Pulsed LEI Peak Shapes and Induced Charge Apportionment

The ability to accurately profile peak shapes not inordinately determined by amplifier response times is only now beginning to emerge (15, 23). Earlier efforts, as shown in Fig. 2.17, were hampered by the all-too-familiar trade-off between response time and sensitivity (5, 24). Analyte concentrations required for adequate signal-to-noise ratio for Fig. 2.17 were high enough to result in distortions produced by the laser-induced charges, causing a significant perturbation to the steady-state field, as illustrated in Fig. 2.12. These types of distortions are predicted by the numerical models discussed earlier, but a good "fit" of theory and experiment is precluded by the many differences between the simple theory and the complexity of the experiment (e.g., the dimensionality).

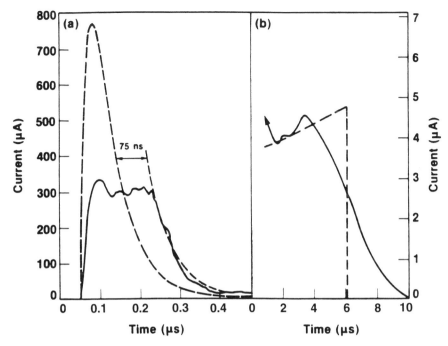

Figure 2.17. Electron component (*a*) and ion component (*b*) of the LEI signal pulse for 100 ppm Na in the acetylene/air flame, with a water-cooled immersed cathode 2 cm above the burner head (anode), -1500 V applied to the cathode, and the laser beam located 0.3 cm below the cathode. The *point charge model* represented by the dashed curves neglects the effects of self-induced charge perturbations and saturated electric fields, as discussed in the text. Reprinted with permission from Havrilla et al. (5).

Both Berthoud et al. (24) and Matveev (23) report "knees" and/or second maxima on electron pulses, which they attribute to electron arrival at the anode. This interpretation is entirely consistent with theory for saturated conditions, under which the field is nonzero at the anode. In the theoretical treatments above, we have not considered this case for either the point charge model or the numerical model. However, it is clear that saturation is readily obtained at moderate voltage, even in the highly seeded air/acetylene flame, as shown in Fig. 2.16. Consequently, the electron pulse would be truncated by arrival at the anode, just as the ion pulse is truncated by arrival at the cathode. It would be entirely feasible to modify the code LEIMODEL (see the Appendix) for the saturated condition using the general form of the Eq. (29) for the base electric field magnitude.

In addition to predicting the peak shape for an idealized case, the point charge model carries with it an implication about the time integral under each of the two (electron and ion) signal pulse components. The relationship can be derived by integrating Eqs. (38) and (40) over the pulse intervals, but it is most easily derived from the induced charge concepts of Eqs. (14). That is, the amount of charge induced in the external circuit should be given by the *change* in the induced charge on either the anode or the cathode between the time the charge was deposited at $x = X_L$ and the arrival of electrons at the *effective* anode ($x = X_s$) or of ions at the cathode ($x = 0$). Taking such differences with Eq. (14) and substituting X_s for W(see below) shows that both the cathode and the anode yield the same result (as they should) and that the integral of the ion pulse, Q_+, and the integral of the electron pulse, Q, are given by

$$Q_+ = Q\left(\frac{X_L}{X_s}\right)$$

$$Q_- = Q\left(\frac{X_s - X_L}{X_s}\right)$$

(48)

assuming that the laser has deposited equal magnitudes, Q, of electrons and ions at the position X_L. Notice that the sum of the two charges gives Q, the total charge magnitude of either the deposited electrons or ions, and that the charge is apportioned between the two pulses according to the fractional distance moved by the corresponding charges.

Equations (48) have been shown to agree with experimental data, validating the "effective anode" concept of taking the anode position to be the sheath edge, X_s, rather than the physical position of the positive electrode, W(5). That is, under subsaturation conditions, for which $X_s < W$, the sheath edge at X_s becomes the effective anode, since the plasma between the sheath and the real anode is simply acting as a conductor. This concept was employed without explanation in the derivation of the point charge model currents.

2.8.3.   Matrix Effects and Recovery Curves

LEI matrix effects are primarily discussed elsewhere in this book, but one in particular deserves mention here, in connection with signal collection theory. That is, the presence of an easily ionizable matrix element can affect the magnitude the LEI analyte signal by modifying the electric field distribution and the effective anode position. Figure 2.18 shows *signal recovery curves* for several types of electrodes (25). These curves are taken for families of binary mixtures with a fixed analyte concentration and varying concentration of a matrix EIE and have been widely employed as a diagnostic in LEI (26–29). The resistance of different electrode configurations to this effect is discussed

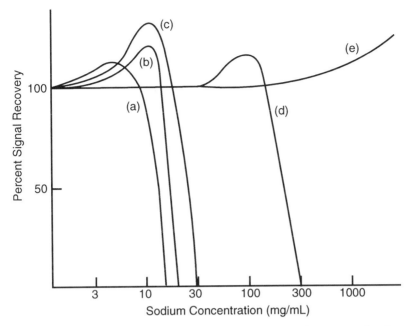

Figure 2.18. LEI signal "recovery" curves as a function of Na matrix concentration for split cathode rods of (*a*) 1, (*b*) 1.5, and (*c*) 2.35 mm diameter; (*d*) for 5 mm wide by 125 mm long split cathode plates; and (*e*) for the water-cooled immersed cathode. Reprinted with permission from Travis et al. (25).

later. However, the general behavior for each type can be interpreted on the basis of the theory discussed above.

Total loss of signal is understood to occur when the exciting laser is outside of the sheath, or $X_L > X_s$. This can be shown by experiments that vary the laser position, though the signal recovery curve experiment does the opposite—varies the sheath edge position. With increasing EIE concentration, the sheath edge withdraws toward the cathode according to Eqs. (30) and (35). Once the sheath is totally restricted to the region between the cathode and the laser beam, no signal is detected.

The reason for an initial enhancement of the signal under some circumstances is harder to understand. However, it is generally attributed to the increase in field strength near the cathode as the volume ionization rate is increased. This local increase arises from the fact that the field *slope* is proportional to the volume ionization rate, but the *integral* under the field gives the applied potential, a constant for the recovery experiments. The field increase less than halfway between the cathode and sheath edge is, of course, offset somewhat by the reduced fractional distance traveled by the electrons as the sheath shrinks with increasing EIE concentration.

2.8.4. Evolution of Electrodes

The earliest LEI experiments were detected using a pair of tungsten welding rods inserted into the flame (30). Experiments exploring the effects of spacing of electrodes and positioning of the laser beam with respect to electrodes soon indicated that the laser beam should be near the cathode and that the anode position was largely irrelevant. This finding led to a one-rod arrangement, with a cathode rod and the burner head acting as anode. The immersed electrode glowed red hot in the flame but was reasonably robust while the flame burned. When the flame was shut off, however, the hot rod formed an oxide layer in air, and these oxide layers eventually flaked off, reducing the diameter of the rod. The rod had to be changed every few days.

The *split-cathode rod arrangement* was designed to get the cathode out of the flame and therefore make it robust (31). Two rods were located on opposite sides of the flame, both biased to the same negative high voltage with respect to the burner head, still acting as the anode. This arrangement worked well enough, although higher voltages were required (than for the single cathode in the flame) to project the cathode sheath to the laser excitation position. Sensitivity loss was minimal, but the arrangement was extremely sensitive to easily ionizable matrix elements, as shown in Fig. 2.18. This was interpreted as being related to the principle of the lightning rod, namely, that sharply curved electrode surfaces yield high local fields. The high fields in the vicinity of a small diameter rod would yield more tightly packed ion sheaths, as well as less sensitive volume for LEI in the flame. This diameter trend was confirmed by an experiment (27), as shown Fig. 2.18, leading to the development of the *split-cathode plate arrangement*, which prevailed for several years (26, 32).

Nevertheless, the development of the one-dimensional theory made it obvious that maximum signal strength and optimal resistance to electrical matrix interference dictated that laser excitation be as close to the cathode as practicable. To meet this condition while accessing the optimal region of the flame for analyte density, Turk (22) developed the stainless steel, *water-cooled, immersed cathode arrangement*, still using the burner body as the anode. This electrode is easy to fabricate, quite robust, and seems to contribute negligible memory effects, since it is located downstream from the active volume in the flame. There is some indication from cw experiments that the electrode cools the flame in its vicinity, but this seems to have little detrimental effect (33). The diameter of the electrode (typically 6 mm) is large enough to avoid the "lightning rod" effect, and it is often somewhat flattened in a vice to better approximate parallel plate geometry between the electrode and the burner head. The matrix resistance of the electrode is compared to the earlier designs in Fig. 2.18.

An additional advantage of the electrode is that the saturation current is reduced with respect to that of the split cathode plate, as shown in Fig. 2.16

(22). This improves the LEI signal-to-noise ratio, since the background current is a primary contributor of the limiting noise.

The water-cooled *coiled cathode* of Szabo et al. (34) has equivalent sensitivity to the immersed cathode and seems to have even greater resistance to high concentrations of EIEs. The working principles of this electrode are not clear, but on one level it seems to continue the "rod diameter" progression of Fig. 2.18 to the point of reversed curvature; that is, the electrode surface facing the laser excitation is concave rather than convex. On another level, the field within an enclosed conductor is zero, so the coil must not behave as an enclosed conductor or no signal would be observed. Two-dimensional modeling will be required to fully understand the operation of the electrode.

2.8.5. Sheath-Displacement Imaging

Figure 2.19 shows an "image" of the laser ionization volume in a flame, generated by translating the sheath edge through the flame by variation of the applied voltage (18). From the low voltage limit, the sheath edge is advanced

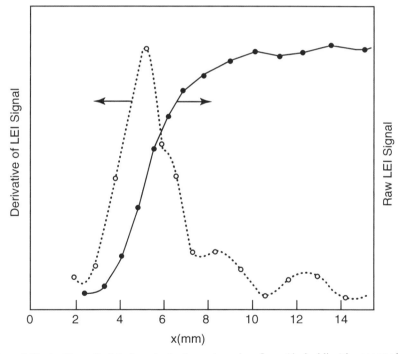

Figure 2.19. An "image" of the laser ionization volume in a flame (dashed line) is generated by translating the sheath edge through the flame by variation of the applied voltage and differentiating the total LEI charge (solid line) as a function of sheath position. Reprinted with permission from Turk (18).

by increasing the voltage until small signals are detected as the edge of the sheath meets the edge of the laser ionization volume. Each increment in voltage then produces a signal increment corresponding to more of the laser ionization volume being included in the sheath, until the entire ionization volume is within the sheath. Therefore, the raw signal as a function of sheath edge position is an integral of the desired image, and the image is obtained using finite differences of adjacent measurements.

Low-fidelity images, which will respond reasonably well to the experimental laser beam position and width, may be obtained very simply. However, higher fidelity images require consideration of both the theory outlined above and the fact that real flames depart from the simple theory. To wit: (i) the sheath does not yield a simple sensitivity step function, due to the electric field slope and the scaling of peak current with field; (ii) the sheath edge position ideally scales as the square root of applied voltage. Thus, even a low-fidelity image should include this assumption, but a high-fidelity image should use experimentation to "calibrate" the voltage vs. position curve for the actual conditions used. This is most readily done by recording the signal onset voltage for a number of equally spaced laser positions in the flame. These considerations are discussed more fully in a report by Turk (18).

2.8.6. Translated-Rod Imaging

A second method of "imaging" the laser ionization event in flames has only been achieved with cw laser excitation and in hydrogen/air flames, though no great effort has been applied to reproducing it in other circumstances. Figure 2.20 shows images obtained by vertically translating a horizontal 1 mm diameter tungsten rod in proximity to a plate electrode at the same potential as the rod, as illustrated in Fig. 2.21 (21). This method was designed to "intercept" ions or electrons (depending on the polarity of the opposite plate) traveling along the nominally parallel electric field "lines of force," assuming minimal perturbation of the field configuration by the presence of the wire. The fact that the experiment worked seems, on the face of it, to be inconsistent with the carefully laid groundwork presented earlier concerning the remote detection of ions through the *induction* of current in the electrodes. Both this experiment and the corollary "shadow" experiment of Fig. 2.22 [see Schenck et al. (21)], for which the signal was taken from the backing electrode rather than the translated rod, actually illustrate that charge induction may be very localized in the presence of a significant electric field.

2.8.7. Modulated Continuous-Wave Current Profiles

Time-response data also published in the paper by Schenck et al. (21) and shown in Fig. 2.23 strongly support the induced current model. That is, signal

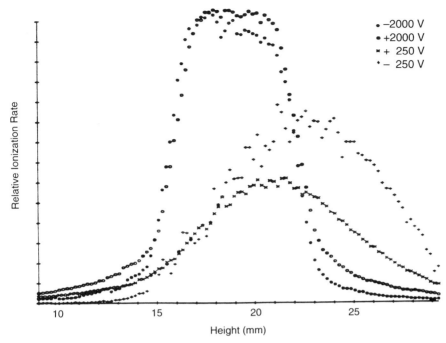

Figure 2.20. Images obtained by vertically translating a horizontal 1 mm diameter tungsten rod in proximity to a plate electrode at the same potential as the rod, as illustrated in Fig. 2.21b. Two magnitudes and two polarities of applied potential were used for the four images shown. At 2000 V, the image faithfully reproduces the laser beam at a height of 19 mm above the burner. At 250 V, the image slips "downstream," due to the flame velocity. Reprinted with permission from Schenck et al. (21).

onset occurred simultaneously with laser irradiation, even at low collection voltages. At low voltages, in fact, the rising and falling edges of signal show a definite exponential time constant related to the applied voltage and consistent with sodium ion mobility, but with the steepest slope at the instant of change in laser intensity, not delayed by any transit time. This behavior was explored empirically in the aforementioned paper but was not explained theoretically.

With the benefit of hindsight, it is possible to provide additional interpretation, while still leaving room for further experimentation (and theory) to thoroughly unravel the puzzle. In the high voltage limit, both the laser turn-on time (with a 40 MHz acousto-optic modulator) and the ion transit time from the center of the flame to an electrode are short with respect to the minimum time interval (10 μs) shown in Fig. 2.23. At lower potentials, the laser turn-on

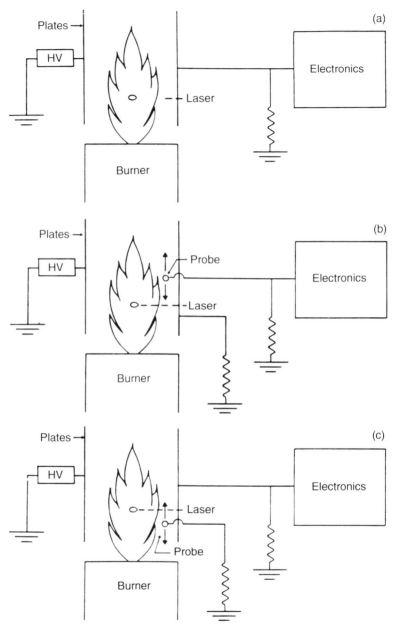

Figure 2.21. Electrode configurations and electrical connections used for (a) V/I curves shown in Figs. 2.14 and 2.15; (b) images shown in Fig. 2.20; and (c) "shadow" image shown in Fig. 2.22. Reprinted with permission from Schenck et al. (21).

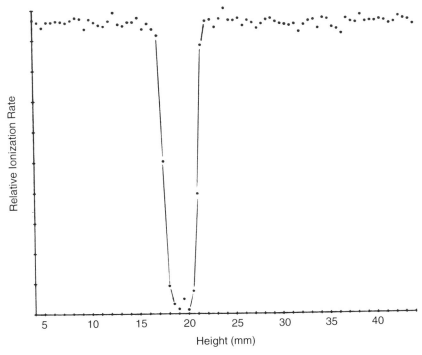

Figure 2.22. "Shadow" image obtained as for Fig. 2.20a, but with the signal taken from the backing plate rather than the translated rod, as shown in Fig. 2.21c. Reprinted with permission from Schenck et al. (21).

time is still negligible, but the ion transit time to the electrode is nominally consistent with the observed time constants.

The exponential leading and trailing edges are not easy to interpret and are seemingly inconsistent with imaging. Assume that a ribbon of charge originates at the laser position at time zero, the leading edge translates toward the appropriate electrode as the ribbon continues to grow, and the ribbon eventually arrives at a steady state when it fills the distance between the laser and the electrode. Further, assume for the moment that even low fields are high enough in the hydrogen/air flame to be nominally uniform across the space, so that the leading edge of the ribbon moves uniformly with time (at constant velocity). Then, the exponential rise implies that the most important charge element in producing signal current is right at the laser position. Simple parallel plate, current induction theory would predict a triangular ramp, extending from laser turn-on to leading edge arrival at the electrode. The exponential character must reflect space charge influence.

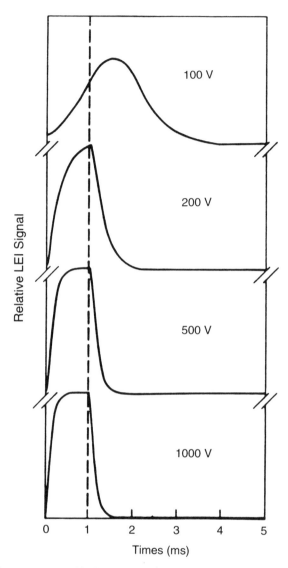

Figure 2.23. Time-response profiles for LEI signals resulting from 1 ppm LEI in a hydrogen/air flame excited by a 1 ms square "pulse" from an amplitude modulated cw laser, for the indicated voltages applied across plate electrodes separated by 1.25 cm. Reprinted with permission from Schenck et al. (21).

Electron transit, by the way, may be governed by space charge influence as well, in the manner of "ambipolar diffusion," by which electrons are constrained to move at the same velocity as ions by the coulombic attraction between the electrons trying to escape and the ions left behind. In the field model based on Poisson's equation, the ions left bare by rapidly escaping electrons perturb the field to the extent that the field at the electron position is vastly reduced. This effect occurs at concentrations large enough for the ion density to significantly perturb the field. In diffusion theory, electron escape is slowed by orders of magnitude, ions are sped up by the tug of the electrons, and both electrons and ions diffuse at nominally twice the rate of ions alone.

2.8.8.   Advantages of Continuous-Wave Lasers

Most of the LEI literature concerns experiments performed with pulsed dye lasers, due to the limited spectral coverage of tunable cw lasers that have been available to date. Thus, it is important to document some very important advantages of cw laser excitation in light of dramatic advances in solid-state laser development. These advantages span two different theoretical arenas of this book: charge production and charge detection. The reader is referred to Chapter 1 for a full understanding of cw advantages for charge production, but some commentary will necessarily be included here as well.

The most obvious point impacting both charge production and collection is simply the duty factor match between the continuous introduction of the sample (which is characteristic of analytical flame spectrometry) and the effective duty factor of the laser. Pulsed lasers have typical optical duty factors of 10^{-7}, for a 10 Hz laser with a 10 ns pulse length. However, if the laser ionizes all analyte atoms within a cylinder of, say 1 mm diameter, then the next laser pulse need not occur until the volume refills at the flame velocity of about 1 m/s. Thus, the *effective* duty factor for efficient pulsed systems is not much less than 10^{-2} at 10 Hz, and approaches unity for kilohertz repetition rate lasers.

An important contrast in signal detection of cw and pulsed LEI is shown in Fig. 2.24 (33). The spatial "map" of pulsed LEI sensitivity as a function of distance from the cathode clearly reflects the electric field distribution and fractional position of the laser beam in the sheath, as predicted by the point charge model. The corresponding cw profile, just as clearly, does nothing of the kind. Indeed, the amplitude variation of the cw profile has been attributed to signal production, rather than detection. Thus, the mild drop-off in signal near the cathode may represent the cooling of the flame by the water-cooled cathode, resulting in a drop in the thermal ionization enhancement rate. At the other extreme, the laser beam enters a flame region where desolvation and

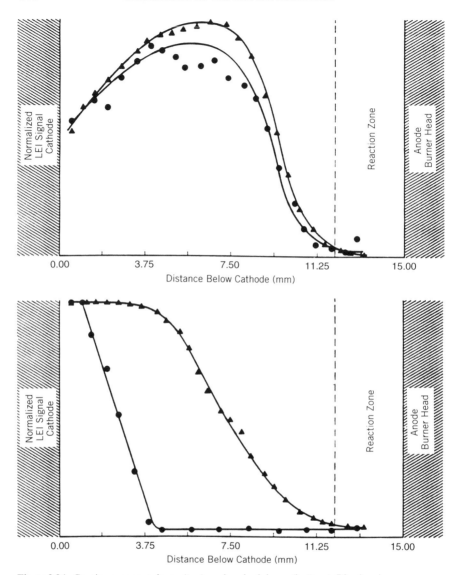

Figure 2.24. Continuous-wave laser (top) and pulsed laser (bottom) Rb signal response as a function of distance of the laser from the immersed cathode for a solution matrix of (▲) distilled water, or (●) 100 ppm potassium, in an H_2/air flame with -1000 V applied to the electrode. Reprinted with permission from Havrilla et al. (33).

vaporization of the sample aerosol is incomplete, and therefore the analyte concentration falls off.

The cw insensitivity to electric field distribution and laser beam position correctly implies an improved tolerance to easily ionizable matrices and may readily be explained. Several important details about Fig. 2.24 have not been given yet. First, the cw laser beam was amplitude modulated by using a 2 kHz chopper, and the signal was detected by using a lock-in amplifier. Second, the "phase" required to maximize to LEI signal was individually adjusted for each laser position in the figure. Finally, the water-cooled, immersed cathode was "downstream" of the laser excitation with respect to the flame convection velocity. Indeed, a study of the lock-in phase as a function of beam position may be used to infer the velocity. Thus, for cw LEI, the laser beam *need not be in the sheath* for successful signal detection. Ions and electrons produced by the laser coexist in a volume that is swept by convection into the sheath, where the charges are then separated by the field and registered in the circuit.

Why can the same approach not be employed for pulsed lasers? It can, of course, but with great loss in sensitivity, which is not the case for cw lasers. The "pulse" created from a 1 mm diameter packet of electrons and ions entering the sheath at the flame velocity of 1 m/s would be about a millisecond wide, determined by the flame velocity and packet diameter rather than the mobility, etc. Thus, the pulse, which should actually be about 100 ns wide but is often slowed in amplification to about 1 μs, would be broadened and reduced in amplitude by several orders of magnitude. The detection gate would be opened accordingly, accepting several orders of magnitude more "noise" current than before. On top of that, the 1 ms pulse would "move around" in time, during the interval between laser pulses, representing the distance between the laser beam and the sheath edge. This is equivalent to the phase change in the cw case, which is done automatically on many modern lock-in amplifiers.

2.9.   UNFINISHED BUSINESS

2.9.1.   Electrode Design

There is not much room for improvement in the *efficiency* with which signal electrons are detected in LEI, since it already approaches 100%. However, background current is detected with similar high efficiency and over a much larger volume than is excited by the laser. Also, with present electrodes, the most sensitive detection zone is nearest the cathode, where the cathode itself may perturb the measurement.

Advances in modern computing suggest the value of modeling electrode/flame systems in at least two dimensions, in order to completely optimize the shape and placement of electrodes.

2.9.2. High-Frequency Detection

The 100 ns "typical" electron pulse length of LEI suggests some interesting possibilities for the use of electric fields oscillating at frequencies above ~ 10 MHz. To begin with, electrons responding to such a field would typically not reach an electrode before being reversed in direction and sent the other way. This implies an ability to continue "sensing" a signal electron until convection and/or recombination removes it from the "field of view." High frequencies also imply that no sheath could form between polarity changes and thus the field should penetrate further into the flame.

This simple extrapolation from the steady-state, one-dimensional model, however, runs counter to the well-known "skin effect" of high-frequency plasmas, such as the analytical inductively coupled plasma (ICP). However, due to the lower electron number density of flames (than the ICP), Berglind and Casparsson (35) have been able to observe the LEI signal via the absorption of a 10 GHz microwave field across a flame. A number of groups have reported rf-detected opto-galvanic spectroscopy in low-pressure discharges (36–39).

2.9.3. Draining of Easily Ionized Elements

Electrode design could improve on the concept of "draining" easily ionizable elements from the flame prior to LEI (28). This concept employs a set of electrodes near the combustion zone to physically extract thermally and chemically produced ions from the flame prior to the region of the LEI experiment, which utilizes a separate set of electrodes and high voltage supply. One can readily conceive of the possibility, inasmuch as many alkalis and alkaline earth elements yield a high natural ion fraction in high-temperature analytical flames, and removing the ions simply tips the equilibrium until the population is exhausted. Indeed, an especially impressive experiment is to aspirate a high concentration of Na in a nitrous oxide/acetylene flame and turn the sodium light on and off with a field across parallel plates bracketing the flame. In reality, the process requires a finite distance in the flame and pushes the LEI excitation volume well above the normal "analytical zone" at which the atomic population is optimum.

2.9.4. Other Plasmas

The analytical flame is nearly optimal for LEI, due to the temperature, the ease of sample introduction, and the amenability of detection with an impressed field. However, the flame is rapidly approaching extinction in analytical spectrometry, with analysts preferring the superior sample consumption, with reduced chemical complications, of the argon ICP. This popular analytical plasma operates at about 6000 K, with rather high ion fractions for most elements of the periodic table. Electron densities of around 10^{16} cm^{-3} compare unfavorably with those of about 10^{10} cm^{-3} for the air/acetylene flame. Initial experiments in the plasma have yielded noncompetitive sensitivities (40–42). However, it has been shown that a dc field can be applied near the excitation coil, by grounding the exciting coil in the same manner as for ICP–mass spectrometry (MS).

The mismatch of excitation temperature and ionization level of the ICP with the optimal values for LEI suggests the separate sampling and excitation approach. Cooling and recombination would be allowed to proceed to a point at which LEI could be performed effectively. Initial efforts with a modulated ICP have been encouraging (41).

2.9.5. Ion Mass Selectivity

Ion-selective LEI is a concept with a great deal of intellectual appeal to analytical chemists, but working with the ion mobilities at atmospheric pressure and relatively high electron densities has proven to be a nontrivial task. Smyth and Mallard (43) clearly distinguished massive ($\sim$ 2300–6100 u) soot precursor ions from sodium ions in a sooting acetylene/air flame. Their success largely was due to the very large mass and mobility difference in the species, and their single welding rod cathode immersed in the flame also probably sharpened the ion pulse with respect to the prediction of the one-dimensional model discussed here.

Similarly, Wu et al. (44) benefited from the use of fine-wire electrodes to sharpen the shape of the pulse resulting from the approach of the ions to the electrode, and they further succeeded in measuring the mobility of a relatively light atom, sodium. This group was working in an $H_2/Ar/O_2$ flame, with considerably lower electron densities (and less general analytical utility) than the conventional analytical hydrocarbon flame. The lower electron density permits an almost uniform field across the flame. These authors give the exact solution for the ion number density as a function of time and position, for an infinitely thin (delta function) deposition of ions by laser ionization at $t = 0$ and a uniform field. They also treat extensions to the method to allow for finite beam size and pulse duration, and for departure of the field from uniformity. Their theoretical treatment would be especially useful in the low-electron-density environment required for successful mobility spectrometry.

Because the ion pulse stretches from time zero to the arrival time, pulses due to ions of different mobility are not distinctly separated, especially with the ion-spreading effect of the normal field gradient. An effective LEI ion mobility spectrometer would need low electron density to eliminate the field gradient (45), and a grid near the cathode to delay current induction until the ions were between the grid and the cathode. (The treatment of current generated in the external circuit by Wu et al. (44) ignores current induction but would be valid for the use of a screen grid.)

Sampling of the flame into a conventional mass spectrometer appears to be the most promising approach to ion-selective LEI (46). This work uses the mass spectrometric end of a commercial ICP–MS device, with an analytical burner substituted for the ICP. The flame is found to be superior to the ICP for selective laser ionization, due to the lower degree of collisional ionization (47).

2.10. CONCLUSION

This chapter has attempted to establish a basis for understanding the transport of electrons and ions through a flame by an impressed electric field, and the induction of current in an external circuit in response to this motion. Many opportunities exist for the extension of the theory to higher dimensionalities and to different geometries and plasma environments. Such development is especially needed in light of the high-speed transient digitizing electronics and high-speed/high-capacity computing capability now available and the alternate atom reservoirs in which LEI is being exploited.

APPENDIX:
A COMPUTER PROGRAM TO MODEL LEI CHARGE TRANSPORT
AND CURRENT GENERATION

```
'LEIMODEL - A model to trace the time history of Gaussian electron or ion

'distributions deposited in an idealized flame by Laser Enhanced Ionization,

'and the current pulse generated by the moving charges.

'Version 6/10/94

DIM n(200)          'Electron number density

DIM j(40)           'Electron current density

DIM e0(201), e(201)     'Nominal field perturbation due to immobile

                    'positive ions; negative charge perturbation of field.

'Initialize parameters, etc.
```

v = 100 'Applied potential, only used for relative diffusion contribution

    'The low value makes diffusion relatively more important to stabilize

    'the computation

k = .2 'Realistic value of kT/e for air/acetylene flame

nskip = 5 ' Thinning factor for output profiles relative to current calculation.

'Keyboard inputs of output filename, normalized laser position and width,

'and the normalized maximum number density of charge deposited by the laser.

5 INPUT "Laser Position, 0.2 < XL < 0.8"; xlas

INPUT "Laser Width, 0.1 < WL < XL/2"; wl

IF wl > xlas / 2 THEN

    PRINT "WL must be < or = XL/2"
    GOTO 5

    END IF

INPUT "Ratio of Deposited Charge to Background Charge, 0.01 < RN < 10"; rn

6 INPUT "Ions (i) or electrons (e)"; chg$

IF chg$ = "e" THEN s = -1

IF chg$ = "i" THEN s = 1

IF chg$ < > "e" AND chg$ < > "i" THEN GOTO 6

'Open output file for charge profiles at 0.05*nskip time intervals,

'and for current profiles as well.

IF chg$ = "e" THEN

    o$ = "neg_dist.csv" 'Output file for electron distributions

    o2$ = "neg_fld.csv" 'Output file for field distributions

    ELSE

```
    o$ = "pos_dist.csv" 'Output file for electron distributions

    o2$ = "pos_fld.csv" 'Output file for field distributions

END IF

OPEN o$ FOR OUTPUT AS #1

'Open output file for field profiles at 0.05*nskip time intervals

OPEN o2$ FOR OUTPUT AS #2

PRINT #1, "Position,Width,Chg_Ratio"

PRINT #1, xlas; ","; wl; ","; rn

PRINT #2, "Position,Width,Chg_Ratio"

PRINT #2, xlas; ","; wl; ","; rn

SCREEN 9: CLS

'Reduced unit spatial interval for sheath edge at x=1

dx = .005: nx% = 1 / dx

'Reduced unit time coordinate where t=1 after one time constant

'Compute dt as an integral subinterval of 0.05, and about 1/5th of dx

i% = 5 * INT(.05 / dx): dt = .05 / i%

PRINT #1, "dx,dt,n": PRINT #1, dx; ","; dt; ","; nx%

PRINT #1, " ": PRINT #1, "t\x"; ",";

PRINT #2, "dx,dt,n": PRINT #2, dx; ","; dt; ","; nx%

PRINT #2, " ": PRINT #2, "t\x"; ",";

FOR i% = 0 TO nx%: PRINT #1, i% * dx; ","; : NEXT i%

PRINT #1, " ": PRINT #1, "0,";

FOR i% = 0 TO nx%: PRINT #2, i% * dx; ","; : NEXT i%

PRINT #2, " ": PRINT #2, "0,";
```

```
norm = 0: elast = 1 + dx

xl = 0: n0 = 0: j(0) = 0

FOR i% = 0 TO nx%

n(i%) = EXP(-4 * (i% * dx - xlas) ^ 2 / wl ^ 2): norm = norm + n(i%)

x = i% * dx

j(0) = j(0) + (1 - x) * n(i%)

'Plot initial charge distribution on screen

LINE (640 * xl, 310 - 100 * n0)-(640 * x, 310 - 100 * n(i%))

'Plot initial field distribution on screen

LINE (640 * xl, 310 - 100 * (1 - xl))-(640 * x, 310 - 100 * (1 - x))

xl = x: n0 = n(i%)

e0(i%) = s * dx * rn * norm'Field perturbation due to stationary charges

e(i%) = -e0(i%) 'Initial field perturbation due to moving charges

IF chg$ = "i" THEN e0(i%) = 0

PRINT #1, n(i%); ",";

PRINT #2, 1 - x; ",";

NEXT i%

PRINT #1, " ": PRINT #2, " "

'Start time loop

tp = .05: jp% = 0' Current calculation interval and index

iskip = 0' Print-thinning counter; prints for every "nskip" values of "tp"

offst = 0' Field offset due to laser-deposited charges

FOR t = 0 TO 2 + dt / 2 STEP dt' Reduced time ranges to 2 units

IF t > = tp - dt / 2 THEN
```

```
        iskip = iskip + 1

        IF iskip = nskip THEN

                PRINT #1, t; ","; n(0); ",";

                PRINT #2, t; ","; e(0) + e0(0) + 1 - offst; ",";

        END IF

        jp% = jp% + 1: j(jp%) = 0

        IF iskip = nskip THEN CLS

END IF

nl = n(0): nc = n(1) ' Left and center densities for derivatives

chsm = 0: vsum = 0: esum = 0: norm = n(0)

FOR i% = 1 TO nx% - 1

'Left, and right fields and right density for derivatives

el = e(i% - 1) + e0(i% - 1) + 1 - dx * (i% - 1) - offst

ec = e(i%) + e0(i%) + 1 - dx * i% - offst

er = e(i% + 1) + e0(i% + 1) + 1 - dx * (i% + 1) - offst

nr = n(i% + 1)

'Field-driven term

dn = s * dt * (er * nr - el * nl) / (2 * dx)

'Diffusion-driven term

dn = dn + dt * (k / (2 * v) * (nr - 2 * nc + nl) / (dx) ^ 2)

n(i%) = n(i%) + dn

nl = nc: nc = nr 'New left and center densities for derivatives

IF t > = tp - dt / 2 THEN

        IF iskip = nskip THEN
```

```
        PRINT #1, n(i%); ",";

        PRINT #2, ec; ",";

    END IF
    xl = dx * (i% - 1): x = dx * i%

    LINE (640 * xl, 310 - 100 * n(i% - 1))-(640 * x, 310 - 100 * n(i%))

    LINE (640 * xl, 310 - 100 * el)-(640 * x, 310 - 100 * ec)

    j(jp%) = j(jp%) + n(i%) * ec: chsm = chsm + n(i%)' Current; charge

END IF

NEXT i% ' End of spatial loop

n(0) = 2 * n(1) - n(2): n(nx%) = 2 * n(nx% - 1) - n(nx% - 2)' Boundary conditions

'Compute new field for next time increment

offst = 0: norm = 0

FOR i% = 0 TO nx%

norm = norm + n(i%)

e(i%) = -s * dx * rn * norm

offst = offst + dx * (e0(i%) + e(i%))

NEXT i%

IF t > = tp - dt / 2 THEN

    tp = tp + .05 ' Increment temporal printout coordinate

    IF iskip = nskip THEN

        PRINT #1, n(nx%)

        PRINT #2, er

        iskip = 0

    END IF

END IF

END IF
```

```
NEXT t ' End of time loop

CLOSE #2

'Print current row-wise

'Time coordinate

PRINT #1, " "

PRINT #1, "Time:,";

FOR t = 0 TO 1.99 STEP .05: PRINT #1, t; ","; : NEXT t

PRINT #1, "2.0"

'Raw current

PRINT #1, "RawCrt:,";

FOR i% = 0 TO 39: PRINT #1, j(i%); ","; : NEXT i%

PRINT #1, j(40): jold = 0: trise = .1

'Current with amplifier time constant of "trise" (above)

PRINT #1, "AmpCrt:,0,";

FOR i% = 0 TO 39

jamp = 0

FOR j% = 0 TO i%

jamp = jamp + .05 / trise * EXP(-.05 * j% / trise) * j(i% - j%)

NEXT j%

PRINT #1, jamp;

IF i% < 39 THEN PRINT #1, ",";

NEXT i%

CLOSE #1

END
```

ACKNOWLEDGMENTS

The authors are especially indebted to many collaborators and coworkers over the years who have contributed to their understanding of charge transport and sensing. An inexhaustive list would certainly include Jim DeVoe, Peter Schenck, Jesse Wen, Tom O'Haver, Gary Mallard, Kermit Smyth, Thierry Berthoud, Rich Simon, George Havrilla, Stephan Weeks, and Bob Green.

REFERENCES

1. G. C. Turk and N. Omenetto, *Appl. Spectrosc.* **40**, 1085 (1986).

2. J. Lawton and F. J. Weinberg, *Electrical Aspects of Combustion*, pp. 316–322. Clarendon Press, Oxford, 1969.

3. I. Magnusson, *Spectrochim. Acta* **42B**, 1113 (1987).

4. J. C. Travis, G. C. Turk, J. R. DeVoe, P. K. Schenck, and C. A. van Dijk, *Prog. Anal. At. Spectrosc.* **7**, 199 (1984).

5. G. J. Havrilla, P. K. Schenck, J. C. Travis, and G. C. Turk, *Anal. Chem.* **56**, 186 (1984).

6. J. C. Travis, in *Analytical Laser Spectroscopy* (S. Martellucci and A. N. Chester, eds.), NATO ASI Ser. B, Vol. 119, p. 213. Plenum Press, New York, 1985.

7. H. S. W. Massey, *Electronic and Ionic Impact Phenomena*, 2nd ed., Vol. 3, Chapter 19. Clarendon Press, Oxford, 1971.

8. E. Nasser, *Fundamentals of Gaseous Ionization and Plasma Electronics,* Sect. 6.2. Wiley (Interscience), New York, 1971.

9. E. W. McDaniel, *Collision Phenomena in Ionized Gases*, Sect. 9–2. Wiley, New York, 1964.

10. E. W. McDaniel, *Collision Phenomena in Ionized Gases*, Chapter 11. Wiley, New York, 1964.

11. E. Nasser, *Fundamentals of Gaseous Ionization and Plasma Electronics*, Sect. 6.8. Wiley (Interscience), New York, 1971.

12. A. B. Gillespie, *Signal, Noise, and Resolution in Nuclear Counter Amplifiers*, Chapter 2, Pergamon, New York, 1953.

13. J. D. Jackson, *Classical Electrodynamics*, Sect. 2.1. Wiley, New York, 1962.

14. P. J. W. Boumans, *Theory of Spectrochemical Excitation*, p. 161. Hilger & Watts, London, 1966.

15. O. I. Matveev and N. Omenetto, Joint Research Center, Ispra, Italy, private communication, September 1993.

16. J. Lawton and F. J. Weinberg, *Electrical Aspects of Combustion*, p. 118. Clarendon Press, Oxford, 1969.

17. J. Lawton and F. J. Weinberg, *Electrical Aspects of Combustion*, pp. 16–19, 232. Clarendon Press, Oxford, 1969.

18. G. C. Turk, *Anal. Chem.* **64**, 1836 (1992).

19. J. Wen, National Institute of Standards and Technology, Gaithersburg, MD, unpublished program LEISIM, December 1982.

20. P. K. Schenck, J. C. Travis, and G. C. Turk, *J. Phys. (Paris)* Colloq. **C7**, 75 (1983).

21. P. K. Schenck, J. C. Travis, G. C. Turk, and T. C. O'Haver, *J. Phys. Chem.* **85**, 2547 (1981).

22. G. C. Turk, *Anal. Chem.* **53**, 1187 (1981).

23. O. I. Matveev, *J. Anal. Chem. USSR (Engl. Transl.)*, **43**, 482 (1988).

24. T. Berthoud, J. Lipinsky, P. Camus, and J.-L. Stehle, *Anal. Chem.* **55**, 963 (1983).

25. J. C. Travis, G. C. Turk, and R. B. Green, *Anal. Chem.* **54**, 1006A (1982).

26. G. C. Turk, J. C. Travis, J. R. DeVoe, and T. C. O'Haver, *Anal. Chem.* **51**, 1890 (1979).

27. R. B. Green, G. J. Havrilla, and T. O. Trask, *Appl. Spectrosc.* **34**, 561 (1980).

28. G. J. Havrilla and R. B. Green, *Anal. Chem.* **52**, 2376 (1980).

29. V. I. Chaplygin, N. B. Zorov, and Yu. Ya. Kuzyakov, *Talanta* **30**, 505 (1983).

30. R. B. Green, R. A. Keller, P. K. Schenck, J. C. Travis, and G. G. Luther, *J. Am. Chem. Soc.* **98**, 8517 (1976).

31. G. C. Turk, J. C. Travis, J. R. DeVoe, and T. C. O'Haver, *Anal. Chem.* **50**, 817 (1978).

32. G. J. Havrilla and R. B. Green, *Anal. Chem.* **52**, 2376 (1980).

33. G. J. Havrilla, S. J. Weeks, and J. C. Travis, *Anal. Chem.* **54**, 2566 (1982).

34. N. J. Szabo, H. W. Latz, G. A. Petrucci, and J. D. Winefordner, *Anal. Chem.* **63**, 704 (1991).

35. T. Berglind and L. Casparsson, *J. Phys. (Paris), Colloq.* **C7**, 329 (1983).

36. C. Stanciulescu, R. C. Bobulescu, A. Surmeian, D, Popescu, I. Popescu, and C. B. Collins, *Appl. Phys. Lett.* **37**, 888 (1980).

37. T. Suzuki, *Opt. Commun.* **38**, 364 (1981).

38. J. R. Brandenberger, *Phys. Rev. A* **36**, 76 (1987).

39. N. V. Denisova and N. G. Preobrazhensky, *Spectrochim. Acta* **49B**, 185 (1994).

40. G. C. Turk and R. L. Watters, Jr., *Anal. Chem.* **57**, 1979 (1985).

41. K. C. Ng, M. J. Angebranndt, and J. D. Winefordner, *Anal. Chem.* **62**, 2506 (1990).

42. G. C. Turk, L. Yu, R. L. Watters, Jr., and J. C. Travis, *Appl. Spectrosc.* **46**, 1217 (1992).

43. K. C. Smyth and W. G. Mallard, *Combust. Sci. Technol.* **26**, 35 (1981).

44. Y.-Y. J. Wu, P. M Hunt, G. E. Leroi, and S. R. Crouch, *Chem. Phys. Lett.* **155**, 69 (1989).

45. M. R. Winchester, National Institute of Standards and Technology, Gaithersburg, MD, private communication, 1993.

46. G. C. Turk, L. Yu, and R. Koirtyohann, *Spectrochim. Acta* **49B**, 1537 (1994).

47. L. Yu, S. R. Koirtyohann, G. C. Turk, and M. L. Salit, *J. Anal. At. Spectrom.* **9**, 997 (1994).

CHAPTER

3

ANALYTICAL PERFORMANCE
OF LASER-ENHANCED IONIZATION IN FLAMES

GREGORY C. TURK

Analytical Chemistry Division, National Institute of Standards and Technology, Gaithersburg, Maryland 20899

3.1. INTRODUCTION

With the fundamental physical mechanisms of laser-enhanced ionization (LEI) and the intricacies of ionization detection having been covered in the previous two chapters, we now begin to describe the strengths and weakness of LEI spectrometry as a technique for quantitative elemental chemical analysis. In this chapter the basic instrumental configurations of LEI that have evolved over the years are described, with reference back to the physical fundamentals to show how instrumental design can affect analytical performance. A summary of the signal-to-noise considerations that ultimately define the sensitivity limits of LEI is given, and experimentally determined limits of detection are compiled. Interferences are characterized, along with techniques used to minimize their effect. Actual applications are described in Chapter 4.

3.2. INSTRUMENTATION

LEI spectrometry can be broken down into the basic steps of atomization, excitation, ionization, detection, and registration. The instrumental components of the LEI spectrometer that perform these functions are the flame (atomization), the laser (excitation), again the flame (ionization), the probe electrodes (detection), and finally the various amplifiers and computers that condition and display the LEI signal (registration). Each of these instrumental components will be discussed in the following subsections.

Laser-Enhanced Ionization Spectrometry, edited by John C. Travis and Gregory C. Turk.
Chemical Analysis Series, Vol. 136.
ISBN 0-471-57684-0 © 1996 John Wiley & Sons, Inc.

3.2.1. Flames

The first LEI measurements were performed in 1976 (1) using the standard premixed flame as the atom reservoir. This was largely a matter of convenience in the laboratories where these first experiments were done, utilizing the flame systems developed for atomic absorption spectrometry. The generally unsuccessful attempts to move LEI into other atom reservoirs, such as the graphite furnace and inductively coupled plasma, lead to the conclusion that the first LEI researchers had perhaps chosen the optimum atom reservoir for their new spectrometric method.

The flame performs a dual function in LEI: atomization of the sample and collisional ionization of the laser-excited atoms. This is analogous to atomic emission, where both atomization and excitation are performed, but differs from atomic absorption and fluorescence, where atomization is the only function. Unlike hotter plasmas, the flame leaves most elements as neutral atoms, leaving the task of ionization to be selectively enhanced by the laser. Low-pressure atom sources have low collision rates. Laser-induced ionization in such sources requires that the laser provide all of the energy for ionization, using the weak process of photoionization. Because photoionization cross sections are so small, very high laser powers must be used. This can result in the distortion of line shapes and increases the likelihood of spectral overlap interferences. For LEI in a flame, the laser is only needed to selectively excite the atoms, which are then efficiently ionized by collision. Much lower laser powers are needed than for photoionization. Very importantly, the flame environment is compatible with probe electrodes for detection of LEI and the level of background ionization is acceptably low.

The flame equipment used for LEI remains the commercially available premixed burner systems designed for use in atomic absorption spectrometry. These systems employ a pneumatic nebulizer with liquid sample uptake rates of 3–6 mL/min. The nebulizer converts the sample solution into a liquid aerosol, which is mixed with the combustion gases in a spray chamber. Roughly 10% of the sample solution reaches the flame, the rest being drained to waste. Water is the sample solvent in the vast majority of LEI applications, but there has been some successful use of organic solvents in LEI for petroleum (2) and chromatographic applications (3–5). The limitation of the use of organic solvents is the formation of soot in the flame, which leads to a non-specific laser-induced ionization background (6). This problem is encountered with aromatic solvents or at higher solution aspiration rates.

Most LEI work has been done using the slot-type burner heads used for atomic absorption. The standard path lengths are 5 and 10 cm. The longer path length has the advantage of allowing a greater laser excitation volume for a single pass of the laser beam through the flame. This advantage is partially offset by a greater flame ionization background. Most of the work in our laboratory has been done using the shorter 5 cm single-slot burner head. This is mostly a matter of the convenience of easier electrode and laser alignment with the smaller flame.

The standard flames used for analytical flame spectrometry have all been used for LEI: air/acetylene, air/hydrogen, and nitrous oxide/acetylene. Desirable flame properties for LEI include those that are sought for the classical flame spectrometries: high atomization efficiency and freedom from chemical interference. These properties roughly improve with increasing flame temperature, with air/hydrogen being the poorest flame and nitrous oxide/acetylene the best in these terms. Only the nitrous oxide/acetylene flame is capable of efficient atomization of the refractory elements.

Since LEI relies on collisional ionization, the higher temperature flames also have the advantage of higher rates of collisional ionization. The usefulness of a higher collisional ionization rate depends on the element being ionized and the energetics of the laser excitation. It is possible to achieve 100% ionization efficiency under some conditions, even for the cooler flames. In this situation there is no advantage to using a hotter flame.

The issue of flame background ionization is very important for LEI. Since LEI must be detected above the normal (non-laser-induced) ionization of the flame, the magnitude of this background determines the ultimate limit to the sensitivity of LEI. It is necessary to distinguish between the natural ionization background of the flame and that caused by the thermal ionization of the sample matrix aspirated into the flame. The natural ionization background is determined by the combustion products of the flame gases. Flames using hydrocarbon fuel have much higher levels of background ionization than those that use hydrogen or other fuels such as carbon monoxide. Ion concentrations in hydrocarbon flames vary greatly with fuel/oxidant ratio and spatially within the flame, but a rough concentration of 10^{10} ions cm^{-3} is a useful estimate. The natural ion concentration in hydrogen flames is less than 10^5 cm^{-3} (7). In fact, trace contamination of alkali metals is often the major source of ionization background in hydrogen flames. In practical situations, it is almost always the thermal ionization of easily ionized elements in the sample matrix that dominates the flame background ionization. Thus the low natural ionization background of the hydrogen flame is rarely a significant advantage, the one exception being that the lower temperature results in lower thermal matrix ionization.

Another characteristic of a flame to be considered is that of spectral background, that is, any laser-induced ionization signal resulting from species naturally present in the flame. In optical flame spectrometry, emission from molecules such as OH, CN, or C_2 can cause background interference. Those particular molecules are of no concern in LEI spectrometry because they have very high ionization potentials and are difficult to ionize by laser-induced processes in a typical LEI measurement. The most common spectral background encountered in LEI is caused by multiphoton ionization (MPI) of NO. This molecule has a relatively low ionization potential (9.26 eV) and is efficiently ionized by a two-photon process throughout wide regions of the ultraviolet (8, 9), or by four-photon ionization between 426 and 454 nm (10). Another MPI background spectrum encountered in LEI is that of PO (ionization potential = 8.23 eV), which is found in acetylene flames between 302 nm and 334 nm as a by-product of phosphine contamination in most industrial sources of acetylene (11, 12). Since MPI is a nonlinear optical process and LEI is generally linear, MPI interference can be minimized by avoiding unnecessarily high laser irradiance.

Most LEI work has been done using the air/acetylene flame. The cooler air/hydrogen flame is appropriate for detection of alkali metals and has been very useful for many fundamental studies of LEI mechanisms. However, years of experience by flame atomic absorption spectrometrists have shown that the air/hydrogen flame is more prone to chemical interferences due to incomplete molecular dissociation, and for this reason this flame should also be avoided for LEI in practical analytical measurements.

There has not been very much experience using the hotter nitrous oxide/acetylene flame for LEI. Messman et al. (13) published limits of detection for six refractory elements at a variety of excitation wavelengths and reported on the observation of LEI signals from nine other refractory elements. There has not neen any study of the use of this flame for real sample analysis by LEI or any study of interferences. It is expected that the hotter temperature and consequent higher degree of thermal ionization of matrix species will amplify interferences caused by matrix ionization.

There has been some exploration of the use of total-consumption burners by LEI (14). These are turbulent flames used in the early days of flame spectrometry before laminar flow premix burners were developed. These older burners had some advantages. Since the liquid sample is introduced directly into the flame, the transfer efficiency is much higher. Also, since there is no premixture of gases, there is no chance for a flashback to occur, allowing the use of faster burning and hotter flame gas mixtures. It was hoped that the non-optical detection of LEI would avoid the many disadvantages of such burners. Reasonable results were obtained only when a desolvation system was used (15), and sensitivity suffered as a result of the short path length of these burners in comparison to slot burners.

3.2.2. Lasers

In 1976, the first measurement of LEI by Green et al. (1) was performed using a coutinuous-wave (cw) dye laser pumped by an argon ion laser. Since that time many different varieties of lasers have been used, generally following the trends of commercially available lasers. These have generally, but not entirely, been tunable dye lasers. The pumping sources for these dye lasers include flashlamps (16), nitrogen lasers (17), Nd:YAG lasers (18), XeCl excimer lasers (19), copper vapor lasers (20), and cw argon and krypton ion lasers (21). Nondye lasers used for LEI include the atomic line laser (22) and the tunable diode laser (23).

These lasers vary widely in peak power and pulse duration, but average laser powers vary much less. The main subdivision of these various laser systems is between the pulsed and the cw lasers. The difference in the time regime of exposure to laser radiation has a great effect on the LEI mechanisms of ion production (as discussed in Chapter 1, Section 1.3.7) and signal detection (as discussed in Chapter 2, Section 2.8.8).

In terms of the fraction of atoms passing through the flame that is exposed to laser radiation, pulsed lasers are at a great disadvantage in comparison to cw lasers. Consider a pulsed Nd:YAG laser with a typical repetition rate of 10 Hz and a beam diameter of 1 mm. If the flame has a typical velocity of 10 m/s, 1 m of flame will go by between laser pulses and only 1 mm of flame will be excited during the pulse, yielding an irradiation efficiency of only 0.1%. However, the pulsed copper vapor laser, which is capable kilohertz repetition rates, can achieve perfect irradiation efficiency.

Laser pulse duration has a controlling influence on the ionization efficiency of LEI — the fraction of irradiated atoms that is ionized. The longer the atoms can be kept in an excited state by laser irradiation, the longer they will be exposed to enhanced rates of collisional ionization and thus the greater will be the fraction ionized. Here again, the cw laser has the advantage, and to a lesser extent so does the relatively long-pulsed (microsecond) flashlamp-pumped dye laser. In order to achieve full ionization efficiency with a Nd:YAG or excimer pumped dye laser, which have pulse durations of only 10–30 ns, the rate of collisional ionization from the excited state must be on the order of $10^8 \, \text{s}^{-1}$. To achieve this rate in an air/acetylene flame requires that the laser-excited state be roughly within 1 eV of the ionization limit. For most elements this requires double-resonance excitation.

Using a cw laser, full ionization depletion has been demonstrated with a much larger energy gap of 3 eV between the ionization limit and the laser-excited state, where the rate of collisional ionization was only about $2 \times 10^4 \, \text{s}^{-1}$ (24). In this experiment, the $3p$ state of Na was populated by excitation with a 200 mW cw dye laser chopped at 2 kHz. A small fraction of this laser beam was split off for use as a low-power probe beam, which was used to monitor absorption by ground state Na atoms at various points in the flame. As a result of LEI, neutral atom depletion was observed as a decrease in the atomic absorption signal in the probe beam, synchronized with the chopping frequency of the LEI pumping beam. By spatially rastering the probe beam, the contour plot shown in Fig. 3.1 was constructed. The maximum depletion at the pump beam was 74% with the effects of diffusion and flame flow filling in the hole in the atomic population above the pump beam. In an accompanying measurement, an acousto-optic modulator was used to produce 20 ms pulses from the cw beam, and time-resolved measurements of the dip in absorption in the probe beam were recorded at various positions in the flame. From these data, the time-resolved contours of Na atom depletion following LEI in the flame were constructed, as seen in Fig. 3.2.

Despite the fact that cw lasers have important advantages for LEI in efficiency of excitation and ionization, the vast majority of LEI spectrometric studies has been done using pulsed lasers. The duty cycle advantage of cw lasers is largely compensated by the use of gated detection with pulsed lasers, so that flame background noise during the long period of time between laser pulses is avoided. Additionally, cw LEI is much more susceptible to sources of low-frequency noise in the flame background current. Particularly damaging are noise spikes from droplets or particles that are difficult to selectively filter from the LEI. Such noise is easily filtered when one is detecting pulsed LEI, where the duration of the pulsed signal is in the range of 100–1000 ns.

The power available from cw lasers is generally insufficient to saturate atomic lines in the flame. This works against the ionization efficiency of the cw laser, since a smaller fraction of atoms can be kept in the excited state during laser excitation. The lower power also makes it much more difficult to extend wavelength coverage of these lasers into the ultraviolet by frequency doubling. This is the biggest practical impediment to the use of cw lasers for LEI, as well as for other varieties of laser spectrometry.

One area where the use of cw lasers shows potential for growth is with tunable diode lasers. These lasers offer great advantage over dye lasers in terms of cost, simplicity, and ruggedness. Their use for LEI in flames has been demonstrated (23). They have also been used for LEI in thermionic diodes (25, 26). In the flame, LEI detection limits for Cs and Rb have been measured using such lasers. These elements have strong ground state transitions in wavelength areas where diode lasers currently are readily available—in the near-infrared. The results obtained were very similar to those obtained with cw dye lasers.

To date, the lasers of choice for LEI are dye lasers pumped by either a pulsed Nd:YAG or XeCl excimer laser. The excimer laser has the edge in its ability to operate at relatively high repetition rates (200–500 Hz) in common commercial systems. It is also somewhat easier to operate with a wide range of laser dyes, whereas the Nd:YAG laser requires the use of different harmonics of the pumping wavelength for different wavelength regions. Both pumping lasers are readily adaptable for stepwise excitation schemes. For stepwise excitation measurements, which are necessary to achieve the best sensitivity

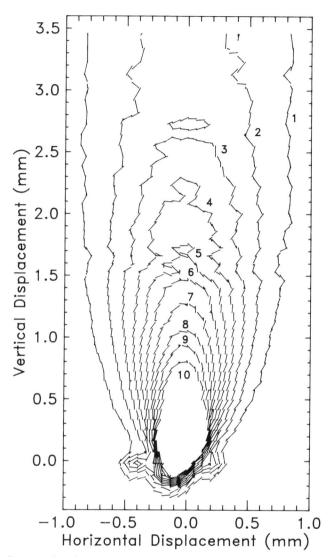

Figure 3.1. Contour plot of steady-state Na atom depletion resulting from cw LEI. Scales are relative to the pump beam position, which is 9.5 mm above the burner. Contours are 1% increments in transmitted light intensity, with an upper limit of 10% to avoid contour bunching. Blank spot in the region between 0 and 0.6 mm in the y axis and -0.2 and 0.2 mm in the x axis is an artifact of the 10% limit. Reproduced from Schenck et al. (24) by permission of Les Editions de Physique, Les Ulis, France.

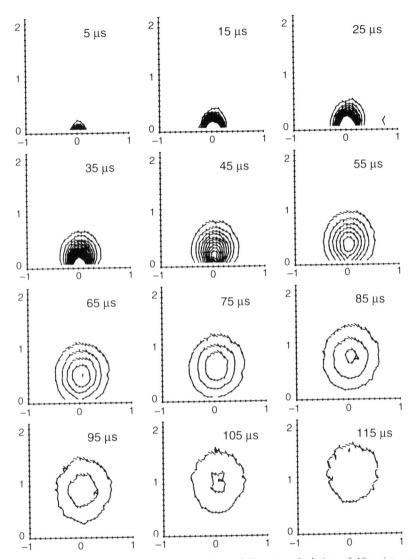

Figure 3.2. Time-resolved nonlinear contour plots of Na atom depletion of 10 μs intervals measured from the beginning of the 20 μs laser pulse. Contours are 1% increments in transmitted light intensity up to a maximum of 15%. Blank region in the center of the 15–35 μs frames is an artifact of the 15% upper limit; 5 μs frame marks pump laser position, and tick marks represent 0.2 mm increments in both vertical and horizontal axes. Reproduced from Schenck et al. (24) by permission of Les Editions de Physique, Les Ulis, France.

and selectivity for LEI, two independently tunable lasers, synchronized in time and space, are required. This is accomplished in a relatively straightforward manner by splitting the pump beam into two beams and pumping two separate dye lasers. In the case of the Nd:YAG laser, it is also possible to pump one dye laser with the second harmonic beam and the other with the third harmonic.

The optimization of laser irradiance for LEI is important not only to maximize the signal strength but also because excessive laser irradiance generally leads to saturation broadening of matrix element lines and increased MPI flame background, thus decreasing elemental selectivity. For a given laser, with the capability of providing a certain laser energy per pulse (for most LEI instruments this would be in the range of $100\,\mu J$ per pulse for the frequency-doubled output of a dye laser), and a fixed laser bandwidth (typically 0.01 nm), the best means of control of the irradiance is through control of the beam diameter by focusing or defocusing. Focusing increases the laser irradiance but decreases the irradiated volume in the flame. Below saturation these two effects cancel. In this case it is usually better to use a larger beam diameter and thus decrease the possibility of spectral interference from matrix elements or flame background. The exception would be in the case where stepwise excitation is being used. In this situation it is important for the first excitation step to achieve as high a fraction of upper level population as possible to produce a high degree of coupling with the second-step transition. In this situation it would be a mistake to defocus the beam bringing the laser irradiance further below the saturation point. The optimum laser irradiance is that which is just sufficient to saturate the transition. Above saturation, focusing will only decrease the irradiated flame volume and the signal will decrease. Obviously defocusing is called for in this situation.

Another issue related to laser instrumentation for LEI is more of a nuisance than a fundamental property, but it can have a dramatic effect on the performance of the total system. It is radio-frequency interference (RFI). All pulsed lasers emit some RFI that is synchronized with the laser output. The source may be the electrical discharge through an excimer or nitrogen laser, a flashlamp, or a Q-switch. Excimer lasers are generally worse than Nd:YAG lasers in this regard, and the presence of high levels of RFI precluded the practical use of a Cu vapor laser as a dye laser pump source for LEI (20). The level of this RFI varies greatly with the quality of shielding and proper design of the high-voltage electronics. Lasers based on the same principles of operation but built by different manufacturers can vary by orders of magnitude in the level of emitted RFI. The probe electrodes and electronics used to detect LEI are very susceptible to this RFI, much more so than photomultipliers used for laser-induced fluorescence.

3.2.3. The Ionization Probe

LEI is detected as an increase in the number of charge carriers in the flame by means of electrical probe electrodes. Chapter 2 has described in detail the many intricacies of this deceptively simple process. It will suffice here to summarize the process for pulsed LEI in a few sentences. Detection of LEI requires the imposition of an electrical field through the laser-irradiated region of the flame in order to separate the ions and electrons created by LEI. The movement of these charged species under the influence of the electric field induces a current in the detection circuit. Electrodes are used to apply the electric field and detect the LEI-induced current.

The most important consideration in evaluating the performance of a particular design of the ionization probe electrodes is the susceptibility to space charge interference. Such interferences occur in the presence of easily ionized elements, when the increased background of thermal ionization results in an increased density of the space charge region surrounding the cathode. This can shield the laser-irradiated region of the flame from the applied electric field, resulting in the loss of detected LEI signal. Chapter 2 deals with this topic in greater detail, including a discussion of the historical evolution of LEI electrode designs and the relationship between electrode design and space charge interference.

3.2.3.1. The Water-Cooled Internal Cathode

The cardinal rule for avoiding space charge interference is to keep the laser beam aligned as close to the cathode as possible. This ensures that LEI will occur in the presence of the electric field of the cathode fall region as long as possible, even as this region compresses into a tighter space surrounding the cathode with increasing concentration of easily ionized elements in the sample matrix. The more-or-less standard configuration for detection of LEI uses a water-cooled cathode immersed inside the flame, just above the laser beam, with the burner head used as the anode. With this system, a matrix containing 3000 μg/mL of sodium could be tolerated without space charge interference. In contrast, earlier electrode designs were subject to space charge interference with only 10 μg/mL of Na present.

Stainless steel tubing is generally used to form the cathode, with a diameter on the order of 6 mm. The cathode is aligned along the long dimension of the slot burner, 1–2 cm above the burner head. This electrode becomes the cathode when a negative high voltage is applied, with the grounded burner head then being the anode.

One detail to be considered when supplying the cooling water for the cathode is the flow of electric current through the cooling water supply from the high-voltage power supply to a grounded water pipe. This can be reduced by using a recirculated supply of deionized water or by using several meters of plastic tubing between the electrode and the water supply. With the latter approach, which is simpler to implement, the current through the cooling water is approximately 0.1 mA. The LEI spectrometers built by SOPRA[1] (Bios-Colombes, France) avoided this problem by using a glass-lined tube electrode, but this system was subject to cracking from thermal shock when the flame was being lit or extinguished.

3.2.3.2. Optimization of Applied Cathode Potential

As the negative potential applied to the cathode is increased, the cathode fall region, or sheath, extends farther from the cathode toward the anode (see Chapter 2 for more detail). As this occurs, more ions and electrons experience the applied electric field and thus are registered by the detection circuit. This applies not only to the LEI-produced charged species but also to the background ions and electrons, and consequently the applied cathode potential affects both the level of the LEI signal and the level of background noise. Figure 3.3. shows the typical threshold and plateau behavior of an LEI signal as a function of applied potential (27). Below the threshold potential the electric field has not reached the laser beam (located 7 mm below the cathode in this example) and no signal is observed. At the plateau potential, the field has passed through the entire laser beam and the LEI signal no longer increases with applied potential. If the background current is observed as a function of applied potential (see, for example, Fig. 2.15), the current increases as the electric field moves through the flame; it finally saturates when the field extends to the anode—the saturation potential. The optimum signal-to-background ratio is thus obtained at the potential where the LEI signal has just reached the plateau. Increasing the potential beyond that point will only increase the background current and the noise associated with it.

One caution is in order regarding the optimized applied potential. When the potential is just to the point where the field has extended through the laser beam, the optimum potential as far as signal-to-background ratio is concerned, the measurement is susceptible to space charge interferences from slight changes in the concentrations of easily ionized elements. For this reason, the optimization should be performed for the one sample in a particular collection of samples and standards to be analyzed that contains the highest concentration of easily ionized elements. This can be easily determined by monitoring the dc background currents for the suite of solutions. In most cases

---

[1] SOPRA: Société de Production et de Recherches Appliquées.

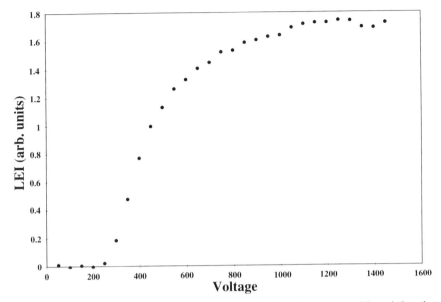

Figure 3.3. LEI signal as a function of applied potential for a laser beam centered 7 mm below the cathode.

it is safer to use an applied potential somewhat higher than the optimum. Typical applied potentials range between -1000 and -1500 V.

3.2.4. Electronics

Figure 3.4 shows a typical configuration of the electronic circuit used to measure a pulsed LEI signal. The connection between the flame and the external electronic circuit is most conveniently made at the burner head anode. This requires isolation of the burner head from ground. A burner premix chamber made of plastic is the most convenient way to accomplish this, but isolation can be simply accomplished with insulating tape for metal premix chambers.

The first order of business is the separation of the pulsed LEI signal from the dc current due to natural thermal ionization, and this is accomplished with the simple RC (resistance–capacitance) combination between the anode and the input to the current preamplifier. The resistor is sometimes mistaken as a load resistor for the LEI current, but the function of this resistor is to provide a path to ground for the natural dc flame current. Without the resistor a charge will build up on the burner and all current will cease to flow. This resistance needs to be larger than the reactance of the capacitor at the frequency of the

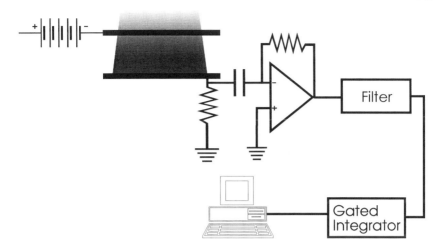

Figure 3.4. Detection electronics for pulsed LEI spectrometry.

LEI signal and the input resistance of the preamplifier, to avoid having the LEI signal current flow to ground. The resistance should be small in comparison to that of the flame, even in the presence of high concentrations of alkali metals, to avoid significant reduction of the potential difference between the anode and cathode. The dc potential at the burner head is a convenient measure of the background ionization current. For this purpose, the resistor does serve as a true load resistor. A typical value for the resistor is $10\,k\Omega$, with a $0.01\,\mu F$ capacitor.

Most laboratories performing LEI measurements have used "homemade" current-to-voltage converters as preamplifiers (16, 28), constructed from operational amplifiers as shown in Fig. 3.4. We have used the National Semiconductor LH0032CG FET operational amplifier, as well as the Analog Devices 50J operational amplifier. Combined with a 1 MΩ feedback resistor, for a gain of 1 V/μA, both of these op amps will have an upper frequency response of approximately 1 MHz. This is too slow to give an undistorted pulse shape for the electron component of the LEI signal. Based on the electron mobility and typical electric fields, the true LEI electron pulse width is on the order of 100 ns, but a 1 μs pulse shape is obtained with the preamplifier described here. This results in some loss of sensitivity, but (as will be discussed later) the inherent pulse integration that occurs when the slower preamplifier is used has an advantage in reducing susceptibility to interference from easily ionized matrix elements.

In most cases it is advisable to perform some additional filtering of the LEI signal following the preamplifier. The purpose is to attenuate lower frequency fluctuations that are associated with the background flame ionization but are not filtered by the input capacitor to the preamplifier. We have found a 10 kHz high-pass filter to be useful, particularly in situations where the level of background current is high due to the presence of easily ionized elements.

As with any pulsed signal with a poor duty cycle, gated detection results in a greatly improved signal-to-noise (S/N) ratio by ignoring the noise generated during the period of time between signal pulses. There are some subtleties to be considered in the optimization of the gate duration. An important consideration in this matter is the duration of the gate in relation to the duration of the electron component of the LEI signal, as well as the often ignored ion component, keeping in mind that the LEI pulse shapes are subject to changes with a changing easily ionized element matrix.

As discussed in Chapter 2, changes in the concentration of easily ionized elements in the flame affects the electric field profile in the flame. When the easily ionized element concentration is increased, the size of the cathode fall region is diminished and thus the same applied potential is dropped over a shorter distance. If the laser beam is properly aligned near the cathode, this will usually have the effect of increasing the magnitude of the electric field at the position of the laser beam. This in turn increases the velocity of the LEI ions and electrons, thus sharpening the LEI ion and electron pulses. An example of this effect is shown in Fig. 3.5, where the LEI pulse from 100 μg/mL of Fe is shown with and without the presence of a matrix containing 300 μg/mL of Na. These measurements were taken with the laser beam aligned 1 mm below the water-cooled cathode, using a preamplifier with a fast enough bandwidth to resolve the LEI electron pulse with minimal distortion (29). A clear sharpening of the LEI electron pulse shape is observed in the presence of the Na matrix. A closer look at the LEI current beyond 400 ns from the laser pulse reveals that the current has not returned to zero in the presence of the Na matrix. This current is the leading edge of the LEI ion pulse, which is observed in this time window because the ions are also moving faster in response to the increased electric field.

Aside from changing the shapes of the LEI electron and ion pulses, changes in easily ionized element concentration can lead to changes in the integrated charge of the electron and ion pulses relative to each other. This effect is also discussed in Chapter 2, Section 2.8.2, under the topic of induced charge apportionment. Since it is the *motion* of the charged species created by LEI that induces the LEI signal current, not simply their arrival at the electrodes, the relative magnitudes of the ion and electron charge are not necessarily equal. An infinitely thin laser beam producing LEI ions and electrons at the surface of the cathode would yield only an LEI electron pulse as the electrons move toward the anode. The ions would never have a chance to induce any current before arriving at the cathode. The opposite would hold if the laser beam were at the opposite end of the cathode fall region, and intermediate

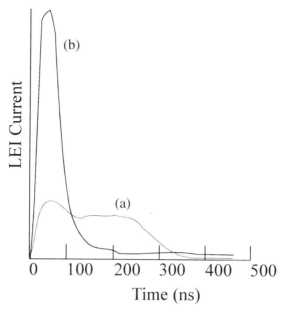

Figure 3.5. LEI signal pulse from 100 µg/mL Fe (*a*) in the absence of a matrix of easily ionized elements and (*b*) with a matrix containing 300 µg/mL Na. The measurements were made with the laser beam centered 1 mm below the cathode and −1500 V applied to the cathode.

positions of the laser between the cathode and the edge of the sheath yield intermediate ratios of electron-to-ion signal. Thus, when the position of the sheath edge moves in response to a change in easily ionized element concentration, the charge apportionment between electron and ion signal changes.

A change in the peak shape or the charge apportionment can result in a change in the fraction of the LEI signal that falls within the gate width of the gated integrator. The only way to avoid this problem completely is to use a gate width that is wide enough to include 100% of the electron and ion signal. However, using a wider gate leads to increased noise in the measurement as more and more background current falls within the gate. In order to keep the gate width as narrow as possible and still detect the electron pulse and the entire ion pulse, it is desirable to have the slower moving ions collected as quickly as possible and to have the charge apportionment such that most of the signal is carried by the faster moving electrons. Both of these conditions can be met by once again ensuring that the laser beam is aligned as close to the cathode as possible. For example, the cathode fall region in an unseeded air/acetylene flame with −1500 V applied to the water-cooled cathode has been measured to extend a distance of 1.32 cm from the cathode (29). Using

Eqs. (2.30) and (2.35), we can calculate that the electric field at a point 0.1 cm from the cathode will be 2100 V/cm under such conditions. Assuming a typical analyte ion mobility of 20 cm^2 V^{-1} s^{-1}, we can use Eq. (2.41) to determine that the ion pulse from a laser beam 0.1 cm below the cathode should end with the arrival of the LEI ions at the cathode 2.3 μs after the laser pulse. Calculation of the charge apportionment from Eq. (2.48) shows that 92% of the total LEI charge is from the electron pulse. In the presence of 1000 μg/mL Na, the sheath width has been measured to decrease to 0.36 cm. In this situation, with the laser beam in the same position, the percentage of electron-induced LEI signal would decrease to 72%, but the electric field 0.1 cm below the cathode would increase to 6020 V/cm and the ion pulse would sharpen to approximately 0.8 μs. Thus using a gate width of 2.3 μs would assure that all ion and electron LEI would be integrated under either matrix condition.

The need to detect the slower ion-induced LEI signal is the reason that the use of a preamplifier with an upper-frequency response that is too slow to resolve the true electron pulse shape is not a disadvantage. However, in situations where changes in easily ionized element concentration are not expected, the optimum ratio of LEI current to background current would be obtained using a preamplifier and gate width matched to the electron pulse, ignoring the ion pulse with its poor S/N "rate of return".

From the output of the gated detector, the only task remaining to complete the LEI measurement is the accumulation, averaging, and storage of the data. Signal averaging of multiple LEI pulses can usually be performed by the gated integrator, or this can be performed after the fact in a computer. Computer-based data collection is preferred, particularly when combined with the ability to control laser wavelength and monitor other variables such as laser power and flame background current.

3.3. NOISE CHARACTERISTICS

In dicussing the sources and characteristics of noise for any analytical measurement technique it is convenient to divide the topic into two categories: additive sources of noise and multiplicative noise. The latter refers to noise that is proportional to the concentration of analyte. Such noise affects the precision of a method but not the detection limit. Additive noise comes from other components of the measured signal besides the analyte response and together with the analyte sensitivity (signal/unit concentration) determines the limits of detection. Table 3.1 outlines a variety of sources of noise in an LEI measurement. In the discussion that follows, a pulsed LEI measurement is assumed. Most of the general principles would also apply to a cw LEI measurement, but obviously in a different frequency range.

Table 3.1. Sources of Noise in LEI

Multiplicative:
  Fluctuations in atomic population
    Nebulizer
    Flame gas flow
  Fluctuations in ionization yield
    Laser properties
      Power
      Wavelength
      Bandwidth
      Spatial profile
    Temperature
  Fluctuations in detection efficiency
    Applied potential
    Flame composition
    Spatial fluctuations of laser beam

Additive:
  Fluctuations in thermal background ionization
    Natural flame ions
    Sample matrix ions
  Fluctuations in laser-induced background ionization
    Natural flame components
      Multiphoton ionization
    Sample matrix
      Molecular bands
      Line wings
  Electronics
    Johnson noise
    Radio-frequency interference (RFI)

3.3.1. Multiplicative Noise

The LEI measurement can be broken into three distinct phases: (*i*) the production of free atoms from the sample solutions; (*ii*) the ionization of these atoms by LEI; and (*iii*) the detection of the LEI ions. Random fluctuations in each of these processes contribute to the total LEI multiplicative noise. Fluctuations in the free atom population can result from the operation of the nebulizer, and the flow, mixing, and combustion of the flame gases as well as the room air. Fluctuations in the LEI yield (the fraction of atoms ionized by LEI) result from the variation in the properties of the laser (power, wavelength, etc.); since LEI is partially a collisional process, temperature fluctuations in the flame could also result in LEI noise. In the detection process, any source of variation of the magnitude and spatial distribution of the applied electric field can lead to LEI noise, as well as any variation in the position of the laser beam within the flame. Variation in the electric field strength could come from noise in the power supply used to apply the field or from variations of the flame composition that affect the size of the cathode fall region.

Of these various sources of multiplicative noise, two have been studied in some detail: noise due to variation in laser power and noise due to variation of the atomic population.

3.3.1.1. Laser Power Fluctuations

The pulse-to-pulse variation in the power of a pulsed dye laser can be substantial. A 4% relative standard deviation (RSD) is typical, but in some circumstances it can be much worse. For this reason it is often desirable to monitor the laser power on a pulse-to-pulse basis. To be more precise, in most cases it is not laser power that is actually measured but rather the integrated

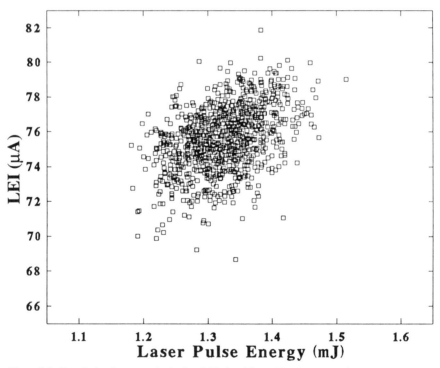

Figure 3.6. Correlation between single-shot LEI signal from 20 μg/mL Na and laser pulse energy at 588.995 nm.

energy of each laser pulse. The conversion from pulse energy to pulse power is done through knowledge of the laser pulse duration.

One obvious approach to reducing the noise in an LEI measurement is normalization of the LEI signal to the measured laser pulse energy. Though often attempted, this approach is rarely successful. A typical result is seen in Fig. 3.6, where the LEI signal from Na is plotted against laser energy on a pulse-by-pulse basis for 1000 laser pulses from a Nd:YAG pumped dye laser (30). Little if any correlation between LEI and laser pulse energy is evident, with a correlation coefficient (r) of only 0.40. The RSD of the laser pulse energy is 4.1%, but the LEI signal varies with an RSD of only 2.4%. The lack of correlation is due to saturation of the absorption process at the high spectral irradiance being used for the measurement. Figure 3.7 shows the saturated laser power dependence under the same conditions as apply to Fig. 3.6, except with the laser power being purposefully varied between the normal operating power and much lower powers (30). The issue of correlation between laser power fluctuations and LEI is further complicated by the variation in laser power across the beam profile. The power is higher and saturation more

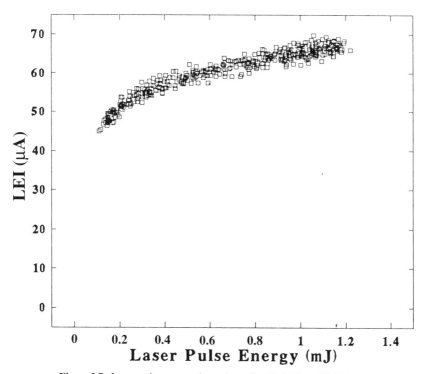

Figure 3.7. Laser pulse energy dependence for Na LEI at 588.995 nm.

pronounced at the center of the laser beam than at the edges. Thus an LEI measurement will be more susceptible to variation in laser power at the edges of the laser beam than in the center. In the measurements shown in Figs. 3.6 and 3.7, the total beam energy is measured. At lower laser irradiances, the correlation between LEI and laser power increases.

3.3.1.2. *Atom Population Fluctuations*

Some interesting insights into the sources of noise in LEI have been gained through simultaneous detection of LEI and laser-induced fluorescence (LIF) (30). These two techniques are based on different responses to the identical process of laser photoexcitation of atoms produced in a flame from a nebulized solution. As such they share many sources of noise and also have some sources of noise that are unique. By studying the degree of correlation between LEI noise and LIF noise, the source of noise can be narrowed to either the category of common noise sources or unique noise sources. Figure 3.8 shows a scatter plot of simultaneously measured LEI and LIF of Na. The data, which were

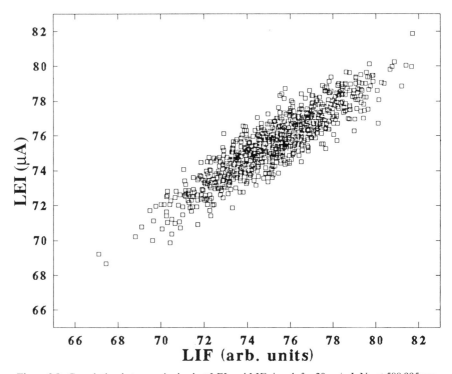

Figure 3.8. Correlation between single-shot LEI and LIF signals for 20 μg/mL Na at 588.995 nm.

collected during the measurement of LEI and laser energy shown in Fig. 3.6, show significant correlation, with r equal to 0.90. The LIF data were also uncorrelated with laser energy at the normal operating power, showing the identical saturated response function when the laser power was varied between normal and lower laser power. Correlation between LEI and LIF noise, without the correlation of either to laser power fluctuations, indicates that the major source of noise for both LEI and LIF in this particular case results from fluctuation of the atomic population within the laser-irradiated volume of the flame.

3.3.2. Additive Noise

As the concentration of an analyte decreases, multiplicative noise decreases accordingly, and it is additive noise, which remains even in the absence of analyte, that ultimately determines the limit of detection. Three categories of additive noise in LEI are listed in Table 3.1.

3.3.2.1. Thermal Background Ionization Noise

The first and most fundamental source of noise listed in Table 3.1 is the fluctuation in the thermal ionization background. Thermal ionization gives rise to the dc background current upon which the LEI signal must be discerned. This background may be the result of natural flame ions, particularly in hydrocarbon flames. Realistically, the size of the background current is usually determined by the concentration of easily ionized elements in the sample matrix. The source of the fluctuations may be many of the same mechanical influences discussed for multiplicative noise, including flame flow fluctuations and nebulizer-induced noise. The more fundamental fluctuations are analogous to photon shot noise familiar to all optical spectroscopists. In this case it is the discrete nature of the number of ions measured in the background current that determines the shot noise (31), which is equal to the square root of the number of ions. Examples of the magnitude of this noise in actual measurements will be given later.

3.3.2.2. Laser-Induced Background Ionization Noise

The next source of noise is again due to background ions, but in this case it is background ionization induced by the laser. Here again the species undergoing laser-induced ionization may be a natural flame constituent or, in real sample situations, an atom or molecule from the sample matrix. The ionization mechanism may be any variety of LEI or multiphoton ionization (MPI), atomic or molecular, resonant or nonresonant. This source of noise is particularly destructive because all of the multiplicative noise sources already discussed in terms of the analyte signal can also apply to the laser-induced background signal. This includes laser power fluctuations, which are usually more significant in this situation because saturation is less prevalent for nonresonant and molecular absorptions. In the case of multiphoton processes, laser power fluctuations have a multiple-power effect. Experience in actual analytical situations has shown this to be the most serious noise source. Examples will be discussed later (in Section 3.5).

3.3.2.3. Electronic Noise

The last source of additive noise listed in Table 3.1 is that from the various electronic components used to measure the LEI current. Ultimately any analog electronic measurement device is going to be limited by Johnson noise, induced by the thermal motion of charged particles in resistors. Of the electronic devices used in the typical LEI measurement, the current preamplifier is the noisiest. Less fundamental in nature but no less significant is electronic noise induced by RFI. To one degree or another, all pulsed lasers broadcast some RFI, resulting from Q-switches, flashlamps, and high-voltage pulsed discharges. In comparison to optically detected spectroscopic techniques, LEI measurements are particularly prone to RFI noise. The LEI electrodes and preamplifier seem to form an excellent antenna and detector for RFI. This is one area where spectroscopic measurements using optical detection procedures and photomultipliers have a significant advantage over LEI, since a photomultiplier can selectively apply gain to the optical signal and not to any RFI. Care in shielding and grounding of the LEI instrumentation is essential. Although the RFI is always synchronized with the laser, and thus the LEI signal, some discrimination can sometimes be accomplished by careful adjustment of the gate position. In some cases the highest RFI precedes the LEI signal. In theory this noise source can always be eliminated. The ultimate cure, although a logistical nightmare, is location of the laser in a room separate from the flame.

3.3.2.4. Measurements of Additive Noise

In most real measurement systems, a combination of noise sources are encountered. Included in this subsection are some measurements of LEI noise under a variety of conditions. These measurements were done specifically to determine the magnitude of additive noise sources, in order to estimate the fundamental limit of detection of LEI. The measurements were done without operation of the laser; thus RFI and any laser-induced background or analyte noise are not measured. The results are given in Table 3.2, which includes the

Table 3.2. Flame Current and Noise Measurements

Flame	Sample	Current (μA)	Root Mean Square (rms) Noise (nA)
None	None	0	1.3
Acetylene	None	7	1.7
Acetylene	SRM Trace Elements in Water	45	2.7
Acetylene	SRM Citrus Leaves	780	23
Hydrogen	None	0.06	1.6
Hydrogen	SRM Trace Elements in Water	2.4	1.7
Hydrogen	SRM Citrus Leaves	47	6.2

dc flame background current, and the rms noise level measured in an air/hydrogen flame and an air/acetylene flame under a variety of sample conditions. The noise measurements were taken from the output of the active high-pass filter, before the input to the gated integrator, using the instrumentation outlined in Fig. 3.4. Two "typical" real samples include a Standard Reference Material synthetic drinking water (Trace Elements in Water, SRM 1643c) and a solution of Citrus Leaves (SRM 1572). The SRM Water contains a moderate level of easily ionized elements, including approximately 12 μg/mL Na, 2 μg/mL K, 37 μg/mL Ca, and 9 μg/mL Mg. The SRM Citrus Leaves were acid digested and diluted to a volume of approximately 1 g/100 mL. The major matrix elements in the diluted solution include 1.6 μg/mL Na, 180 μg/mL K, 315 μg/mL Ca, and 58 μg/mL Mg. Table 3.2 also includes a measurement made with the flame off, as a measure of the contribution of the electronics to the total noise.

Three observations can be made through comparison of the results seen in Table 3.2: (*i*) noise increases with increasing background current; (*ii*) the background current is always lower in the cooler air/hydrogen flame; and (*iii*) in the absence of a significant easily ionized element matrix, most of the noise can be attributed to electronic noise rather than flame current noise.

When the noise level is high enough that any contribution of electronic noise is negligible, the noise magnitude is consistent with what would be expected from ion shot noise. Consider the case of the water SRM in the air/acetylene flame. The background current is 45 μA, which corresponds to 2.8×10^{14} ions/s. The frequency bandwidth of the measurement system at the point where the measurement was made is approximately 1 MHz. This is equivalent to an integration time of 0.5 μs (32), and during an effective

Table 3.3. LEI Noise Measurements at the Output of Various Devices in the System

	Preamp (nA)	High-Pass Filter (nA)	Gated Integrator (nA, 10 Hz, 10-sample average)	Gated Integrator (nA, 300 Hz, 300-sample average)
Flame off	1.4	1.4	0.27	0.04
Flame on	3.4	1.7	0.29	0.06
SRM Trace Elements in Water	9.4	2.7	0.45	0.07
SRM Citrus Leaves	130	23	3	0.52

integration period 1.4×10^8 ions are measured. Since shot noise increases with the square root of the number of ions measured, the predicted level of shot noise is 12,000 ions/0.5 μs, or 3.8 nA. This is within reasonable agreement with the measured noise level of 2.7 nA, with the difference likely due to inaccuracy of the estimated measurement bandwidth.

The noise levels given in Table 3.2 are without the benefit of signal averaging and thus are higher than what would be encountered in an actual analysis situation. Table 3.3 shows some noise measurements made for the same samples in an air/acetylene flame using a gated integrator with a 1 μs gate width. The latter table includes results obtained using a pulse repetition rate of 10 Hz with 10-sample exponential averaging, as well as data collected using a 300 Hz repetition rate with 300-sample exponential averaging. These two settings roughly represent the situation of a low repetition rate Nd:YAG laser and a higher repetition rate excimer laser. Table 3.3 also shows the noise level encountered directly from the output of the preamplifier and after the high-pass filter. The beneficial effects of the high-pass filter and signal averaging with the gated integrator are clearly evident.

Frequency spectra of the noise measurements just described were generated by performing a Fourier transform on the digitized noise measurements. Figure 3.9 shows a collection of such noise frequency spectra collected with the flame off and the flame with a distilled water sample, SRM Trace Elements in Water, and the SRM Citrus Leaves solution. These noise measurements were taken at the output of the preamplifier, before the high-pass filter, and show noise out to the upper-frequency limit of the preamplifier at 1 MHz. A comparison of the frequency spectra with the flame off and with the unseeded flame on shows no significant difference above 100 kHz, indicating that most of the noise encountered in the absence of an easily ionized element matrix is due to electronic noise, except at the lower frequencies. As the easily ionized element

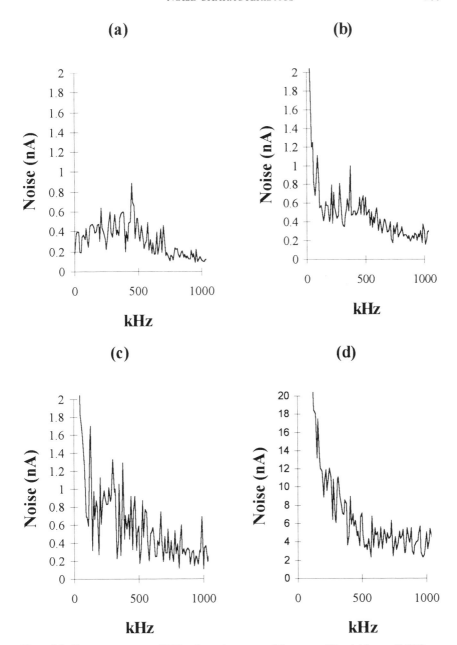

Figure 3.9. Frequency spectra of LEI noise at the output of the preamplifier: (*a*) flame off; (*b*) flame on with distilled water aspirated; (*c*) Trace Elements in Water (SRM 1643c) aspirated; (*d*) solution of Citrus Leaves (SRM 1572) aspirated.

level increases in the above SRM Water and SRM Citrus Leaves solution, noise at all frequencies is seen to increase, but especially at the lower frequencies. The $1/f$ frequency nature of the noise indicates that the noise is due to more than just the increased ion shot noise in the presence of easily ionized element matrix, since pure shot noise should show a white noise spectrum. The excess noise at the lower frequency is due to the mechanical variety of fluctuations seen in the nebulizer and gas flow.

3.4.   LIMITS OF DETECTION

The primary advantage LEI offers over other analytical methods is high sensitivity, and consequently the topic of the detection limit is a critical one. Before going on to a compilation of experimentally determined detection limits, it is useful to consider the ultimate detection capability of LEI from a more theoretical point of view (31, 33–35). This topic will be explored in the following subsections, using the background current and noise measurements described above as a connection to experimental reality.

3.4.1.   Theoretical Detection Limits

The most optimistic—and unrealistic—estimate of the detection power of an LEI measurement is one in which a 100% ionization yield is assumed (not at all unrealistic) and the only additive source of noise encountered is due to shot noise in the natural ion background current of the flame (very unrealistic). For this theoretical measurement, let us consider the use of a pulsed laser operating at 300 pps, a laser beam diameter of 0.5 cm, a 5 cm path length slot burner, an air/hydrogen flame, and a total measurement time of 1 s. In addition, let the bandwidth of the preamplifier be fast enouth to resolve the true LEI electron pulse shape, lasting 100 ns. The ion pulse will be ignored, so a 100 ns gate width can be used on the gated integrator.

From Table 3.2, the background current in an air/hydrogen flame, with only distilled water being aspirated, was measured to be 0.06 µA. This corresponds to 4×10^{11} background ions per second. During the 1 s measurement time, the 100 ns gate on the integrator will open 300 times, for a total integration time of 0.03 ms. During this time approximately 10^7 background ions will be integrated, and the shot noise level will be equal to the square root of 10^7, or 3×10^3 ions.

On the signal side of the equation, a rule of thumb often used to estimate atom density in a flame is that a microgram per milliliter of an analyte element in the aspirated solution will yield an atom density of 10^{10} atoms/cm^3 in the flame (36). To put it in more optimistic terms, 1 pg/mL corresponds to

10^4 atoms/cm^3. If we assume a 100% LEI yield and a laser-irradiated flame volume of 1 cm^3, a 1 pg/mL solution will result in the formation of 10^4 atoms per laser shot, and 3×10^6 during the total measurement. The S/N ratio for the 1 pg/mL solution is therefore 10^3, and extrapolating to an S/N ratio of 3 yields a detection limit of 3 fg/mL. This is only about 300 analyte atoms in the laser beam per laser pulse. Between laser pulses, with the flame flowing at a rate of 10 m/s, 6.7 beam volumes of atoms will pass without "seeing" the laser.

A more realistic detection limit can be derived from the real noise measurements found in Table 3.3. In this case let us consider the air/acetylene flame, using the gated integrator, with 300-sample exponential averaging, which gave a measured noise level of 0.06 nA. The same dimension laser beam and repetition rate used in the previous example will be assumed, as will a 100% LEI ionization yield. In this case the frequency bandwidth of the preamplifier was 1 MHz, and a 1 μs gate width was used on the gated integrator. As before, we will assume that a 1 pg/mL solution results in the formation of 10^4 LEI ions and electrons per laser pulse. The average current of these 10^4 charges over the 1 μs duration of the gated integrator is 1.6 nA. With the additive noise level at only 0.06 nA, the detection limit is 0.1 parts per trillion.

3.4.2. Compilation of Measured Detection Limits

Table 3.4 (37–61) lists more than 200 experimentally measured limits of detection for LEI in flames for 34 elements. In the best cases, these measured detection limits are within a factor of 3 of the predicted best case detection limit of 0.1 pg/mL. In many cases, the measured detection limits are much poorer, owing to poor atomization fractions for some elements, high contamination levels in blanks for some elements, RFI, low repetition rate lasers, etc. Even among measurements made for the same element, wide variations in the limit of detection are seen. In some cases this is due to differences in the flames and lasers utilized or to differences in experimental procedure, i.e., the time constant. Of more fundamental interest is the variation seen among different laser excitation wavelengths for the same element.

Wherever possible, Table 3.4 includes information regarding the time constant of the signal averaging system used for the measurement. In cases where a linear integration method was used, the equivalent exponential time constant is listed in the table. In most cases, the detection limit listed is that computed on the basis of an S/N ratio of 3. Published detection limits calculated on an $S/N = 2$ basis have been recalculated. In situations where the coverage factor used for calculation of the detection limit is not published, the values are presented as they appear in the original publication.

Table 3.4. LEI Limits of Detection

	First-Step Wavelength (nm)	Second-Step Wavelength (nm)	Laser[a]	Flame[b]	Time Constant (s)	Detection Limit (ng/mL)	Ref.
Ag	328.068	421.094	E	AA	—	0.075	(37)
Ag	328.068	546.549	Y	AA	—	0.3	(38)
Ag	328.068	547.155	Y	AA	—	0.4	(38)
Ag	328.068	—	Y	AA	—	2	(38)
Ag	328.068	—	F	AA	1.1	1	(39)
Al	265.248	—	E	AA	1	3	(40)
Al	266.039	—	E	AA	1	2	(40)
Al	308.215	—	F	AN	1	0.2	(13)
Al	309.271	—	F	AN	1	0.2	(13)
As	278.022	—	E	AA	1	3000	(40)
As	286.044	479.266	E	AA	1	50000	(40)
Au	242.795	—	Y	AA	1.4	1	(18)
Au	242.795	—	Y	AA	1.4	1000	(18)
Au	264.148	—	E	AA	1	4	(40)
Au	267.595	—	E	AA	1	1.2	(40)
Au	274.825	—	E	AA	1	200	(40)
Ba	270.263	—	E	AA	1	0.6	(40)
Ba	307.158	—	F	AA	1.1	0.2	(39)
Bi	302.464	—	E	AA	3	45	(19)
Bi	306.772	—	F	AA	1.1	2	(39)
Ca	272.165	—	E	AA	1	0.4	(40)
Ca	300.686	—	F	AA	1.1	0.1	(39)
Ca	422.673	468.527	N	AA	1	0.05	(41)

188

El	λ₁	λ₂					Ref
Ca	422.673	468.527	N	MA	1	0.5	(41)
Ca	422.673	585.745	Y	AA	1.5	0.03	(42)
Ca	422.673	518.885	N	AA	1	0.02	(41)
Ca	422.673	518.885	N	MA	1	0.1	(41)
Ca	422.673	518.885	N	AA	1	0.02	(43)
Ca	422.673	—	Y	AA	1.5	15	(42)
Ca	422.673	518.885	N	HA	3	100	(17)
Ca	422.673	—	Kr cw	AA	0.3	1	(21)
Ca	422.673	—	N	HA	3	30000	(17)
Ca	422.673	—	N	AA	1	1	(41)
Ca	422.673	—	N	MA	1	10	(41)
Cd	228.802	466.235	Y	AA	1.4	0.1	(18)
Cd	228.802	—	Y	AA	1.4	100	(18)
Co	252.136	591.680	Y	AA	1.4	0.08	(18)
Co	252.136	—	Y	AA	1.4	10	(18)
Co	273.112	—	E	AA	1	50	(40)
Co	274.046	—	E	AA	1	25	(40)
Co	276.419	—	E	AA	1	6	(40)
Co	281.556	—	E	AA	1	7	(40)
Co	304.400	—	E	AA	3	6	(19)
Co	315.878	531.678	Y	AA	—	0.2	(38)
Co	315.878	534.339	Y	AA	—	0.2	(38)
Co	315.878	—	Y	AA	—	2	(38)
Co	321.915	515.405	Y	AA	—	0.3	(38)
Co	321.915	—	Y	AA	—	4	(38)
Cr	272.651	—	E	AA	1	0.9	(40)

(Contd.)

189

Table 3.4. (*Contd.*)

	First-Step Wavelength (nm)	Second-Step Wavelength (nm)	Laser[a]	Flame[b]	Time Constant (s)	Detection Limit (ng/mL)	Ref.
Cr	278.070	—	E	AA	1	1.5	(40)
Cr	298.600	—	F	AA	15	2	(44)
Cr	301.492	—	E	AA	3	36	(19)
Cr	301.757	—	F	AA	15	2	(44)
Cs	455.531	—	N	PBA	1	0.004	(45)
Cs	455.531	—	N	PBA	1	0.004	(46)
Cs	455.531	—	N	PBA	0.1	0.1	(35)
Cs	455.500	—	N	AA	—	0.002	(47)
Cs	852.124	—	Diode	HA	1	0.25	(23)
Cu	276.637	—	E	AA	1	50	(40)
Cu	282.437	—	F	AA	15	100	(44)
Cu	282.437	—	E	AA	1	40	(40)
Cu	296.116	—	E	AA	1	600	(40)
Cu	324.754	453.078	E	AArod		0.2	(48)
Cu	324.754	453.078	Y	AA	1.4	0.07	(18)
Cu	324.754	—	Y	AA	1.4	3	(18)
Cu	324.754	—	F	AA	15	100	(16)
Cu	324.754	—	Y	AA	1	2	(14)
Cu	324.754	—	Y	AAtc	1	2	(14)
Cu	510.600	453.078	N	HA	3	500	(17)
Eu	459.404	564.021	N	AA	0.1	4000	(49)
Fe	271.902	—	E	AA	1	0.1	(40)

Element	Wavelength		Source	Mode			Ref
Fe	273.358	—	E	AA	1	2	(40)
Fe	273.548	—	E	AA	1	3	(40)
Fe	274.698	—	E	AA	1	30	(40)
Fe	278.810	—	E	AA	1	1.5	(40)
Fe	281.329	—	E	AA	1	5	(40)
Fe	298.357	—	F	AA	15	4	(44)
Fe	302.064	—	E	AA	3	0.12	(19)
Fe	302.064	—	F	AA	15	2	(44)
Fe	318.490	—	Y	AA	—	100	(38)
Fe	319.166	—	Y	AA	—	4	(38)
Fe	319.323	—	Y	AA	—	3	(38)
Fe	321.440	—	Y	AA	—	200	(38)
Fe	364.784	538.337	N	HA	3	100	(17)
Fe	364.784	—	N	HA	3	2000	(17)
Ga	265.987	—	E	AA	1	0.1	(40)
Ga	271.965	—	E	AA	1	0.04	(19)
Ga	287.424	—	F	AA	15	0.07	(44)
Ga	287.424	—	E	AA	1	0.06	(40)
Ga	294.364	—	E	AA	3	0.06	(19)
Ga	294.364	—	F	AA	15	0.1	(44)
Ga	417.200	—	Kr cw	AA	0.3	60	(21)
In	271.026	—	E	AA	1	0.001	(40)
In	271.394	—	E	AA	1	0.008	(40)
In	275.388	—	E	AA	1	0.005	(40)
In	293.263	—	E	AA	1	0.03	(40)
In	303.936	532	Y	AA	75	0.0004	(50)
In	303.936	—	F	AA	1.1	0.006	(39)
In	303.936	—	F	AAtc	1	0.1	(14)

(Contd.)

Table 3.4. (*Contd.*)

	First-Step Wavelength (nm)	Second-Step Wavelength (nm)	Laser[a]	Flame[b]	Time Constant (s)	Detection Limit (ng/mL)	Ref.
In	303.936	—	F	AA	1	0.02	(14)
In	303.936	—	Y	AAtc	1	0.1	(51)
In	303.936	—	Y	AA	1	0.02	(51)
In	303.936	—	F	AA	15	0.008	(44)
In	303.936	—	Y	AA		0.007	(47)
In	303.936	—	E	AA	3	0.03	(19)
In	410.176	—	Kr cw	AA	0.3	20	(21)
In	451.131	571.0	E	AArod		0.0004	(48)
In	451.131	501.8	N	HA	3	0.6	(17)
In	451.131	501.8	N	AA		0.007	(47)
In	451.131	502.3	N	AA		0.03	(47)
In	451.131	525.4	N	AA		0.003	(47)
In	451.131	526.3	Y	AA		0.01	(47)
In	451.131	571.0	N	AA		0.001	(47)
In	451.131	572.8	N	AA		0.03	(52)
In	451.131	572.8	N	AA		0.003	(47)
In	451.131	—	Kr cw	AA	0.3	0.1	(21)
In	451.131	—	N	HA	3	100	(17)
Ir	266.479	562.004 + 642.0	E	AA	1	0.3	(53)
K	294.268	—	F	AA	15	1	(44)
K	296.321	—	E	AA	1	1.5	(40)
K	404.414	—	N	PBA	—	0.1	(45)
K	766.490	—	Kr cw	HA	0.3	0.1	(21)

Li	274.119	—	E	AA	1	0.005	(40)
Li	460.286	—	Kr cw	AA	0.3	20	(21)
Li	610.362	—	F	AA	1.1	0.01	(39)
Li	639.146	639.146	F	AA	1.1	0.4	(39)
Li	670.784	460.286	E	AA	—	0.0003	(37)
Li	670.784	610.362	N	HA	3	0.04	(17)
Li	670.784	610.362	Y	AA	0.1	0.03	(35)
Li	670.784	—	F	AA	1.1	0.001	(39)
Li	670.784	—	N	HA	3	4	(17)
Mg	285.213	470.299	N	AA	0.1	0.4	(34)
Mg	285.213	—	F	AA	15	0.1	(16)
Mg	285.213	—	E	AA	3	0.005	(19)
Mn	279.482	—	E	AA	1	0.04	(40)
Mn	279.482	521.482	Y	AA	1	0.02	(54)
Mn	279.482	—	F	AA	15	0.3	(16)
Mn	279.827	—	E	AA	1	0.05	(40)
Mn	279.984	—	F	AA	15	5	(16)
Mn	280.106	—	E	AA	1	0.08	(40)
Mn	292.557	—	E	AA	1	3	(40)
Mn	292.557	—	E	AA	3	3	(19)
Mn	403.076	602.180	N	HA	3	5	(17)
Mn	403.076	—	N	HA	3	30	(17)
Mo	267.985	—	F	AN	1	30	(13)
Mo	306.428	—	F	AN	1	400	(13)
Mo	307.437	—	F	AN	1	500	(13)
Mo	308.562	—	F	AN	1	500	(13)
Mo	311.212	—	F	AN	1	900	(13)
Mo	313.259	—	F	AN	1	70	(13)

(Contd.)

Table 3.4. (*Contd.*)

	First-Step Wavelength (nm)	Second-Step Wavelength (nm)	Laser[a]	Flame[b]	Time Constant (s)	Detection Limit (ng/mL)	Ref.
Mo	315.816	—	F	AN	1	70	(13)
Mo	317.035	—	F	AN	1	20	(13)
Mo	319.397	—	F	AN	1	10	(13)
Mo	320.883	—	F	AN	1	50	(13)
Na	268.037	—	E	AA	1	0.1	(40)
Na	268.046	—	E	AA	1	0.1	(40)
Na	285.281	—	E	AA	3	0.0015	(19)
Na	285.301	—	F	AA	15	0.05	(16)
Na	540	540	Y	PBA	—	70	(45)
Na	550	550	Y	PBA	—	3	(45)
Na	578.732	578.732	E	AA	—	0.001	(55)
Na	578.732	578.732	Y	PBA	—	0.9	(45)
Na	588.995	568.266	E	AA	—	0.003	(37)
Na	588.995	568.821	N	HA	3	0.04	(17)
Na	588.995	568.821	Y	AA	—	0.012	(56)
Na	588.995	568.821	Y	PBA	—	0.002	(45)
Na	588.995	568.821	N	AA	—	0.0006	(47)
Na	588.995	616.075	Y	PBA	—	0.01	(45)
Na	588.995	—	Kr cw	AA	0.3	0.03	(21)
Na	588.995	—	N	HA	3	6	(17)
Na	588.995	—	E AL	HA	10	0.3	(22)
Na	588.995	—	E	AA	—	0.02	(55)
Na	588.995	—	F	AA	—	0.01	(57)
Na	588.995	—	F	AA	—	20	(58)

Na	588.995	—	Y	HA	1	0.8	(59)
Na	588.995	—	Y	AA	1	0.6	(59)
Na	589.592	568.263	N	AA		0.001	(47)
Na	589.592	568.263	N	AA	0.1	0.005	(34)
Na	588.995	—	F	AA	15	0.1	(16)
Ni	269.649	—	E	AA	1	24	(40)
Ni	279.865	—	E	AA	1	0.4	(40)
Ni	282.129	—	E	AA	1	0.3	(40)
Ni	300.249	—	F	AA	1.1	7	(39)
Ni	300.249	—	F	AA	15	8	(44)
Ni	300.249	576.755	Y	AA	1.4	0.08	(18)
Ni	300.249	—	Y	AA	1.4	8	(18)
Ni	301.200	—	E	AA	3	1.5	(19)
Ni	324.846	—	Y	AA		2	(38)
Pb	280.199	—	E	AA	1	0.4	(40)
Pb	280.199	—	F	AA	15	0.6	(16)
Pb	282.320	—	E	AA	1	0.5	(40)
Pb	282.320	600.193 + 1064	Y	AA	75	0.0007	(50)
Pb	282.320	—	F	AA	15	3	(16)
Pb	283.306	—	Y	AA	1.4	0.09	(18)
Pb	283.306	600.193	E	AA		0.3	(37)
Pb	283.306	600.193	Y	AA	1.4	3	(18)
Pb	283.306	—	E	AA	1	0.2	(40)
Pb	287.331	—	E	AA	3	3	(19)
Pb	287.331	—	E	AA	1	0.6	(40)
Rb	420.185	—	Kr cw	HA	0.3	0.7	(21)
Rb	420.185	—	N	PBA		0.1	(45)

(Contd.)

Table 3.4. (*Contd.*)

	First-Step Wavelength (nm)	Second-Step Wavelength (nm)	Laser[a]	Flame[b]	Time Constant (s)	Detection Limit (ng/mL)	Ref.
Rb	420.185	—	N	AA	1	0.0006	(43)
Rb	780.023	—	Kr cw	HA	0.3	0.09	(21)
Rb	780.023	—	Diode	HA	1	0.3	(23)
Sb	276.995	—	E	AA	1	90	(40)
Sb	287.792	—	E	AA	1	50	(40)
Si	288.158	—	F	AN	1	40	(13)
Sn	266.124	—	E	AA	1	30	(40)
Sn	270.651	—	E	AA	1	8	(40)
Sn	270.651	—	F	AN	1	2	(13)
Sn	283.999	597.028	Y	HA	1.4	0.3	(18)
Sn	283.999	—	Y	HA	1.4	8	(18)
Sn	283.999	—	E	AA	1	2	(40)
Sn	283.999	—	F	AN	1	0.4	(13)
Sn	283.999	—	F	AA	15	6	(44)
Sn	286.333	—	F	AN	1	2	(13)
Sn	286.333	—	F	AA	15	10	(44)
Sn	286.333	—	E	AA	3	20	(19)
Sn	286.333	—	E	AA	1	3	(40)
Sn	300.914	—	F	AN	1	10	(13)
Sn	303.412	—	F	AN	1	6	(13)
Sn	317.505	—	F	AN	1	3	(13)
Sn	326.234	—	F	AN	1	2	(13)
Sr	293.183	—	E	AA	1	0.01	(40)
Sr	459.513	—	E	AA	—	15	(60)

			Kr cw				
Sr	460.733	—	Kr cw	AA	0.3	0.4	(21)
Sr	460.733	—	Y	HA	1	3	(59)
Sr	460.733	—	Y	AA	1	1	(59)
Sr	460.733	554.336	E	AA	—	0.3	(60)
Ti	294.200	—	F	AN	1	10	(13)
Ti	294.826	—	F	AN	1	8	(13)
Ti	295.613	—	F	AN	1	6	(13)
Ti	300.087	—	F	AN	1	20	(13)
Ti	318.645	—	F	AN	1	1	(13)
Ti	319.199	—	F	AN	1	1	(13)
Ti	319.992	—	F	AN	1	1	(13)
Ti	331.442	—	F	AN	1	3	(13)
Ti	334.188	—	F	AN	1	2	(13)
Ti	335.469	—	F	AN	1	3	(13)
Ti	337.145	—	F	AN	1	4	(13)
Tl	276.787	377.572	E	AA	1	0.008	(61)
Tl	276.787	—	E	AA	10	0.02	(61)
Tl	276.787	—	E	AA	1	0.006	(40)
Tl	291.832	—	E	AA	3	0.02	(19)
Tl	291.832	—	E	AA	1	0.008	(40)
Tl	291.832	—	F	AA	15	0.09	(44)
Tl	377.572	—	E AL	HA	10	3	(22)
V	292.362	—	F	AN	1	20	(13)
V	305.633	—	F	AN	1	6	(13)
V	306.046	—	F	AN	1	4	(13)
V	306.638	—	F	AN	1	3	(13)

(Contd.)

Table 3.4. (Contd.)

	First-Step Wavelength (nm)	Second-Step Wavelength (nm)	Laser[a]	Flame[b]	Time Constant (s)	Detection Limit (ng/mL)	Ref.
V	318.398	—	F	AN	1	0.9	(13)
V	318.540	—	F	AN	1	0.9	(13)
W	283.138	—	E	AA	1	300	(40)
Yb	267.198	—	E	AA	1	1.7	(40)
Yb	555.647	581.2	Y	AA	—	0.1	(47)
Zn	213.856	396.545	Y	AA	1	1	(9)
Zn	213.856	—	Y	AA	1	3	(9)
Zn	307.590	472.216	Y	AA	1	15	(9)

[a] E = excimer pumped dye laser; E AL = excimer pumped atomic line laser; F = flashlamp pumped dye laser; Kr cw = krypton ion pumped cw dye laser; N = nitrogen pumped dye laser; Y = Nd:YAG pumped dye laser.
[b] AA = acetylene/air; AN = acetylene/nitrous oxide; HA = hydrogen/air; MA = methane/air; PBA = propane/butane/air; rod = graphite rod in flame; tc = total consumption burner.

198

3.5. SPECTRAL INTERFERENCES

The harsh reality of analytical chemistry is that not all samples find the analyte element alone in a distilled water matrix. There are three major sources of matrix interferene encountered in LEI spectrometry: (i) increased flame background current; (ii) space charge interference; and (iii) spectral interference. The first two of these interferences have already been discussed, but spectral interferences have proved to be the greatest impediment to the application of LEI. A spectral interference occurs when laser-induced ionization of any species other than the analyte occurs simultaneously with analyte LEI.

Flame background and space charge interferences are exclusively associated with easily ionized matrix elements. Spectral interferences can be caused by any matrix element, but are again most often caused by easily ionized elements. A low ionization potential generally is indicative of a high LEI sensitivity. Consequently the presence of such elements, often at high concentrations, while attempting the measurement of a low-concentration analyte element, with a higher ionization potential, often leads to spectral interference. Spectral interferences can result from the overlap of atomic lines but are more often the result of an overlap between the analyte line and some broadband spectral feature of a matrix component.

3.5.1. Line Overlaps

Such interferences are rare, and they are easy to deal with when double-resonance LEI is used. An example of such an interference is the determination of cobalt at 252.136 nm in the presence of indium, which has an overlapping line at 252.137 nm. If single-wavelength excitation were used and equal concentrations of Co and In were present in the sample, a 13-fold error would occur in the Co determination (18). By using double-resonance excitation, excitation of the Co from the ground state to $39649 \, cm^{-1}$ induced by the 252.136 nm excitation line can be continued up to a level located at $56546 \, cm^{-1}$ by a second-step excitation line at 591.680 nm. This second step of excitation enhances the Co sensitivity by a factor of 170 without affecting the interfering In signal. This enhancement provides a measurement of Co concentration that is completely free from the In interference. This is illustrated in Fig. 3.10 (62), which shows the LEI spectrum in the vicinity of the second-step line with the first-step line fixed at 252.14 nm for solutions of 100 ng/mL Co, with and without the presence of 100 ng/mL of In in the matrix. The presence of indium in the mixed sample results in an increased baseline signal due to the In line overlap on the first excitation step and consequently increased baseline noise, but 100% recovery of the Co signal is obtained in the In matrix by measuring from the baseline to the peak.

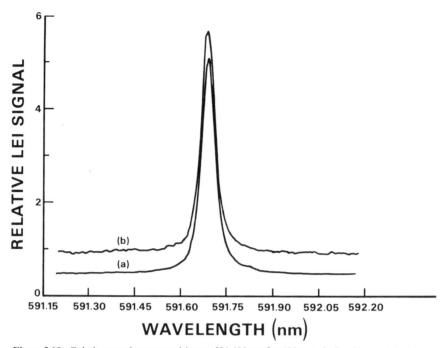

Figure 3.10. Cobalt second-step transition at 591.680 nm for 100 ng/mL Co. Curves: (*a*) without indium present (*b*) with 100 ng/mL indium present. First-step Co transition at 252.136 nm overlaps In line at 252.137 nm. Reproduced from Travis et al. (62) by permission of Elsevier Science Ltd., The Boulevard, Langford Lane, Kidlington, OX5 1GB, UK

One disadvantage of using double-resonance excitation is that with the use of two excitation wavelengths some type of spectral interference can occur at either, or even both, wavelengths. In addition, the use of two laser beams together can lead to rather unusual two-photon transitions, which can be an unexpected source of line overlap interference. This is illustrated in the dual-wavelength LEI spectrum shown in Fig. 3.11 for a barium double-resonance line, depicted as a three-dimensional projection of the first-step line, the second-step line, and the LEI signal (63). In addition to the peak observed when both lasers are on resonance for Ba, three ridge features are observed. Two of these ridges are orthogonal to a wavelength axis and are the result of single-resonance LEI. The third ridge follows a diagonal course through the spectrum, intersecting the double-resonance peak. This is the result of two-photon transitions, resonantly enhanced by the intermediate level that is the upper level of the first-step wavelength and the lower level of the second-step wavelength. Along this diagonal ridge one laser wavelength is detuned to the blue of the resonance wavelength while the other is equally detuned to the red,

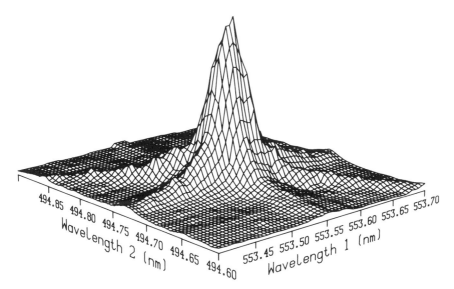

Figure 3.11. Dual-wavelength LEI spectrum near the barium double-resonance at 553.548 nm + 494.733 nm. Reproduced from Turk et al. (63) by permission of the Society for Applied Spectroscopy, Frederick, Maryland.

such that the sum of the two energies always equals that of the upper level of the second-step wavelength. The appearance of such two-photon ridges is very hard to predict, and the measurement of such three-dimensional spectra, while time consuming, is a useful diagnostic. An example in a real sample situation is shown in Fig. 3.12, where a three-dimensional LEI spectrum of a solution of a high-alloy steel sample (SRM 1289a) is shown (63). A near miss occurs between a cobalt double-resonance peak and a diagonal ridge that extends from an iron double-resonance peak outside the range of the recorded spectrum.

3.5.2. Broadband Spectral Interferences

Overlaps between analyte lines and broadband spectral features of matrix components are much more likely to be encountered, simply because the interferences cover a larger wavelength range. Such interferences include line wings, molecular bands, and thermionic ionization of particles. Correction for such interferences requires scanning of the laser wavelength across the analyte line and performing the appropriate background correction. Often this can be accomplished by slew scanning the laser wavelength between the peak of the analyte line and one or two background correction points (64). Wavelength

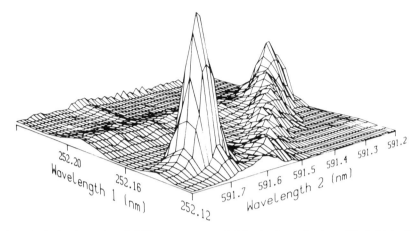

Figure 3.12. Dual-wavelength LEI spectrum recorded for the determination of Co in high-alloy steel (SRM 1289a). The double-resonance peak for Co is seen in the foreground corner at 252.136 nm + 591.680 nm. The concentration of Co in the solution is 250 ng/mL. The diagonal ridge is caused by two-photon excitation of Fe. Reproduced from Turk et al. (63) by permission of the Society for Applied Spectroscopy, Frederick, Maryland.

modulation has also been used to deal with this interference (4). The more serious problem caused by such interferences is the degradation in detection limits and measurement precision caused by the fluctuations of the interfering signal.

3.5.2.1. Line Wing Interferences

The wings of atomic lines are rarely an important issue in analytical spectrometry when conventional light sources are used, but they are easily observed in laser spectrometry, including LEI. The intensity of a line wing relative to the peak is exaggerated at higher laser powers if saturation of the transition occurs. Saturation will generally occur at or near the peak of the transition, but not in the far wings, thus decreasing the peak-to-wing ratio. It is not uncommon to observe LEI signals from the wings of lines several nanometers from the peak wavelength, as seen in Fig. 3.13. Here the wings of the sodium D-lines from a solution containing 20 μg/mL of Na are seen throughout the range of the laser dye, from 581 to 596 nm. The peak of the transition is off the scale of the figure, at approximately 70 μA. The laser irradiance for this measurement was not unusually high, at 200 MW/cm^2-nm (1.6 mJ laser energy in a 7 mm diameter beam with a 0.0025 nm bandwidth), but high enough to saturate the strong Na transitions.

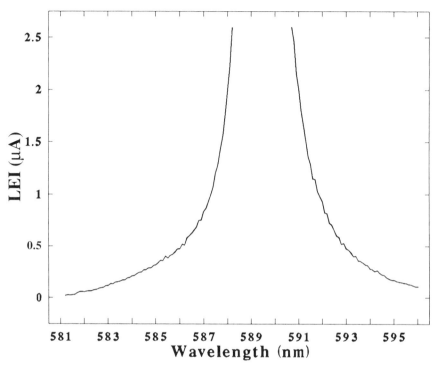

Figure 3.13. LEI spectrum for 20 μg/mL Na, showing the wings of the 3s–3p transitions. The line peaks are off-scale.

The effect of another line wing interference, again from Na, on the determination of Ni by LEI is seen in Fig. 3.14. Here a stepwise excitation procedure is being used (300.249 nm + 561.479 nm), and the second-step wavelength is being scanned across the transition while the first-step wavelength is fixed on the line center. The figure shows the Ni second-step line for a 100 ng/mL Ni solution with and without the presence of 100 μg/mL of Na in the solution. A background interference signal, more than twice as intense as the analyte signal, is encountered in the Na matrix, and the S/N ratio is clearly degraded. The source of the background interference can be seen by scanning the laser wavelength through a wider range. Figure 3.15 shows the line wing of the Na 3p–4d transition, which peaks at 568.3 nm, seen here between 554 nm and 568 nm for a 100 μg/mL Na solution. For comparison the figure also shows the stepwise excitation 100 ng/mL Ni signal.

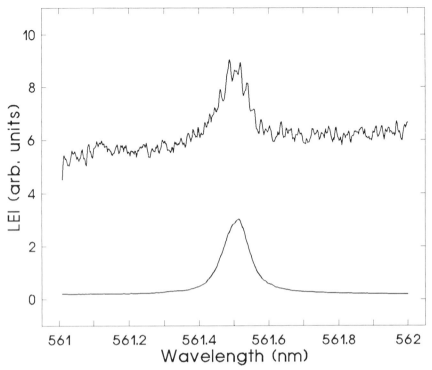

Figure 3.14. Second-step LEI spectra for 100 ng/mL Ni: (*top*) with and (*bottom*) without 100 μg/mL Na present in the solution. The first-step wavelength is fixed on the Ni transition at 300.249 nm.

3.5.2.2. *Molecular Band Interferences*

The most common molecular band interference encountered in LEI analysis is due to LEI of CaOH. The prevalence of Ca in many sample matrices, the incomplete dissociation of CaOH in the air/acetylene flame, the low ionization potential of CaOH (5.7 eV), the broad spectrum from green to red wavelengths, and the location of many second-step LEI stepwise excitation lines in this wavelength range combine to make this a common problem. The LEI spectrum of a 10 μg/mL Ca solution over the range of three laser dyes is shown in Fig. 3.16. The relative LEI signal has been divided by the relative laser power at each wavelength in an attempt to compensate for the variation in laser power encountered over the range of these three dyes. The shape of this spectrum corresponds approximately to the LIF spectrum for CaOH in the air/acetylene flame (65).

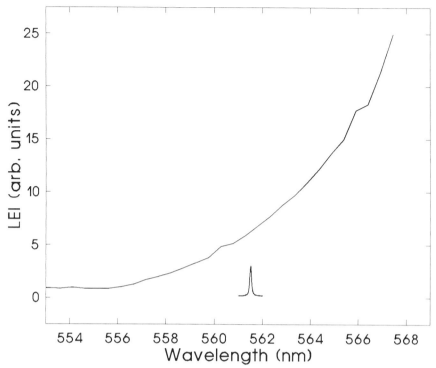

Figure 3.15. Wing excitation of the $3p$–$4d$ resonance of Na from a 100 µg/mL solution of Na, shown with the Ni second-step resonance line from a 100 ng/mL solution of Ni.

3.5.2.3. Laser-Induced Particle Ionization

Another variety of broadband spectral interference occurs when laser light interacts with particles in the flame to induce ionization of the particles. Such interferences are spectrally featureless and occur when incandescent soot particles are visible or when a sample matrix contains high concentrations of refractory elements. Soot formation occurs when a fuel-rich flame is being used or when some organic solvents are aspirated. The mechanism of the ionization is thought to be thermionic in nature (6).

3.5.2.4. Amplified Stimulated Emission

One problem that is easy to mistake for a broadband spectral interference is a signal induced by amplified stimulated emission (ASE). In most dye lasers,

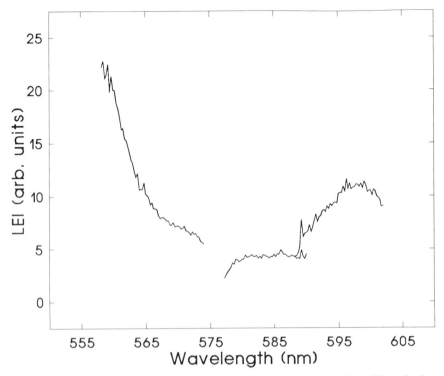

Figure 3.16. LEI spectrum for CaOH, recorded over the range of three laser dyes while aspirating a solution of 10 µg/mL Ca. The LEI signal has been divided by the relative laser power in an attempt to compensate for laser power variation.

the output beam is accompanied by a beam of ASE, which is usually of much lower power, broadband, and not wavelength tunable except by changing the laser dye. ASE competes with the true laser emission and is more prevalent at the edges of the tuning curve of a dye, where the laser output is weaker. Occasionally the ASE wavelength will overlap matrix absorption lines and can be of sufficient power to induce an LEI signal. Because the ASE wavelength does not change as the laser wavelength is tuned, any signal induced by ASE appears to be wavelength independent. One telltale sign of ASE-induced background is an increase in the magnitude of the background at the edges of the dye tuning curve, where ASE intensity is greatest. Such interference can be minimized by adjusting the laser cavity for less ASE, or in some cases through the use of a filter.

3.6. SUMMARY AND FUTURE DIRECTIONS FOR LEI SPECTROMETRY

A one-sentence summary of LEI spectrometry that might be given in a review article or a textbook would state that LEI is among the most sensitive methods available for trace metal analysis but is generally more susceptible to interferences than other methods. Another important drawback of LEI that would need to be mentioned is the reliance on expensive and complex laser technology, and as a consequence the practical restriction of LEI to single-element analysis. Future development of LEI should exploit the strengths—detection limits have not yet reached the fundamental limits—and work to minimize the weaknesses.

The performance of any analytical method is evaluated in relationship to other analytical methods that might accomplish the required measurement task. In the early years of LEI, the most significant competition for LEI as a method of trace metal analysis was from graphite furnace atomic absorption spectrometry. This method also offers very low limits of detection at a much lower cost. On the other hand, it is also susceptible to interferences and is a single-element technique. The discrete sampling nature of the graphite furnace, in comparison to continuous solution aspiration in a flame, offers an advantage in terms of sample size requirements, but also some disadvantage in measurement time.

In more recent times, the most obvious competition for LEI comes from inductively coupled plasma–mass spectrometry (ICP–MS). Although the first publications of ICP–MS follow those of LEI by four years, ICP–MS has seen rapid development and commercialization. In ICP–MS, sample solutions are aspirated into an argon ICP at atmospheric pressure and the resulting ions are sampled through a nozzle into a differentially pumped quadrupole mass spectrometer. Most of the periodic table can be rapidly detected in solutions at concentrations below 1 part per trillion. The instrumentation is commercially available from several manufacturers, and hundreds of laboratories now routinely utilize ICP–MS for a variety of applications.

In light of such strong competition, it is fair to question where a method such as LEI "fits in." At the present state of development, most analytical chemists would agree that the use of any dye-laser-based spectrometric method is sensible only if the measurement cannot be accomplished by one of the more conventional analytical mehods. This is a consequence of the complexity of the laser instrumentation and the considerable skill and training needed to operate such an instrument. Analytical measurements requiring the use of laser spectrometry do occur but are rare, and consequently laser-based spectrometric methods are available only at a few research institutions.

One special area that has found a need for laser spectrometry, including LEI, is the certification of SRMs at the National Institute of Standards and

Technology (Gaithersburg, Maryland). In such work, an important technique used to ensure the accuracy of the certification is the use of independent methods of analysis for all certified analytes. The situation often arises where a particular analytical determination can be performed using one of the more common analytical methods but more exotic methods must be used for independent verification. For example, manganese was determined in SRM 1598, Inorganic Constituents in Bovine Serum, by graphite furnace atomic absorption, but neutron activation analysis and LEI were used as second and third independent measurements. LEI measurements have been used for the certification of many SRMs, including metal alloys, sediments, water, and biological fluids.

The unique nature of LEI, an optical spectroscopic method with ionization-based detection, assures it a place among analytical methods. One feature of LEI that is only recently being realized is the fundamental physical dimension of the detection system. Integration of an LEI pulse leads to determination of the absolute number density of the ions through the Coulomb. This procedure can be considered a flame-based variety of coulometry.

Another direction for the future development of LEI can be found in the success of ICP–MS. Combinations of LEI with mass spectrometric detection have been demonstrated in flames (66) and in the ICP (67). This combination addresses a primary weakness of LEI, the nonspecific nature of the conductivity-based detection procedure. It is this lack of elemental specificity that makes LEI so susceptible to spectral interference. The combination of LEI with MS also can be seen as a solution to isobaric interferences encountered in conventional mass spectrometry, through the combination of optical spectroscopic selectivity with mass spectrometric selectivity.

This chapter has shown that LEI has reached a level of considerable maturity as an analytical method. The fundamentals of signal production and detection are well understood, and this understanding has led to optimization of analytical performance. Applicaions of LEI to elemental analysis are presented in Chapter 4.

REFERENCES

1. R. B. Green, R. A. Keller, P. K. Schenck, J. C. Travis, and G. G. Luther, *J. Am. Chem. Soc.* **98**, 8517 (1976).

2. G. C. Turk, G. J. Havrilla, J. D. Webb, and A. R. Forster, in *Analytical Spectroscopy* (*Proceedings of the 26th Conference on Analytical Chemistry in Energy Technology*) (W. S. Lyon, Ed.), pp. 63–68. Elsevier, Amsterdam, 1984.

3. T. Berglind, S. Nilsson, and H. Rubinsztein-Dunlop, *Phys. Scr.* **36**, 246 (1987).

4. K. S. Epler, T. C. O'Haver, G. C. Turk, and W. A. MacCrehan, *Anal. Chem.* **60**, 2062 (1988).

5. K. S. Epler, T. C. O'Haver, and G. C. Turk, *J. Anal. At. Spectrom.* **9**, 79 (1994).

6. K. C. Smyth and W. G. Mallard, *Combust. Sci. Technol.* **26**, 35 (1981).

7. R. Mavrodineanu and H. Boiteux, *Flame Spectroscopy*, p. 556. Wiley, New York, 1965.

8. W. G. Mallard, J. H. Miller, and K. C. Smyth, *J. Chem. Phys.* **76**, 3483 (1982).

9. G. J. Havrilla and K. J. Choi, *Anal. Chem.* **58**, 3095 (1986).

10. B. H. Rockney, T. A. Cool, and E. R. Grant, *Chem. Phys. Lett.* **87**, 141 (1982).

11. K. C. Smyth and W. G. Mallard, *J. Chem. Phys.* **77**, 1779 (1982).

12. G. C. Turk, *Anal. Chem.* **63**, 1607 (1991).

13. J. D. Messman, N. E. Schmidt, J. D. Parli, and R. B. Green, *Appl. Spectrosc.* **39**, 504 (1985).

14. J. E. Hall and R. B. Green, *Anal. Chem.* **55**, 1811 (1983).

15. J. E. Hall and R. B. Green, *Anal. Chem.* **57**, 431 (1985).

16. G. C. Turk, J. C. Travis, J. R. DeVoe, and T. C. O'Haver, *Anal. Chem.* **50**, 817 (1978).

17. G. C. Turk, W. G. Mallard, P. K. Schenck, and K. C. Smyth, *Anal. Chem.* **51**, 2408 (1979).

18. G. C. Turk, J. R. DeVoe, and J. C. Travis, *Anal. Chem.* **54**, 643 (1982).

19. O. Axner, I. Lindgren, I. Magnusson, and H. Rubinsztein-Dunlop, *Anal. Chem.* **57**, 776 (1985).

20. M. J. Rutledge, M. E. Tremblay, and J. D. Winefordner, *Appl. Spectrosc.* **41**, 5 (1987).

21. G. J. Havrilla, S. J. Weeks, and J. C. Travis, *Anal. Chem.* **54**, 2566 (1982).

22. D. J. Ehrlich, R. M. Osgood, Jr., G. C. Turk, and J. C. Travis, *Anal. Chem.* **52**, 1354 (1980).

23. R. W. Fox, C. S. Weimer, L. Hollberg, and G. C. Turk, *Spectrochim. Acta Rev.* **15**, 291 (1993).

24. P. K. Schenck, J. C. Travis, and G. C. Turk, *J. Phys. (Paris), Colloq.* **C7**, 75 (1983).

25. K. Niemax, J. Lawrenz, and A. Obrebski, in *Resonance Ionization Spectroscopy 1986* (*Proceedings of the 3rd International Symposium on Resonance Ionization Spectroscopy and Its Applications*) (G. S. Hurst and C. G. Morgan, Eds.), pp. 45–50. IOP Publishing, Bristol, 1987.

26. A. Obrebski, L. Lawrenz, and K. Niemax, *Spectrochim. Acta* **45B**, 15 (1990).

27. G. C. Turk, *Anal. Chem.* **64**, 1836 (1992).

28. G. J. Havrilla and R. B. Green, *Chem. Biomed. Environ. Instrum.* **11**, 273 (1981).

29. G. J. Havrilla, P. K. Schenck, J. C. Travis, and G. C. Turk, *Anal. Chem.* **56**, 186 (1984).

30. G. C. Turk and J. C. Travis, *Spectrochim. Acta* **45B**, 409 (1990).

31. J. C. Travis, *J. Chem. Educ.* **59**, 909 (1982).

32. H. V. Malmstadt, C. G. Enke, S. R. Crouch, and G. Horlick, *Electronic Measurements for Scientists*, p. 822. Benjamin, Menlo Park, CA, 1974.

33. O. I. Matveev, *J. Anal. Chem. USSR (Engl. Transl.)* **42**, 1121 (1987).

34. O. I. Matveev, *J. Anal. Chem. USSR (Engl. Transl.)* **43**, 944 (1988).

35. N. B. Zorov, Yu. Ya. Kuzyakov, O. I. Matveev, and V. I. Chaplygin, *J. Anal. Chem. USSR (Engl. Transl.)* **35**, 1108 (1980).

36. J. D. Winefordner and T. J. Vickers, *Anal. Chem.* **36**, 1939 (1964).

37. N. Omenetto, B. W. Smith, and L. P. Hart, *Fresenius' Z. Anal. Chem.* **324**, 683 (1986).

38. G. J. Havrilla and C. C. Carter, *Appl. Opt.* **26**, 3510 (1987).

39. G. C. Turk, J. C. Travis, and J. R. DeVoe, *Anal. Chem.* **51**, 1890 (1979).

40. O. Axner, I. Magnusson, J. Petersson, and S. Sjöström, *Appl. Spectrosc.* **41**, 19 (1987).

41. A. A. Gorbatenko, N. B. Zorov, S. Yu. Karopova, Yu. Ya. Kuzyakov, and V. I. Chaplygin, *J. Anal. At. Spectrom.* **3**, 527 (1988).

42. G. C. Turk and M. De-Ming, *NBS Spec. Publ. (U.S.)* **260-106**, 30–33 (1986).

43. Yu. Ya. Kuzyakov, N. B. Zorov, V. I. Chaplygin, and A. A. Gorbatenko, in *Resonance Ionization Spectroscopy 1988 (Proceedings of the 4th International Symposium on Resonance Ionization Spectroscopy and Its Applications)* (T. B. Lucatorto and J. E. Parks, Eds.), pp. 179–182. IOP Publishing Bristol, 1989.

44. J. C. Travis, G. C. Turk, and R. B. Green, *ACS Symp. Ser.* **85**, 91–101 (1978).

45. V. I. Chaplygin, Yu. Ya. Kuzyakov, ánd O. A. Novodvorsky, *Talanta* **34**, 191 (1987).

46. V. I. Chaplygin, N. B. Zorov, and Yu. Ya. Kuzyakov, *Talanta* **30**, 505 (1983).

47. I. V. Bykov, A. B. Skvortsov, Yu. G. Tatsii, and N. V. Chekalin, *J. Phys. (Paris), Colloq.* **C7**, 345 (1983).

48. N. V. Chekalin, V. I. Pavlutskaya, and I. I. Vlasov, *Spectrochim. Acta* **46B**, 1701 (1991).

49. N. B. Zorov, Yu. Ya. Kuzyakov, and O. I. Matveev, *Zh. Anal. Khim.* **37**, 520 (1982).

50. A. G. Marunkov and N. V. Chekalin, *J. Anal. Chem. USSR (Engl. Transl.)* **42**, 506 (1987).

51. R. B. Green and J. E. Hall, *J. Phys. (Paris), Colloq.* **C7**, 317 (1983).

52. L. E. Salcedo Torres, N. B. Zorov, and Yu. Ya. Kuzyakov, *J. Anal. Chem. USSR (Engl. Transl.)* **36**, 1016 (1981).

53. O. I. Matveev, P. Cavalli, and N. Omenetto, in *Resonance Ionization Spectroscopy 1994 (Proceedings of the 7th International Symposium on Resonance Ionization Spectroscopy and Its Applications)* (H. J. Kluge, J. E. Parks, and K. Wendt, Eds.), pp. 269–272. AIP Press, New York, 1995.

54. G. C. Turk, J. C. Travis, and J. R. DeVoe, *J. Phys. (Paris), Colloq.* **C7**, 301 (1983).

55. L. C. Chandola, P. P. Khanna, and M. A. N. Razvi, *Anal. Lett.* **24**, 1685 (1991).

56. A. S. Gonchakov, N. B. Zorov, Yu. Ya. Kuzyakov, and O. I. Matveev, *Anal. Lett.* **12**, 1037 (1979).

57. Y. Yan, D. Ding, Z. Zhang, S. Jin, and Q. Jin, *Guangpuxue Yu Guangpu Fenxi* **4**, 1 (1984).

58. P. Zhang, J. Du, J. He, and H. Li, *Fenxi Huaxue* **10**, 66 (1985).

59. M. D. Seltzer and R. B. Green, *Appl. Spectrosc.* **43**, 257 (1989).

60. L. P. Hart, B. W. Smith, and N. Omenetto, *Spectrochim. Acta* **40B**, 1637 (1985).

61. N. Omenetto, T. Berthoud, P. Cavali, and G. Rossi, *Anal. Chem.* **57**, 1256 (1985).

62. J. C. Travis, G. C. Turk, J. R. DeVoe, P. K. Schenck, and C. A. VanDijk, *Prog. Anal. At. Spectrosc.* **7**, 199 (1984).

63. G. C. Turk, F. C. Ruegg, J. C. Travis, and J. R. DeVoe, *Appl. Spectrosc.* **40**, 1146 (1986).

64. G. C. Turk and H. M. Kingston, *J. Anal. At. Spectrom.* **5**, 595 (1990).

65. S. J. Weeks, H. Haraguchi, and J. D. Winefordner, *J. Quant. Spectrosc. Radiat. Transfer* **19**, 633 (1978).

66. G. C. Turk, L. Yu, and S. R. Koirtyohann, *Spectrochim. Acta* **49B**, 1537 (1994).

67. L. Yu, S. R. Koirtyohann, G. C. Turk, and M. L. Salit, *J. Anal. At. Spectrom.* **9**, 997 (1994).

CHAPTER

4

APPLICATIONS OF LASER-ENHANCED IONIZATION SPECTROMETRY

ROBERT B. GREEN

Associated Western Universities, Inc., Northwest Division, Richland, Washington 99352

4.1. INTRODUCTION

One measure of the success of a new technique for instrumental analysis is the extent to which it has been successfully applied. In spite of the intrinsic value and satisfaction of understanding physical phenomena and the resulting advancement of scientific knowledge, chemical analysis of real samples is the *raison d'être* for analytical chemistry.

The evolution of laser-enhanced ionization (LEI) spectrometry has followed a predictable course. The physical phenomenon was discovered in a research laboratory in 1976 (1) and since then has progressed through the developmental stages common to any analytical method: genesis, growth, and maturation. The high sensitivity and simplicity of the detection technique and the selectivity afforded by tunable dye lasers have been the drivers for the continued efforts to exploit LEI spectrometry as an analytical method. Today we can begin to assess its value for chemical analysis.

A thorough literature search limited to applications of LEI spectrometry for the first 16 years after the discovery resulted in 40 citations. Figure 4.1 illustrates the growth of papers on the subject of the applications of LEI spectrometry to real samples. Many of the papers counted in this survey, particularly the early ones, do not concentrate on the application of LEI but merely include it as evidence of the potential value of continued investigation of the technique.

Since its discovery, LEI spectrometry has been embraced by the worldwide scientific community. Almost 60% of the publications on applications of LEI spectrometry originated outside the United States. Although a complete

Laser-Enhanced Ionization Spectrometry, edited by John C. Travis and Gregory C. Turk.
Chemical Analysis Series, Vol. 136.
ISBN 0-471-57684-0 © 1996 John Wiley & Sons, Inc.

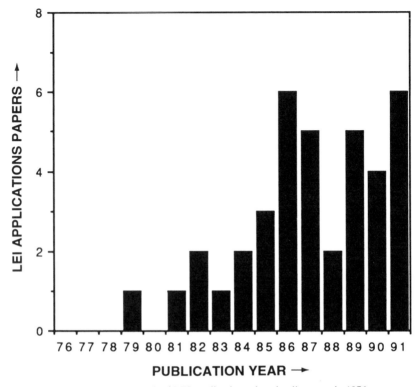

Figure 4.1. The growth of LEI applications since its discovery in 1976.

bibliography of applications of LEI spectrometry is given at the end of this chapter, only those topics that should be of particular interest to the analyst will be discussed in detail.

4.2. LEI SPECTROMETRY OF REAL SAMPLES

If every sample was composed of a metal salt dissolved in distilled and deionized water, this chapter would be very brief and "LEI spectrometers" would be rolling off the assembly line by the hundreds. Unfortunately, most samples of any interest contain traces of analytes in a complicated matrix. The inherent ease of collecting and sensing ions that contributes to the simplicity of the LEI detection scheme also makes it vulnerable to interferences, primarily due to easily ionizable elements (EIEs) (2).

LEI depends on collisional processes in the atom reservoir to increase the rate of ionization of the laser-excited analyte atoms. Even though laser excitation can drive the rate of thermal ionization into saturation, the collisional ionization of the atoms and molecules can significantly perturb ion collection and add noise to the signal. Simply stated, the selectivity provided by the laser source is limited in the amount of resolution it can provide when coupled with an undiscriminating detector. In many cases, the extremely high sensitivity intrinsic to LEI spectrometry will be sacrificed without effective methods of dealing with EIEs.

Simple dilution of samples is the first defense against interferences. Because of the high sensitivity of LEI spectrometry, it is often possible to bring interfering concomitants into a range where no further treatment is necessary prior to analysis. The use of an immersed electrode provides a good antidote to loss of LEI signal due to ion collection interferences (3) but the dc background current in the flame can only be reduced by removing the EIEs from the sample. A nonresonant background can limit sensitivity if a sample contains substituents with low-lying energy levels. EIEs in the sample matrix will be thermally excited to levels from which photoionization is probable via single or multiphoton processes. If ultraviolet (UV) light is used and the EIE concentration is high, even elements with low photoionization cross sections will make a major contribution to the gross ionization signal.

The applications discussed in this chapter will be categorized on the basis of the complexity of the sample, from the simplest case to the most difficult. Because of the different goals for analysis, the following discussion of applications will be divided into two parts: Section 4.2.1 will consider general analysis; Section 4.2.2 will review diagnostic measurements.

4.2.1. General Analysis

4.2.1.1. Determinations Without Interferent Removal

LEI spectrometry is particularly suited to samples of high purity with small amounts of EIEs. In these cases, high sensitivity and precision are at a premium and relatively little accommodation of interferences is necessary. Alloy samples are particularly amenable to LEI spectrometry because they typically contain low levels of sodium and potassium. The determination of indium in nickel-based, high-temperature alloys (4) was an early example of application of LEI spectrometry to a difficult analytical problem. Because of the low levels of EIEs, external plate electrodes produced satisfactory results for an acetylene/air flame supported on a slot burner. Similar samples routinely present spectral background interferences for conventional furnace atomic absorption methods and require time-consuming extraction procedures to "clean up" the samples prior to determinations. When LEI flame spectrometry was used, alloy samples were successfully analyzed "as is," avoiding the extraction procedure. The LEI determinations of indium in the alloy samples were in close agreement with values obtained by the more tedious furnace atomic absorption method. Lowering the temperature of the atom reservoir is a potential remedy for interferences from EIEs as long as significant atom fractions of the analyte are maintained. Using a solid stainless-steel rod immersed in a low temperature propane-butane-air flame was sufficient to accurately determine low concentrations (nanograms per milliliter) of cesium in tap water samples by LEI spectrometry, even with tens of milligrams per milliliter of sodium, potassium, and calcium present (5). Both of these applications employed single-step excitation. An expression has been derived for the prediction of the optimum transition for the strongest LEI signal by one-step excitation (6). The most sensitive one-step transitions are reported for several elements.

Natural water samples also lend themselves to analysis by LEI spectrometry. The concentration of several elements at picogram per milliliter levels was validated in a simulated rainwater Standard Reference Material (SRM 2694) at the National Bureau of Standards, now known as the National Institute of Standards and Technology (NIST), using LEI spectrometry (7). As is the practice, this was one of several unrelated analytical methods used to certify the elemental content of the SRM for accuracy and precision. Some spectral interferences due to excitation in the wings of nearby peaks were corrected for by standardization using matrix matching, whereas others required two-wavelength excitation.

Two-step or two-wavelength excitation of connected, bound atomic transitions into a high-lying excited state improves selectivity in addition to boosting sensitivity (8). The probability of both analyte transitions coinciding with spectral features of an interferent is very small. A broad spectral background may overlap either analytical line, but the LEI signal will be enhanced only by excitation at both wavelengths, so a correction can be made without scanning the laser wavelength. Two-step excitation has also been used to determine zinc in SRM 1643a, trace elements in water, in the presence of a background interference (9). The experimental value for zinc was slightly high, but no attempt was made to remove potential interferences beyond using an immersed electrode and sample dilution. Matrix matching of the standards may have improved the accuracy of the measurement.

Decreasing the incident laser power density can also minimize spectral interferences if the signal for the interfering element decreases linearly with laser power while the analyte signal remains unchanged (8). This may result because the signal *enhancement* due to the second laser excitation step increases with decreasing laser intensity in the first step when the first-step analyte transition is optically saturated and the interferent transition is not. Background signals from sodium have been thoroughly investigated (10). It was concluded that an optimum laser intensity exists for the most efficient detection of a particular analyte in a sodium matrix. Background signals from intrinsic flame species such as NO can also be mitigated by two-step excitation (8).

As part of an environmental monitoring program, lead was determined in unpolluted waters from mountainous regions and compared with results for natural waters impacted by industrial development (11). Many spectroscopic methods have insufficient sensitivity to make accurate measurements where species exist at very low natural background levels, e.g., in rain, snow, and ice samples from pristine areas. Efficient thermal ionization of lead in the acetylene/air flame required a two-step excitation scheme, with the first step involving a UV photon. The pristine natural water samples presented no interference difficulties since they contained few minerals and the low (nanograms per milliliter) lead content was well above the detection limit for LEI spectrometry. In the case of river waters impacted by development, concentrations of calcium, potassium, sodium, and magnesium impurities exceeded lead concentrations by 4–5 orders of magnitude and produced broad background signals. Further study related the background principally to the absorption of CaOH molecules at *both* the UV and visible excitation wavelengths. By tuning off the resonance lines, it was possible to use background subtraction successfully.

Several elements have been determined in rock samples by LEI spectrometry (12). Most other analytical methods require the use of complicated procedures prior to analysis unless the sample is preconcentrated or interferences are removed. The dissolved geological reference materials were aspirated into a propane/butane/air flame, and a water-cooled immersed electrode was used for the LEI measurements. Pure aqueous solutions of the elements of interest were used for calibration. A broad ionization background from laser excitation at analyte wavelengths was attributed to the formation of CaOH in the flame and its subsequent multiphoton ionization. By careful selection of one- and two-step excitation schemes and laser power, it was possible to optimize the signal-to-background ratio. The loss of signal due to the lowering of incident laser power had to be balanced with reduction of the ionization background over the excitation wavelengths available. Detection limits in rock of 0.002, 0.001, and 0.5 µg/g were determined for cesium, lithium, and rubidium, respectively. Comparison with certified values was good.

Since the levels of dopants and impurities in modern semiconductor materials approach the lower limits of detection for conventional methods of metal determination, this presents a challenge to the analyst. The presence of minute quantities of metals, whether intentional or not, markedly affects the semiconductor's performance, so the detection and quantification of these species is very important. Two-step LEI spectrometry was demonstrated as a viable approach for detecting dopants and impurities in acid-dissolved bulk gallium arsenide (13). The added spectral selectivity of the two-step excitation scheme permitted an effective means of background correction, which, in turn, improved accuracy. Trace amounts of chromium, iron, nickel, indium, manganese, and cobalt were detected, the first three elements being quantified. Background ionization due to the matrix elements and the flame molecules was accounted for by subtracting the signals for one-step and two-step excitation. This approach was shown to be as effective for background correction as scanning the laser wavelengths and less time consuming. Two-step LEI spectrometry has also been used to determine sodium in semiconductor silicon (14).

The determination of trace metals in petroleum products by LEI spectrometry has provided a means for monitoring elemental nickel that poisons catalysts used in petroleum processing (15). Samples of both heavy-oil flash distillate and an oil-based SRM were diluted with a xylene/n-butanol solvent mixture and aspirated into an acetylene/air flame. Figure 4.2a shows the single-step LEI spectrum for the heavy-oil flash distillate. There are two nickel lines at 300.249 and 300.363 nm, and two iron lines at 300.95 and 300.303 nm. An unidentified line at approximately 300.28 nm partially overlaps the nickel analytical line. The results of stepwise excitation of this sample are shown in Fig. 4.2b. The first-step wavelength was fixed at 300.25 nm, and the second-step wavelength was scanned. When the fixed and scanned wavelengths were reversed, some nearby lines were observed at much lower intensity but still no overlap was present. A small background signal that included the single-step nickel signal was subtracted. Nickel determination in the SRM was in good agreement with the NIST certified value. Because of the high sensitivity of LEI spectrometry, it was possible to dilute the samples considerably. This helped to minimize interferences resulting from viscosity differences between samples and standards and possibly eliminated the need for matrix matching of standards.

Electrothermal atomization has been coupled with LEI spectrometry by inserting a resistively heated graphite rod in the premixed flame of a slot burner. Propane/butane/air or acetylene/air fuel/oxidant mixtures were used. A determination of indium in a CdHgTe alloy was accomplished in both liquid solutions and solid samples without sample preparation (16, 17). In the latter, a 10 mg sample of the alloy powder was introduced into a groove on the graphite rod surface. There was no signal suppression by the matrix, and calibration was possible using standard aqueous solutions. The nonresonant background signal, which was accounted for by calibration in the solution determination, decreased for the powder samples. It was suggested that the matrix species responsible for the background must evaporate as different molecular compounds with differing ionization efficiencies, depending on

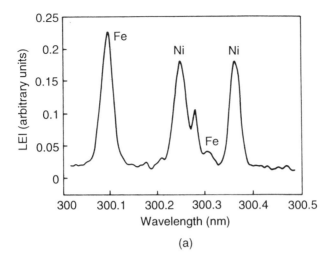

(a)

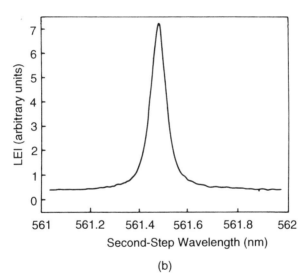

(b)

Figure 4.2. (a) Single-step LEI spectrum near nickel analytical line at 300.249 nm. (b) Stepwise excitation LEI with the first wavelength fixed at 300.25 nm and the second wavelength scanned. Adapted from Turk et al. (15) by permission of Elsevier Science Publishers, Amsterdam.

whether the sample is liquid or solid. A good correlation between results for liquid and solid samples was indicative of analytical accuracy and the absence of analyte losses for solid sampling.

Some metals of interest form refractory oxides in acetylene/air and cooler flames. A mixture of acetylene with nitrous oxide provides a higher temperature flame with a reducing atmosphere the decomposes these refractory oxides, rendering them accessible to atomic spectrometry. A nitrous oxide/acetylene flame supported on a premix burner has been used for LEI spectrometry (18). An immersed cathode was necessary because of the high level of electrical noise generated by the flame and the profound signal suppression due to space charge formation. The detection limits determined for six refractory-forming elements in aqueous solution were comparable to or better than detection limits reported for flame atomic absorption, plasma emission, and atomic fluorescence techniques. Only graphite furnace atomic absorption spectrometry exhibited superior detectability, although at the expense of speed, convenience, and the ease of continuous sample introduction. Unfortunately, IA and IIA concomitants cause signal interferences in acetylene/nitrous oxide flames, even when an immersed electrode is used.

4.2.1.2. Determinations with Interferent Removal

Since EIEs seem to be the burr under the saddle of LEI spectrometry, a simple solution, at least conceptually, would be to remove them. As a result, a variety of strategies have been implemented to eliminate interfering elements.

A creative remedy for spectral interferences due to sodium has been demonstrated by using preionization to remove the interferent (19). Magnesium was selected as the test analyte because its most useful transition is located within 0.07 nm of the strongest sodium transitions. Several preionization schemes were investigated using up to three photons of different energies to achieve as much as 83% sodium depletion in the flame. The probe laser then interrogated the preionized "hole" with 285 nm photons to enhance thermal ionization of the analyte. This technique is complementary to the two-step excitation of magnesium reported by Magnusson et al. (8). Although it produces satisfactory results, the preionization approach will probably not become popular because of the additional expense and complexity of using two separate laser systems, one for removing the interferent and one for determination of the analyte. Perhaps more important, signal collection interferences are related to the bulk flame environment and are not mitigated by laser preionization. Therefore, matrix concentrations of EIEs must be kept in the tens of milligrams per milliliter or lower to avoid analyte signal suppression.

Using the *rod–flame system* described in the previous subsection, Chekalin and colleagues determined copper and sodium in concentrated orthophosphoric acid by introducing the solution directly on the graphite rod (16, 17). The sodium as an interferent was removed by selective volatilization from the dried sample at 1000 °C. When the temperature was raised to 2000 °C the copper signal could be detected in the absence of noise. The detection limits were determined by the purity of the rod material.

More generally applicable solutions to interferent removal involve chemical treatment prior to determination by LEI spectrometry. Solvent extraction has been shown to be effective for the determination of trace amounts of manganese using a single-step excitation scheme (20). The manganese was complexed in water with sodium diethyldithiocarbamate and extracted into diisobutyl ketone. The extraction process resulted in a 10-fold concentration of manganese in addition to interferent removal. Because of the effectiveness of the removal of EIEs, including sodium, potassium, calcium, and magnesium, more complex stepwise ionization schemes with the requirement for an additional laser were avoided. The proposed methods was successfully applied to the determination of nanograms per milliliter of manganese in groundwater, river and lake waters, seawater, tap water, and wastewater. An extraction method in combination with a two-step excitation scheme made it possible to determine 0.001% calcium in aluminum alloys (21). The depression of the calcium LEI signal due to aluminum was documented, and the difference in solubilities of calcium and aluminum chlorides in methanol were exploited for separation of the analyte from the interferent. The limit of detection was determined by the level of calcium contamination in the methanol solution rather than the sensitivity of LEI spectrometry. Noise and signal suppression considerations required a dissolution procedure to partially remove germanium as GeF_4 before determination of copper using a rod–flame system was possible (16, 17). Copper determinations in liquid and solid samples were in sufficient agreement to suggest that the use of aqueous samples for calibration was possible.

Chromatography has been used effectively to remove interfering elements prior to LEI spectrometry. Trace amounts of copper were determined in a sulfate plating solution and seawater, both high salt matrices, using a chelating resin to separate the interfering elements from the analyte (22). Both signal collection and photoionization interferences were encountered with these samples. Seawater is arguably the "worst case" sample matrix for LEI spectrometry. Two resonance lines and five stepwise excitation schemes for copper were investigated: 324 nm plus 453 nm photons produced the lowest detection limits. A microsampling cup coupled to the premix burner (acetylene/air) permitted absolute determinations of copper as low as 0.05 ng. A water-cooled, immersed electrode was also used. Silver, cobalt, iron, and nickel were detected within the same dye tuning range. The separation of EIEs from the analytes was accomplished using Chelex 100 (23). In this process, transition and heavy metals are chelated in the 5.2–5.6 pH range while ammonium acetate is used to selectively elute the alkali and alkaline earth metals by ion exchange. In the final step, the trace metals are eluted with nitric acid and can be introduced into a burner for LEI spectrometry.

Alkali and alkaline earth elements occur in biological and environmental samples at high levels. Turk and Kingston combined automated chelation chromatography with computer-controlled LEI spectrometry to determine a large number of elements in a wide range of NIST SRMs (24). Chelex 100 resin was capable of separating the IA and IIA elements from the other metals, producing samples that were amenable to LEI spectrometry without further treatment, using a water-cooled, immersed electrode in an acetylene/air flame. The chromatography process also provided an opportunity to matrix modify the samples of widely varying composition. Matrix-matched standards eliminated many of the multiplicative interferences common to flame spectrometry. After preliminary work, the separation procedure was automated with a laboratory robot. The robotic apparatus performed the operations in a clean room, reducing contamination of blanks. In most cases, double-resonance stepwise excitation was used. Computer control of the wavelengths of both lasers permitted the appropriate background correction for each sample. The elements determined and the reference materials analyzed include the following: cadmium, cobalt, copper, manganese, nickel, and lead in Trace Elements in Water (SRM 1643b); manganese and nickel in Inorganic Constituents in Bovine Serum (SRM 1598); nickel and lead in Buffalo River Sediment (SRM 2704); copper, manganese, and nickel in Total Diet (SRM 1548); manganese and nickel in Apple Leaves (SRM 1515); manganese and nickel in Peach Leaves (SRM 1547). The concentrations of the elements determined in SRM 1643b ranged from 20 to 47 ng/g; in SRM 1598, they were 3.87 and 0.76 ng/g; in SRM 2704, 45.7 and 154.4 mg/g; in SRM 1548, 2.43, 5.04, and 0.41 mg/g; in SRM 1515, 54.9 and 0.97 mg/g; and in SRM 1547, 98.4 and 0.74 mg/g. Precisions ranged from 0.8% relative standard deviation (RSD) for cobalt in SRM 1643b at a concentration of 26.0 ng/g to 36% RSD for nickel in SRM 1598 at a concentration of 0.76 ng/g. This impressive demonstration of technology should go a long way toward establishing the credibility of LEI spectrometry as a practical analytical method.

In several related efforts, LEI spectrometry has been utilized as a selective detector for liquid chromatography in a synergistic relationship. The analyte detection after chromatography is enhanced by the sensitivity of LEI spectrometry while chromatography separates interfering ions, permitting the lowest possible LEI detection limits. The feasibility of coupling high-pressure liquid chromatography to a LEI spectrometer was demonstrated by connecting the waste outlet of a conventional UV detector of a high-pressure liquid chromatograph to the aspirator of a premixed burner (25). Flow rates and typical mobile phases for liquid chromatography are compatible, and LEI provides detection limits that are generally 100 times more sensitive than flame atomic absorption spectrometry. Organometallic forms of iron and chromium were separated and determined in this study.

Alkyltins were determined in sediment after separation by ion-exchange chromatography with LEI detection (26, 27). The sediment samples were collected from a river used for recreational purposes and a more pristine bay. Tributyltin, a primary toxicant in marine antifouling paints, was extracted from sediment into 1-butanol. Double-resonance excitation of the tin using 284.0 nm and 603.8 nm laser radiation enhanced thermal ionization in the acetylene/air flame. Tap water samples spiked with organotins contained high enough concentrations of sodium and calcium that a broadband spectral interference required correction. This was accomplished under computer control by stepping the visible laser wavelength on and off the tin resonance. The sediment samples contained both sodium and calcium, but they could be separated at sufficient resolution to avoid background correction. Whereas the direct detection limits for inorganic tin are similar for LEI spectrometry and graphite furnance atomic absorption, the chromatography required for the determination of the speciated alkyltins seriously degrades the latter's limits of detection. This occurs because the heating cycle for the furnace is relatively slow compared with the 10 Hz laser, which samples the elutant 500 times more frequently. This necessitates a weaker eluting mobile phase and slower flow rates to broaden chromatographic peaks for improved compatibility of the furnace detector with the separation process. The detection limit determined with the LEI detector was 3 ng/mL tin as tributyltin or 0.06 ng of tin.

An interesting attempt was made to improve the response of a flame ionization detector (FID) using laser enhancement (28). The FID is widely used for the gas chromatography of hydrocarbons. The rate of reaction leading to chemi-ionization in the FID flame is much greater for excited-state CH radicals than for ground-state species. Theoretically, laser population of the excited states of these radicals should lead to an enhanced FID response. Although a small signal enhancement was observed, the ultimate benefit was limited by the subsaturation irradiance provided by the continuous wave dye laser laser used for this experiment and the inability to irradiate a larger cross section in the flame. It was anticipated that with optimization of the experiment, enhancements of flame ionization approaching a factor of 100 could be obtained.

4.2.2. Diagnostic Measurements

In a typical analytical measurement, standards are used to calibrate the system. The analyte signal is then compared with standard signals for quantification. Rarely is there a need to determine the absolute concentration of the analyte. So-called diagnostic measurements seek quantitative information about the concentration of native as well as introduced species in the sample reservoir, i.e., the flame or plasma. The term *diagnostic* also includes measurements of dynamic processes occurring within the sample reservoir, such as the measurement of flame velocity.

Precursors and products of combustion, such as H_2, O_2, OH, and CH, are not typically measured by LEI spectrometry. Because of the high ionization potentials of these species, laser-induced multiphoton ionization is necessary and has been used effectively for their determination. Multiphoton processes leading to ionization may occur simultaneously with LEI, sometimes interfering with analyte signals. For example, radical OH has been observed in an acetylene/air flame (29). A characteristic band structure in the vicinity of 283.0 nm is indicative of multiphoton ionization and could potentially interfere with analytes absorbing in this wavelength region. LEI spectrometry finds its main application in the measurement and understanding of dynamic processes in the flames and plasmas. Early applications of LEI spectrometry to combustion systems have been reviewed by Schenck and Hastie (30).

Lifetime measurements of the metastable levels of thallium and lead in an acetylene/air flame have been measured by LEI spectrometry (31). The procedure is easily understood by considering a two-step excitation process that shares an intermediate level, i.e., the lower level for the second excitation is the same as the upper level for the first excitation. If the second photon is delayed with the respect to the first, the LEI signal will diminish because the common level will be depopulated by spontaneous radiative and nonradiative processes. Several experiments at different delay times will describe the decay of the common excited state and permit the calculation of its lifetime. The two-step excitation schemes for thallium and lead in this work do not share common levels. The thallium energy level diagram is shown in Fig. 4.3. Although the excited state from the first excitation and the initial state for the second excitation for both thallium and lead are not the same, they are strongly coupled. Therefore, the procedure described for the simpler case holds equally well for a two-step excitation with a metastable level. In this experiment, two excimer lasers pumping separate dye lasers were used for excitation. One excimer laser was delayed with respect to the other, and the resulting LEI signal was measured to determine lifetimes for the metastable P states in both thallium (81 ns) and lead (360 ns).

LEI spectrometry has been applied by Su et al. (32) to the determination of atomization efficiencies for lithium and sodium in an acetylene/air flame. A mathematical relationship between the time-integrated LEI signal and the total number density of free atoms in the flame was derived. The development of this expression involved the solution of rate equations for the population densities at relevant atomic energy levels and the relevant ionic state of the atom. The atomization efficiency was defined as the quotient of the *free atom number density* and the *total number density* of the element actually nebulized. The experimental determination of the free atom number density required integrating the LEI pulse over a time interval and combining the result with the known electron charge, the amplifier gain factor, the probing volume, and

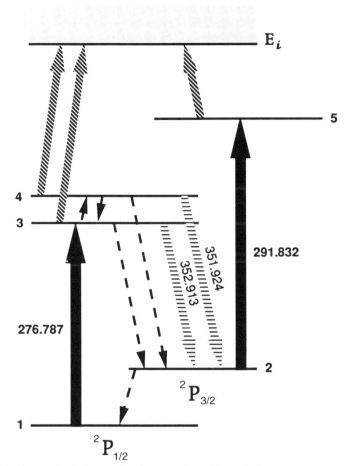

Figure 4.3. Energy level diagram for thallium: the solid, vertical arrows correspond to laser excitation at 276.787 and 291.832 nm; dashed arrows refer to nonradiative transitions; horizontally hatched arrows indicate radiative transitions at 352.913 and 351.924 nm; thermal ionization from levels 3, 4, and 5 is denoted by diagonally hatched arrows.

the determined collection efficiencies. The total number density was evaluated from measured experimental values. The LEI measurements were made with both rod and plate electrodes. When the results obtained for atomization efficiencies for lithium and sodium were compared with values in the literature, there were discrepancies with some published results for atomic absorption measurements and good agreement with others. Su et al. (32) suggested that the variation was related to differences in flame composition. The authors also pointed out that interferences with collection efficiency may account for differences between rod and plate electrodes and differences in the agreement of lithium and sodium values with published studies. As long as the critical collision quenching and ionization rate coefficients are known or can be estimated, measurements of time-integrated LEI signal will allow calculation of the total free atom number density in a flame.

As many analysts have observed, the LEI signal for an analyte is extremely sensitive to the concentration of concomitants in the flame. The relationship of the LEI signal and the total ion concentration in an acetylene/air flame was exploited for the measurement of relative ion fractions for IA and IIA elements (33). Indium was used as a "pilot" analyte and was an indirect measure of the concomitant's presence in the flame. By comparing the relative signal suppression of indium when various concentrations of alkali and alkaline earth metals were aspirated, ionization fractions could be determined. These measurements were duplicated within experimental error for lithium, sodium, and potassium by means of atomic absorption spectrometry.

Ionization efficiencies have been determined for excited lithium and sodium in acetylene/air flames by relating LEI signals from various excited states to laser-induced photoionization signals (34). The ionization efficiency is defined as the probability that the excited atoms ionize instead of returning to the ground state. In this work, the determination of ionization efficiencies of atoms seeded into flames does not require any estimates of flame- or nebulizer-dependent parameters or atom densities.

Single-step and two-step laser excitation, followed by collisional ionization, have been used to measure the ionization yield of lithium atoms produced in a separated acetylene/air flame (35). The ionization yield is defined as the number of ions (electrons) produced during the laser excitation time divided by the total number of atoms present in the laser volume. Two dye lasers pumped by an excimer laser are tuned to two connected transitions of the atom. Atomic fluorescence is monitored simultaneously with the LEI signal. The results obtained were incorporated into a simplified excitation–ionization model to derive order-of-magnitude estimates of the effective ionization rate coefficients. It was found that if the level excited by a single laser pulse lies several electron volts below the ionization potential and the pulse is as narrow as a few nanoseconds, the ionization yield will be insignificant. This has implications for flame and plasma diagnostics because it means that, under optically saturated conditions, a resonance fluorescence measurement can be related to the total number atom density in the levels connected by laser excitation.

Atomic ion mobilities in acetylene/air and carbon monoxide/oxygen flames have been measured by detecting the arrival times of the ions produced by laser enhancement (36). The time from the production to the collection of the ions was measured as a function of the distance from the laser beam to the collecting electrode for a series of applied voltages. At the excitation wavelength, 539.9 nm, there is a discrete two-photon sodium transition and a broadband uranium transition. The arrival of the electron pulse and the ion signals due to sodium and uranium were both visible. The mobility measurements for 10 ions with a wide variety of atomic weights were compared to Langevin theory. The model predicted the mobility of some of these ions to within 10%, provided that the dielectric constant of the flame gases was considered. In general, the larger, more polarizable ions were subject to the largest deviation, sometimes as much as 50% larger than predicted. For atomic ions, Langevin theory is accurate and provides an upper limit for estimating absolute ion mobility.

The temporal and spatial evolution of the depleted neutral atom density following laser enhancement of ionization has been used to characterize the flow velocity of the flame gases in a laminar flow flame (37). Continuous-wave dye laser radiation tuned to the 589.0 nm sodium transition was split into two beams as illustrated in Fig. 4.4. The lower beam was acousto-optically modulated and periodically depleted the sodium atom population within the beam cross section in the flame via LEI. The upper probe beam was directed counter to the modulated beam through a sodium D-line filter onto a photomultiplier tube. The output of the photomultiplier that monitored the sodium absorption was sent to a signal averager. The averaged signal was processed by a computer to fit it to a model for arrival time. When this measurement was

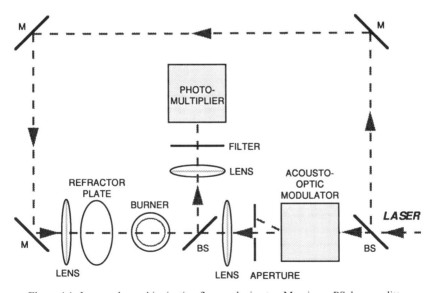

Figure 4.4. Laser-enhanced ionization flame velocimeter: M, mirror; BS, beam splitter.

repeated for two probe beam heights, the difference in arrival times and the probe beam height difference yielded the flame velocity at the average probe beam height above the burner with 2% precision. This approach is limited to flames with reasonably laminar flows. Otherwise turbulence will mix the depleted neutral atom region before it can be registered in the probe beam. Also the background electron density must be low enough to prevent ion–electron recombination and loss of the depleted region. A method for determination of flame velocities using two-step LEI has also been reported (38). The mobility of gases in propane/butane/air and acetylene/air flames was measured.

The application of LEI to the determination of temperatures in an acetylene/air flame has been demonstrated (39). A three-level model of rate equations was developed to define flame temperature in terms of Boltzmann population of the ground fine structure of the analyte gallium. Using published values for spontaneous transition probabilities, it was possible to extract the flame temperature from the power dependence of LEI measurements for two gallium transitions that terminated in the same excited state. Flame temperatures were also determined for excitation to different intermediate states of gallium to confirm that the model was applicable to any three-level system. The determined flame temperatures were consistent with each other and previously reported values measured using atomic absorption, emission, and fluorescence.

The excellent spatial resolution provided by intersecting laser beams lends itself to the profiling of both native and introduced species in flames. A high repetition rate copper vapor laser-pumped dye laser system was used to excite two-step LEI and laser excited atomic fluorescence (LEAF) for crossed-beam spatial diagnostics in an inhomogeneous, separated acetylene/oxygen/nitrogen flame (40). Detailed mapping of the analyte concentration was obtained throughout the flame. Particular care must be taken with LEI measurements because the placement of the electrode(s) may distort the flow field and electric field variances, and collection effects can complicate the spatial results. LEI and LEAF measurements both yielded similar spatial information and high sensitivity, but the latter produced lower detection limits. This information will permit the optimization of experimental conditions for flame-based spectrometric techniques.

Two reports have appeared that investigate the use of LEI spectrometry as a diagnostic tool for laser-generated plumes. At first reading, the results of these studies seem contradictory. The first publication, by Coche et al. (41), explored the possibility of LEI detection in the plasma created by the laser ablation of a solid metal target at atmospheric pressure. A second laser tuned to the element of interest was used to probe the plasma originating at the target surface some time after the ablation laser pulse. The detection electrodes were positioned on either side of the plasma. According to theoretical expressions developed in this work, the ratio of collected electrons to ablated atoms was very small. This led the authors to the conclusion that the laser-produced plasma was strongly ionized but the signal collection was weak. Therefore, the charges created by resonant excitation would not be detectable by electrical means. In fact, the second laser did not produce a detectable LEI signal in their experiment. The second study, by Pang and Yeung (42), was completed independently and published in the same year. It employed a single excimer laser-pumped tunable dye laser for both the evaporation and ionization of the analyte. The samples were sodium tungstate and copper metal, mounted in an evacuated chamber. The dye laser wavelength was insufficiently energetic to cause multiphoton processes to occur. One-centimeter-square plate electrodes separated by 1 cm collected the ion current. Acoustic signals monitored during the ablation process showed that when the dye laser was tuned to the sodium resonant wavelength, ion formation increased significantly while the same amount of material was generated on or off resonance. This was true for copper also. Linear correlations between the acoustic wave and the ion signal at different laser energies and focusing conditions was obtained. Paug and Yeung (42) suggested that the different results compared to those of Coche et al. (41) were due to the differences in the experimental procedure: Coche et al. had used two lasers with different vaporization and ionization beams. The smaller pulse energies of the ionizing beam and the longer time delays between the two lasers could explain the variance. The aforementioned work was also completed at atmospheric pressure. Pang and Yeung suggested that the LEI spectrometry may be one of the few techniques that can be used to probe very dense plasmas with good spatial and temporal resolution.

The electric field used to detect LEI is sufficient to induce Stark broadening of the high-quantum-number Rydberg states of atoms in some cases (43). The Stark effect causes mixing of states and consequent redistribution of transition probabilities among several levels. Although the integrated probability of a transition to the high-lying overlapped states remains relatively constant, the peak sensitivity at any given wavelength may be reduced significantly due to "smearing out" of the broadened levels. At excessively high applied voltages, the advantage of populating high-lying states is negated to some extent by the effects of Stark broadening. The experimental conditions that lead to the Stark effect should be avoided for analytical measurements because the reduced peak height of the LEI signal decreases sensitivity. The use of the Stark effect to profile the electric field distribution in flames was demonstrated. An expression was developed that allowed calculation of the full Stark width in the flame and the corresponding electric field strength from an experimental measurement of the full-width half-maximum LEI signal.

4.3. CONCLUSIONS

While it is possible to overinterpret statistical data, particularly where small samples are involved, it is worth revisiting Fig. 4.1. The data indicate that the published applications of LEI spectrometry have tripled since 1985. These recent papers are more sophisticated and less an afterthought than earlier publications, but a leveling off of published applications after 1985 is apparent. One might argue that publications related to applications are less likely to result as LEI spectrometry becomes more mature, so perhaps the number of publications is a poor indicator. In any case, LEI spectrometry has proved itself in the arena of real sample analysis.

Important advances have been made in adapting LEI spectrometry·to more routine analyses. For LEI to become more widely used and usable, it will require continued evolution of software and hardware to facilitate *routine* computer control of laser wavelength for excitation wavelength surveys and selection, background correction, and implementation of two-step excitation processes. Chromatographic pretreatment of samples to remove interferences also needs automation, perhaps using flow injection analysis technology. Making LEI spectrometry "user friendly" is well within the scope of current technology, as several papers reviewed here illustrate.

Even taking the most optimistic view, there are unlikely to be hundreds of "LEI spectrometers" rolling off the assembly line ever. The outlook for a commercial, multiple-capability "laser spectrometer" is considerably better. Why would any analyst want to limit him- or herself to LEI spectrometry, or LEAF for that matter? With relatively minor modifications, the whole battery of laser-based analytical techniques, including LEI, LEAF, and multiphoton ionization techniques, could be accessible to an instrument operator. Perhaps the best strategy for an instrument manufacturer would be to provide modules that could be coupled to the user's laser rather than building a fully integrated system.

There is little doubt that laser-induced ionization techniques, including LEI spectrometry (44), will continue to play a significant role in chemical analysis because of their high sensitivity, high accuracy, and high precision. Because of the relative simplicity of ionization detection, these methods complement purely optical schemes. In some cases, no other instrumental technique will provide the outstanding figures of merit yielded by laser-induced ionization methods. Diagnostic measurements using LEI spectrometry are some of the most compelling applications and should continue to provide unique and important information about dynamic chemical systems.

ACKNOWLEDGMENTS

The author wishes to acknowledge M. D. Seltzer for critically reading the manuscript of this chapter and those scientists who kindly provided reprints of their work.

REFERENCES

1. R. B. Green, R. A. Keller, P. K. Schenck, J. C. Travis, and G. G. Luther, *J. Am. Chem. Soc.* **98**, 8517 (1976).
2. O. Axner, M. Lejon, M. Norberg, and M. Persson, *Conf. Ser.—Inst. Phys.* **114**, 315 (1991).
3. G. C. Turk, *Anal. Chem.* **53**, 1187 (1981).
4. G. C. Turk, J. C. Travis, J. R. DeVoe, and T. C. O'Haver, *Anal. Chem.* **51**, 1890 (1979).
5. V. I. Chaplygin, N. B. Zorov, and Y. Y. Kuzyakov, *Talanta* **30**, 505 (1983).
6. O. Axner, *Spectrochim. Acta* **45B**, 561 (1990).
7. G. C. Turk and M. De-Ming, *NBS Spec. Publ. (U.S.)* **260-106**, 30 (1986).
8. I. Magnusson, O. Axner, and H. Rubinsztein-Dunlop, *Phys. Scr.* **33**, 429 (1986).
9. G. J. Havrilla and K. J. Choi, *Anal. Chem.* **58**, 3095 (1986).
10. O. Axner, M. Norberg, and H. Rubinsztein-Dunlop, *Appl. Spectrosc.* **44**, 1124 (1990).
11. A. G. Marunkov, T. V. Reutova, and N. V. Chekalin, *Zh. Anal. Chim.* **41**, 681 (1986).
12. N. V. Chekalin, A. G. Marukov, V. I. Pavlutskaya, and S. V. Bachin, *Spectrochim. Acta* **46B**, 551 (1991).
13. O. Axner, M. Lejon, I. Magnusson, H. Rubinsztein-Dunlop, and S. Sjöström, *Appl. Opt.* **26**, 3521 (1987).
14. J. Du, H. Li, L. Pan, H. Chen, J. Lian, C. Jin, C. Wang, and P. Zhang, *Fenxi Huaxue* **18**, 607 (1990).
15. G. C. Turk, G. J. Havrilla, J. D. Webb, and A. R. Forster, *Anal. Chem. Symp. Ser.* **19**, 63 (1984).
16. N. V. Chekalin, V. I. Pavlutskaya, and I. I. Vlasov, *Conf. Sef.— Inst. Phys.* **114**, 283 (1991).
17. N. V. Chekalin and I. I. Vlasov, *J. Anal. At. Spectrom.* **7**, 225 (1992).
18. J. D. Messman, N. E. Schmidt, J. D. Parli, and R. B. Green, *Appl. Spectrosc.* **39**, 504 (1985).
19. O. Axner, M. Norberg, M. Persson, and H. Rubinsztein-Dunlop, *Appl. Spectrosc.* **44**, 1117 (1990).
20. A. Miyazaki and H. Tao, *J. Anal. At. Spectrom.* **6**, 173 (1991).
21. A. A. Gorbatenko, N. B. Zorov, S. Y. Karpova, Y. Y. Kuzyakov, and V. I. Chaplygin, *J. Anal. At. Spectrom.* **3**, 527 (1988).
22. G. J. Havrilla and C. C. Carter, *Appl. Opt.* **26**, 3510 (1987).

23. H. M. Kingston, I. L. Barnes, T. J. Brady, T. C. Rains, and M. A. Champ, *Anal. Chem.* **50**, 2064 (1978).

24. G. C. Turk and H. M. Kingston, *J. Anal. At. Spectrom.* **5**, 595 (1990).

25. T. Berglind, S. Nilsson, and H. Rubinsztein-Dunlop, *Phys. Scr.* **36**, 246 (1987).

26. K. S. Epler, T. H'Haver, G. C. Turk, and W. A. MacCrehan, *Anal. Chem.* **60**, 2062 (1988).

27. G. C. Turk, W. A. MacCrehan, K. S. Epler, and T. C. O'Haver, *Conf. Ser.—Inst. Phys.* **94**, 327 (1989).

28. T. A. Cool and J. E. M. Goldsmith, *Appl. Opt.* **26**, 3542 (1987).

29. Y. Yan, D. Ding, H. Liu, Y. Ren, G. Bing, and Q. Jin, *Fenxi Huaxue* **19**, 1141 (1991).

30. P. K. Schenck and J. W. Hastie, *Opt. Eng.* **20**, 522 (1981).

31. N. Omenetto, T. Berthoud, P. Cavalli, and G. Rossi, *Appl. Spectrosc.* **39**, 500 (1985).

32. K. D. Su, C. Y. Chen, K. C. Lin, and W. T. Luh, *Appl. Spectrosc.* **46**, 1370 (1992).

33. J. E. Hall and R. B. Green, *Anal. Chem.* **57**, 16 (1985).

34. O. Axner and T. Berglind, *Appl. Spectrosc.* **43**, 940 (1989).

35. B. S. Smith, L. P. Hart, and N. Omenetto, *Anal. Chem.* **58**, 2147 (1986).

36. G. W. Mallard and K. C. Smyth, *Combust. Flame* **44**, 61 (1982).

37. P. K. Schenck, J. C. Travis. G. C. Turk, and T. C. O'Haver, *Appl. Spectrosc.* **36**, 168 (1982).

38. Y. Y. Kuzyakov, O. I. Matveev, and O. I. Novodvorskii, *Zh. Prikl. Spektrosk.* **40**, 145 (1984).

39. K. D. Su, C. Y. Chen, K. C. Lin, and W. T. Luh, *Appl. Spectrosc.* **45**, 1340 (1991).

40. M. J. Rutledge, M. Mawn, B. Smith, and J. D. Winefordner, *Appl. Spectrosc.* **41**, 1398 (1987).

41. M. Coche, T. Berthoud, P. Mauchien, and P. Camus, *Appl. Spectrosc.* **43**, 646 (1989).

42. H. M. Pang and E. S. Yeung, *Anal. Chem.* **61**, 2546 (1989).

43. O. Axner and T. Berglind, *Appl. Spectrosc.* **40**, 1224 (1986).

44. O. Axner, H. Rubinsztein-Dunlop, and S. Sjöström, *Mikrochim. Acta* **3**, 197 (1989).

CHAPTER

5

NONFLAME RESERVOIRS FOR LASER-ENHANCED IONIZATION SPECTROMETRY

NIKITA B. ZOROV

*Department of Chemistry, Moscow State University,
119899 Moscow, Russia*

5.1. INTRODUCTION

The outstanding performance of laser-enhanced ionization (LEI) in trace element analysis has been demonstrated mostly in flames, since the determination of elements in flames is fast and simple and has good reproducibility. As flames provide an excellent collisional thermal medium, they are the most commonly used atom reservoirs in LEI spectrometry. The use of flame atomizers often permits the reduction or elimination of interference from the sample matrix and thus permits direct analysis of multicomponent samples.

Yet the flame atomizer also has the following serious drawbacks that limit its application in LEI spectrometry for trace element analysis: (i) dilution of sample vapors by flame gas combustion products; (ii) limited range of temperatures used; (iii) a small (0.1–0.15) sample utilization factor; (iv) the impossibility of separating the processes of sample evaporation and atomization; (v) problems associated with handling microvolumes of liquids and solid samples; and (vi) the fact that the combustion products of flames may hinder some of spectral regions for successful implementation of LEI.

These limitations of flames have prompted consideration of alternate atom reservoirs for LEI spectrometry.

5.2. ATMOSPHERIC PRESSURE ELECTROTHERMAL ATOMIZERS

Since graphite furnace atomization improved the sensitivity of atomic absorption spectrometry by 1–2 orders of magnitude over flame atomization (1, 2),

---

Laser-Enhanced Ionization Spectrometry, edited by John C. Travis and Gregory C. Turk.
Chemical Analysis Series, Vol. 136.
ISBN 0-471-57684-0 © 1996 John Wiley & Sons, Inc.

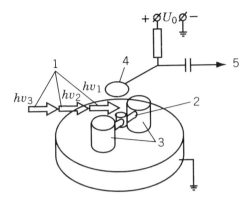

Figure 5.1. Experimental setup for LEI spectroscopy in a graphite cup: *1*, laser beams; *2*, atomizer–graphite cup; *3*, water-cooled terminal blocks, between which support electrodes and cup are clamped; *4*, LEI electrode detecting of electrons; *5*, LEI signal output.

the use of electrothermal atomizers in LEI was investigated in the early years of the development of this technique.

The first attempts to use a graphite tube atomizer at atmospheric pressure for LEI determinations of elements failed (3). Then Gonchakov et al. (4) reported the successful application of electrothermal atomization at atmospheric pressure for determination of small amounts of Na whose atoms were ionized by the three-step scheme (Fig. 5.1). Two dye lasers were pumped by radiation of the second harmonic of a Nd:YAG laser. A graphite cup in an argon atmosphere was used as an atomizer, and a tungsten loop was placed 2 cm above the graphite cup as an electrode. The estimated detection limit of Na was 1×10^{-15} g. It was found that the best signal-to-noise (S/N) ratio was obtained when the atomization temperature of the cup reached about 1900 °C. Higher temperatures caused considerable interference due to thermionic emission by the high-temperature graphite cup surface. When a negative potential was applied to the electrode, the noise signal from thermionic emission was reduced drastically.

Experiments in LEI with a Varian-Techtron CRA-90 electrothermal tube atomizer were described by Salsedo Torres (5). The electrode detecting charged particles and the graphite tube were placed end to end on the same axis. The graphite tube was placed horizontally (Fig. 5.2). The laser beams passed through the graphite tube without touching the wall. The atomizer was used for LEI determination of Cs. The estimated detection limit of Cs was 5×10^{-12} g. However, the determination of some other elements with this atomizer was not feasible because the temperature required for efficient atomization was more than 2500 °C. At these temperatures LEI signal was not observed because of a high noise level from thermionic emission of the

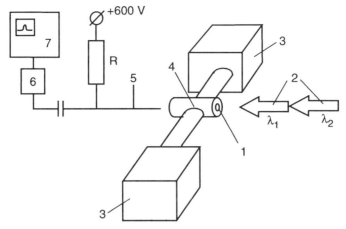

Figure 5.2. Experimental arrangement for laser-enhanced ionization with Varian-Techtron CRA-90 atomizer: *1*, graphite tube; *2*, laser beams; *3*, water-cooled terminal blocks; *4*, support electrodes; *5*, LEI electrode; *6*, amplifier; *7*, oscilloscope.

atomizer material. Salsedo Torres further found that this technique sometimes produces arcing between the ion-detecting electrode and the surface of the graphite atomizer.

Bykov et al. (6) used the graphite tube to determine the presence of sodium and indium at the level of 10^{-14} to 10^{-15} g. A tungsten wire placed axially and centered inside the graphite tude served as a voltage electrode (Fig. 5.3). The experiments were carried out at atmospheric pressure directly in air or in nitrogen. Determination of levels of cesium and yttebium in aqueous solutions were also made. The LEI signal was 10–100 times stronger than when a flame was used as an atomizer, but the reproducibility of the determination was not very good. The main problem in this procedure was arcing between the electrode and the tube when the temperature of the atomizer increased.

The most detailed investigations into the application of graphite furnaces to LEI spectrometry were made by Swedish scientists (7–9). The electrode used for establishing an electric field inside the furnace was a tungsten wire, 0.5 mm in diameter, mounted axially and centered inside the graphite tube. The laser light was directed into the graphite furnace, illuminating almost the whole of the interior. If two laser beams were used, they were aligned parallel, with good spatial and temporal overlap in the furnace. The interior diameter of the graphite tube was 9 mm; the diameter of the overlapping area of the beams was approximately 5 mm. The LEI signal was studied with both positive and negative voltage relative to the grounded tube.

The Swedish scientists were faced with the same problems described in earlier reports (4–6). Severe interferences originated from electrons thermally

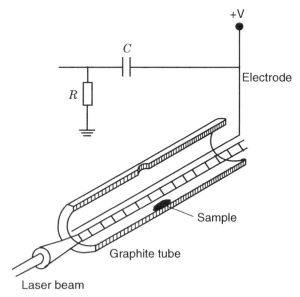

Figure 5.3. LEI detection in the graphite furnace (Perkin-Elmer HGA-72 model).

emitted from the heated graphite tube and the electrode, making it impossible to measure the LEI signal for elements which atomize slowly.

Despite the fact that in the zone of interaction with laser radiation the concentration of atoms was 2–3 orders of magnitude higher than that in a flame, there was not a drastic enhancement of limits of detection. The sensitivity of LEI determination of Mn and Pb was approximately 2 orders of magnitude higher than that in the flame; however, the limits of detection of these elements in furnaces and flames for LEI experiments were of the same order of magnitude (8).

An alternative modified electrothermal atomizer at atmospheric pressure was suggested by Magnusson (9) and demonstrated by Sjöström et al. (10) to solve this problem and to spatially separate the regions of atomization and detection. In this T-shaped furnace, atomization of the analyte occurred inside a Perkin-Elmer HGA-72 graphite furnace. A flow of argon was used to transport the atoms to an external cavity adjacent to the tube, in which laser excitation and LEI detection were performed. Disturbances from electrons emitted from the heated graphite were not observed with a negative potential on the outer plate. The plate served as a wall of the external cavity and as a high-voltage LEI electrode. A positive voltage, however, resulted in a very high noise level due to a large dc current carried by thermionically emitted electrons. Detection limits in the picogram range were obtained for manganese and strontium when LEI experiments were performed in this atomizer in a relatively cool environment.

Despite the low detection limits, this construction of the T-furnace has some drawbacks. As a result of a temperature gradient between the center of a graphite tube and the LEI detection region, relatively few of the atoms in the sample actually reach the detection region: the atoms strike the walls and form molecules. The matrix effects in this T-shaped furnace would be a real problem for real sample analysis because of the pressure of vapor phase interferences.

Considering the applicability to trace element analysis of graphite furnace LEI spectroscopy, Magnusson (9) discussed optimal experimental conditions for this technique. The choice of sheath gas and applied voltage were investigated to obtain an optimum S/N ratio. He also discussed the origin of double peaks for LEI signals in an HGA-72 furnace with wall atomization and a tungsten wire electrode. This event was observed first by Magnusson et al. (7), who postulated that the walls of the tubular furnace heated earlier than the radiatively heated electrode; consequently, the analyte atoms inside the tube condensed on the cold tungsten and finally revaporized upon the radiative heating of the electrode.

Butcher et al. (11) reported graphite furnace LEI that employed modern furnace technology with probe atomization. A graphite probe was used for both sample introduction and as the high-voltage electrode (-50 V). In the probe atomization approach, the atoms vaporize off the radiatively heated probe into an already isothermal furnace environment. When the probe reaches a temperature high enough to vaporize the atoms, the temperature of the probe is also sufficiently high to prevent any atom condensation. The graphite probe was also used to avoid excessive heat deformation of the tungsten wire electrode used in other experimental configurations (6–8). Finally, the probe atomization reduces matrix interferences as the sample is vaporized into an isothermal environment.

Butcher et al. (11) investigated in detail, both theoretically and experimentally, the optimization of the S/N ratio in the probe atomization technique for graphite furnace LEI. For elements such as thalium, indium, and lithium, detection limits in the 0.7–2 pg range were obtained. Lead, magnesium, and iron had detection limits between 10 and 60 pg. The linear dynamic range for each of the elements was between 3 and 4 orders of magnitude; the precision for aqueous standards was between 12 and 16%.

Most of the publications on graphite furnace LEI did not investigate the effects of sample matrices upon the LEI signal. However, a number of papers (12–24) have focused upon the effect of matrices upon the flame LEI signal. Only Butcher et al. (11) have evaluated the effects of easily ionized elements and mineral acids for graphite furnace LEI with probe atomization.

The effect of sodium upon the indium graphite furnace LEI signal was investigated. There was no suppression when the mass ratio of sodium to indium (100 pg) was less than 5. On the other hand, when the mass ratio of sodium to indium exceeded 5, the indium furnace LEI signal was suppressed, with a maximum suppression by a factor of 3 when the ratio of Na to In was between 2000 and 100,000. These results indicate that easily ionized elements suppress the graphite furnace LEI signal in the same manner as the suppression observed in flames, so it would be difficult to use graphite furnace LEI for the analysis of the substances containing appreciable amounts of easily ionized elements. Nitric acid and sulfuric acid had little effect upon the indium graphite furnace LEI signal. These data differed from those obtained by Trask and Green (17) for the flame. However, the indium signal was suppressed at high concentrations of hydrochloric acid (the ratio of HCl to In exceeded 2000). Butcher et al. (11) attribute this suppression to the formation of diatomic indium chloride. Molecular absorption and molecular fluorescence from indium chloride have been observed in a graphite furnace.

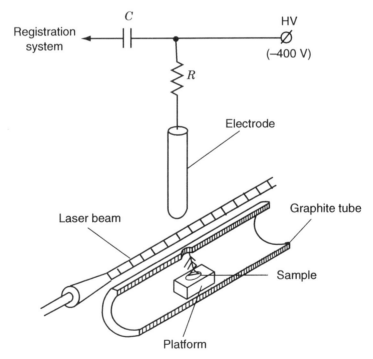

Figure 5.4. Design of the graphite furnace atomizer Reproduced from Chekalin and Vlasov (25) with permission.

Table 5.1. Comparison of Graphite Furnace LEI Detection Limits in Aqueous Solution with Other Methods

| | Limits of Detection (pg/mL) | | |
| | | | |
Element	Graphite Furnace LEI[a] (25)	Flame LEI (26, 27)	Graphite Furnace AAS[b] (28)
In	0.008	0.4	50
Yb	10	1.5	—

[a] Sample volume 100 μL.
[b] Atomic absorption spectrometry.

In almost all experiments with a graphite furnace (7, 8, 11) the LEI electrode was inserted inside the graphite tube at the atomization zone of analyte. The main problem when working with systems of this type is the thermionic emission of electrons from either the heated graphite tube or the LEI electrode placed in it, resulting in the suppression of the analytical signal and an electrical breakdown (9).

A novel design for a graphite furnace atomizer–ionizer was examined by Chekalin and Vlasov (25). This design (Fig. 5.4), with an outer electrode, excludes completely the emission from the cooled electrode; the emission from the graphite tube is suppressed by the electrode voltage selection. Laser excitation and ionization of atoms were performed inside and outside the graphite tube, above the sample injection hole. This graphite furnace atomizer–ionizer permitted reproducible signals of the analyte and low detection limits in aqueous solutions to be obtained (Table 5.1) (25–28).

The detection limit of indium on the femtogram level was 2 orders of magnitude better than for LEI spectrometry in flames. The much lower enhancement factor for ytterbium can be explained by the low atomization efficiency in the graphite furnace due to possible ytterbium carbide formation. These preliminary experiments demonstrate that this atomizer–ionizer graphite furnace system is very promising for the analysis of high purity materials.

5.3. HYBRID COMBINATIONS OF FLAME AND ELECTROTHERMAL ATOMIZERS

The electrothermal atomizers described above are unable to separate the following processes: sample evaporation and atomization; nonselective ionization background and thermoelectron emission from the surface of graphite; matrix interferences. In order to solve these problems, hybrid combinations of

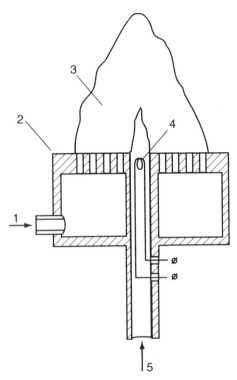

Figure 5.5. Atomizer with separate vaporization and atomization of a sample: *1*, introduction of fuel and air; *2*, burner body; *3*, flame; *4*, electrically heated wire loop; *5*, argon introduction.

flame and electrothermal atomizers have been developed. Despite the fact that such systems have not received wide recognition in atomic absorption spectrometry (29–33), for LEI analysis their use looks very promising (34–38).

This type of atomization system for LEI was first proposed by Chaplygin et al. (34). The atomizer was a specially designed cylindrical burner (Fig. 5.5). The central channel of the burner contains an electrical heating wire loop or filament upon which the sample is deposited. An argon flow in the central channel of the burner was used for the introduction of a sample vapor into a flame and to protect the loop against rapid burining. This burner solves several problems. Depostion of small quantities of a substance being analyzed into the filament means that the sample matrix will contain only a small absolute quantity of easily ionized atoms that will interfere with the determination (12, 13, 19).

The use of a Varian-Techtron CRA-90 programmed temperature-conrol unit for heating the filament not only allows separation of signals at various

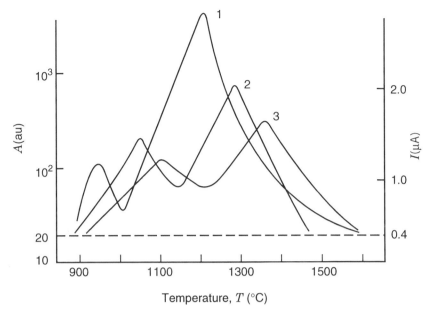

Figure 5.6. Temperature dependence of LEI signal of cesium (A) and background signals (I) of sodium and rubidium. Curves: *1*, cesium signal ($m_{Cs} = 5 \times 10^{-9}$ g); *2*, background current of rubidium ($m_{Rb} = 5 \times 10^{-7}$ g); *3*, background current of sodium ($m_{Na} = 5 \times 10^{-7}$ g). The noise level is 20 arbitrary units; flame conductance 0.4 μA for evaporation of 5 μL of high-purity water.

temperatures but also permits separation by the appearance times of different elements (namely, the analyte element and the sample matrix elements) due to the difference in volatilization temperatures of the compounds of these elements. The results of application of this atomizer are presented in Fig. 5.6.

This atomization system reduces considerably the thermoelectron emission background from a hot filament. The flame region irradiated by laser beams was located much higher than the combustion zone and afforded not only a maximum LEI signal but also an optimal S/N ratio.

The detection limit for cesium was 5×10^{-13} g; reproducibility was better than 5–6%. This type of burner, providing separate vaporization and sample atomization, facilitates not only the reduction of the reciprocal influence of elements but also the analysis of microsamples. The calibration graph obtained for Cs solutions was linear within a concentration range of more than 4 orders of magnitude (5×10^{-4} to 10 μg/mL). It was shown (36) that even a 10-fold excess of rubidium does not influence the LEI signal of cesium (Fig. 5.7). The LEI determinations of impurities of cesium at the level of 10^{-4} wt% in RbCl and Rb_2CO_3 salts were carried out with the help of this atomization system.

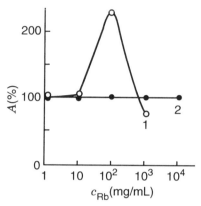

Figure 5.7. Dependence of LEI signal for Cs (percentage of nominal signal A) on the concentration of Rb. Curves: *1*, sample nebulization into the flame ($c_{Cs} = 0.01$ µg/mL); *2*, atomizer with electrothermal sample vaporization into the flame ($c_{Cs} = 1$ µg/mL).

A Micro Sample system for inductively coupled plasma–atomic emission spectrometry (ICP–AES) (Seiko Instruments Inc., Tokyo) was used by Miyazaki and Tao (39) as the electrothermal vaporizer into a fuel-lean air/acetylene flame. The exit of a glass chamber for the tungsten boat was connected to a burner unit with a water-cooled immersed electrode via a poly(tetrafluoroethylene) (PTFE) tube. For ICP–AES, the electrothermal vaporizer uses argon mixed with hydrogen as the carrier gas. Hydrogen prevents the oxidation of the tungsten boat on heating. However, only argon was used in this experiment, because introduction of hydrogen into an air/acetylene flame would be dangerous. Over 200 heating cycles were possible before the boat required replacement.

The applicability of the electrothermal vaporizer to LEI spectrometry in flames was confirmed in the determination of Tl in several types of natural water samples. Interferences from other elements were studied. At 276.79 nm, only alkali and alkaline earth elements interfered at 10-fold excess or more. To eliminate the interferences, Tl was extracted into 2,6-dimethyl-4-heptanone and 20 µL of the extract was used for the measurements. With a 200-fold concentration factor, the detection limit of Tl ($S/N = 3$) was 0.043 ng/mL. The analytical results measured by LEI spectrometry were in agreement with results measured by ICP–AES.

Smith et al. (40) have interfaced a graphite furnace vaporization step with a flame LEI detection step in order to acquire the advantages of both graphite furnace vaporization and flame ionization detection while maintaining independent control of each process.

A graphite furnace was used to vaporize samples, which were then transferred in an argon carrier gas to a low-noise miniflame LEI detection system that

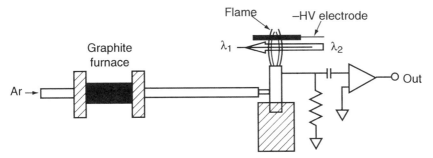

Figure 5.8. Experimental setup for graphite furnace vaporization with miniflame LEI detection system. Reproduced with permission from Smith et al. (40). Copyright 1993, American Chemical Society.

had been previously developed for use as a resonance ionization detector (41). Two-step excitation of analyte atoms was provided by dual Nd:YAG pumped dye lasers operating at 30 Hz. Figure 5.8 schematically shows the experimental system. A temporal probing efficiency of the system was determined, and it was shown that more than 97% of the flame were not probed. The main reason was the low repetition rate of the laser.

Magnesium, thallium, and indium have been studied by injecting aqueous solutions into the furnace. All of these elements exhibited good linearity over at least 5 orders of magnitude. The absolute detection limits for Mg, Tl, and In, limited by the poor probing efficiency and losses in sample transport, ranged from 1 ro 260 fg.

Marunkov (42) performed the first LEI experiments with sample vapors being introduced into a flame by an electrically heated graphite rod and by diffusion through the wall of an electrically heated closed graphite tube. Detailed experiments were not carried out, but detection limits (5×10^{-15} g) confirm that the use of such atomization systems in LEI spectrometry is very promising.

A hybrid "rod–flame" arrangement combining the advantages of flame and electrothermal atomizers was proposed by Chekalin et al. (25, 38, 43, 44). In this system (Fig. 5.9) the sample is evaporated by an electrically heated graphite rod into the flame, where the analyte is atomized, laser excited, and ionized, whereupon the produced charged particles are detected. The sample atomization occurs in a flame where most elements have high atomization efficiencies and matrix interference is considerably reduced.

The temperature for evaporation is usually much lower than that for atomization and the regions of evaporation and atomization are separated from one another. Therefore the problem arising from the ionization background is substantially reduced. In comparison with flames, this atomization

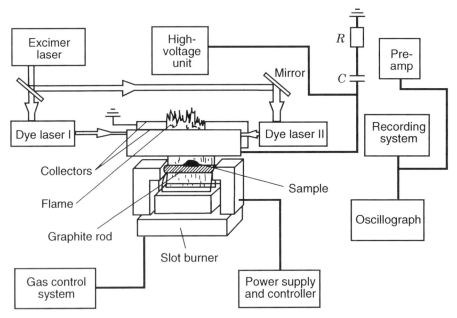

Figure 5.9. Block diagram of the LEI spectrometer with a rod–flame system. Reproduced with permission from Chekalin et al. (43).

system provides the following: (i) an increase of sample utilization factor; (ii) analysis of microsamples; and (iii) direct analysis of solid samples.

The pulsed evaporation of the sample and minimization of matrix interferences due to the use of the flame enable one to obtain detection limits that are low, even in comparison with those typical of AAS in the furnace (Table 5.2) (44–46). The calibration graphs obtained were linear within a concentration range of 4–5 orders of magnitude.

Working with such low levels of element concentrations, Chekalin and Vlasov (25) faced other problems, one of these being the laser-enhanced nonselective background ionization from the compounds evaporating from the heated graphite rod. It was observed that for all carbon-based materials at temperatures above 2000 °C, the background ionization increased rapidly with rising temperatures; the exciting laser wavelength also exerted a weak influence on the background ionization. For the detection of the volatile elements (indium, sodium, and copper) in a rod–flame LEI system, this factor is insignificant, but for the detection of gold this factor becomes dominant. An isotropic pyrographite (PGI TUG-02-595-85) was used with the lowest nonselective background among the materials tested.

Table 5.2. Comparison of Rod–Flame LEI Detection Limits in Aqueous Solutions with Other Methods

Element	Limits of Detection (pg/mL)		
	Rod–Flame LEI (25, 44)	Flame LEI (44, 45)	Graphite Furnace AAS (46)
Au	2	1000	100
Co	100	50	10
Cr	20	2000	10
Cu	2	70	20
In	0.04	0.4	50
Mn	30	20	10
Na	0.02	0.6	50
Ni	8	4	100

The purity of graphite materials posed the next problem. For widespread elements (e.g., copper) the value of the blank signal from the analyte in the rod is significant. To minimize the influence of this factor, kinetic studies were made of changes in signals from copper, both in the sample and in the rod itself, occurring as the temperature and the rate of heating changed (38). Although careful selection of the heating conditions permitted this signal to be minimized, it remains desirable to use high-purity graphite rods when working with lower concentrations of analyte.

The potential of the LEI method with the rod–flame atomizer was demonstrated by Chekalin and colleagues (25, 38, 43, 44) for the analysis of high-purity substances, such as orthophosphoric acid, germanium, Cd–Hg–Te alloy, and silver nitrate, and fluorine-containing materials for optical fibers. Samples in the form of a solution, powder, or solid, having a weight of 1–20 mg, were placed in a groove of the pyrographite rod surface. The experiments began with stepwise heating of the rod for drying and, if necessary, sample preheating. When the experimenters switched to the evaporation mode for the analyte compounds, the rod was inserted into the flame.

5.3.1. Analysis of Orthophosphoric Acid

Concentrated (86%) orthophosphoric acid was introduced directly onto the graphite rod and heated to 200 °C without intensive boiling and splashing of the sample. A vitreous mass formed on the rod; it was removed without losing the analyte in the next stage, heating to 1000 °C. The process was accompanied by a 2 orders of magnitude increase in the flame noise. In the subsequent stage,

Table 5.3. Analysis of High-Purity Substances by LEI Spectrometry with a Rod–Flame System

Sample	Element	Content (10^{-8} wt%)	Detection Limit (10^{-8} wt%)
H_3PO_4	Na	2200 ± 200	3
	Cu	84 ± 8	2
Cd–Hg–Te (No. 1)			
Solution ($10\,g/L$, $10\,\mu L$)	In	32 ± 3	3
Solid ($10\,mg$)	In	34 ± 3	0.01
Cd–Hg–Te (No. 2)			
Solution ($10\,g/L$, $10\,\mu L$)	In	29 ± 2	3
Solid ($10\,mg$)	In	26 ± 3	0.01
Ge (No. 1)			
Solution ($10\,g/L$, $10\,\mu L$)	Cu	790 ± 60	100
Solid ($10\,mg$)	Cu	710 ± 70	5
Ge (No. 2)			
Solution ($10\,g/L$, $10\,\mu L$)	Cu	370 ± 60	100
Solid ($10\,mg$)	Cu	300 ± 40	5
NH_4F			
Solution ($100\,g/L$, $10\,\mu L$)	Cr	700 ± 50	20
Solid ($10\,mg$)	Cr	710 ± 60	2
Solution ($100\,g/L$, $10\,\mu L$)	Co ⎫	$\leqslant 100^a$	100
Solid ($10\,mg$)	Co ⎭		10
Solution ($100\,g/L$, $10\,\mu L$)	Mn ⎫	880 ± 60^a	30
Solid ($10\,mg$)	Mn ⎭		3
Solution ($100\,g/L$, $10\,\mu L$)	Ni	560 ± 60	8
Solid ($10\,mg$)	Ni	575 ± 55	0.8
NaF			
Solution ($20\,g/L$)	Cr	2900 ± 700	150
Solution ($20\,g/L$)	Mn	400 ± 220	220
Solution ($20\,g/L$)	Ni	22000 ± 4000	60
Solution ($20\,g/L$)	Co	$\leqslant 700$	700

[a] Average of solution and solid results.

when the temperature was elevated to 2000 °C for copper, the noise disappeared and the LEI signal of copper was detected. The more volatile sodium required a more careful temperature selection in order to separate the acid and sodium signals. The absence of matrix interferences in both instances was shown by the standard additions method. The sample analysis results and the detection limits obtained are presented in Table 5.3. The detection limits were determined by the purity of the rod material used.

5.3.2. Determination of Indium in Cd–Hg–Te Alloy

5.3.2.1. Analysis of Sample Solutions

When the standard additions method was used, no suppression of the indium LEI signal by the matrix was observed. The laser-generated nonselective ionization background from the sample matrix was small, its fluctuations being of the same order of magnitude as the recording system noise. On heating the rod in the working temperature region, no signal from indium in the rod was detected; in other words, the blank signal was negligible. This background fluctuation resulted in a detection limit of 3×10^{-8} wt% (Table 5.3). The independence of the analytical signal from the matrix permitted calibration with standard aqueous solutions to be used.

5.3.2.2. Analysis of Solid Samples

A 10 mg amount of Cd–Hg–Te alloy powder was introduced into a groove on the rod surface. During preheating to $1000\,^\circ$C the alloy sample partially decomposed, yielding mercury evaporation and fume liberation, and then melted and converted to a droplet. During the working cycle (3–4 s) the sample residue evaporated together with the indium contained therein.

Since no dilution of the sample into a solution occurs, LEI signals from solid samples are enhanced accordingly. No background signals from thermal matrix ionization were detected. As a result, the detection limit of indium in Cd–Hg–Te alloy was 1×10^{-10} wt% for a 10 mg sample. Solid samples did not show signal suppression by the matrix; therefore determination of indium in these samples was achieved using standard aqueous solutions.

5.3.3. Determination of Copper in Germanium

5.3.3.1. Analysis of Sample Solutions

The analysis of this sample is more complicated because germanium is an easily ionizable matrix. Strong thermal ionization of the matrix occurred during the determination of copper in germanium samples containing less than 1 μg/g. To analyze solutions of these samples, a dissolution procedure of solid germanium was used with partial removal of the matrix in GeF_4 form. When the greater part of the sample matrix was separated, the flame noise level considerably decreased. Investigation of the copper evaporation rate from the sample solutions when they were introduced on the surface of the rod showed that the current fluctuations in the operating cycle could be reduced and the analysis could be performed using the standard additions and calibration graph methods. No influence of the matrix on the analytical signal was detected, as with the Cd–Hg–Te alloy.

5.3.3.2.   Analysis of Solid Samples

The rod heating conditions were selected, and the influence of the thermal ionization of the matrix in the operating cycle was completely excluded. It follows that as the solid sample matrix does not effect the analytical signal, calibration can be performed with aqueous standard solutions. The copper detection limit in germanium decreased to the level determined by the copper content in the graphite rod and was equal to 5×10^{-8} wt%.

The comparison of the results for analysis of the same Ge samples in solution and solid form (Table 5.3) shows close correspondence. This demonstrates accuracy of the analysis and the absence of matrix influence when working with solid samples.

5.3.4.   Determination of Gold in Silver Nitrate

The determination of gold in silver nitrate demonstrated one of the main advantages of the LEI method: stepwise excitation makes possible a considerable improvement in spectral selectivity. This analysis proved to be the most complicated, as the matrix behaves in the same way in both solution and solid samples—it decomposes, yielding silver metal. Both gold and silver atoms evaporate into the flame practically at the same temperature, that is, close to 2000 °C.

The laser light with wavelengths used for gold atom excitation will not excite silver atoms from their ground state. However, silver has a large number of transitions into Rydberg and autoionization states from the $5p$ level, 3.66 eV higher than the ground state. The population of this level in a flame at 2500 °C is only 2×10^{-8}, but the excitation and ionization efficiency within the range 265–420 nm is high. Even with tuning far away from the resonance transitions, the superposition of the wings of the absorption lines yields a silver background signal comparable to the analytical signal of gold at a 0.1 ppm concentration.

The two-step excitation of gold atoms ($\lambda_1 = 267$ nm; $\lambda_2 = 294$ nm) led to a detection limit of 10^{-6} wt% for the determination of gold in $AgNO_3$; although the background signal is only doubled by use of the second excitation step, the analytical signal increased by 2 orders of magnitude. However, for the determination of gold in silver nitrate, the analytical conditions adopted were far from optimal. More suitable schemes may be selected according to both efficiency and spectral range; the possible reduction of the silver $5p$ state population by working with colder atomizers also needs to be examined.

5.3.5.   Analysis of NH₄F and NaF

The trace content of Co, Cr, Mn, and Ni in two fluorine-containing materials (NH₄F and NaF) was measured by Chekalin et al. (44) using the LEI technique either with a flame as the atomizer or employing a rod–flame system for atomization. It was shown that the rod–flame system is more versatile and allows direct analysis of both sample solutions and solid samples. This system minimizes the matrix influence of NH₄F.

The analysis of NaF samples was a more complicated task for the LEI technique. With conventional aspiration of the NaF solutions into a flame, the aforementioned authors obtained a substantially larger ionization background as well as a strong suppression of the signal due to the space charge effect, even when analyzing solutions diluted to sample concentrations as low as 0.2 g/L. It was found that it was possible to determine the Co and Ni content in the NaF sample by using an air/acetylene flame without preconcentration or matrix removal.

With the rod–flame system, these easily ionized matrix influences were much less pronounced when these authors worked with diluted sample solutions (2%). Increasing the matrix concentration up to 10% resulted in strong signal suppression and introduced large fluctuations in the signal due to an increased current through the flame. It was impossible to reduce these effects either by a longer ash stage time or with a higher ash stage temperature since these resulted in considerable loss of the elements under study. Therefore the rod–flame system was used for analysis of NaF samples presented only in the form of solutions. The results of the element determinations in the NH₄F and NaF samples are presented in Table 5.3.

5.4.   ELECTROTHERMAL ATOMIZATION IN A LOW-PRESSURE NOBLE GAS ATMOSPHERE

Another LEI method using electrothermal atomization in a low-pressure noble gas atmosphere and two-step excitation with continuous-wave (cw) lasers has been developed by Niemax et al. (47–49), who used Doppler-free spectroscopy and thermionic diode detection of ions. Thermionic diodes are known to be extremely sensitive detectors (50) for ions, having a large dynamic range (over 4 orders of magnitude) and a very high gain (10^6 or even larger). In its simplest form the setup is a cylindrical anode with an axially mounted cathode filament that is heated directly by a dc current. Normally no bias is applied to the diode running in the space-charge-limited mode. If ions are created, e.g., by LEI, they are trapped within the negative space charge. The presence of ions reduces the space charge, and as a result the diode current increases. The physical processes leading to ionization of the analyte atom are the same in thermionic diode spectroscopy and LEI. By one- or two-step laser excitation the atom is excited to a high-lying level, allowing further ionization by collision with other surrounding atoms.

If the excitation of analyte atoms takes place in a low-pressure noble gas atmosphere, there are crucial advantages to the use of single-mode cw lasers instead of pulsed laser systems. Theoretically (50) it was shown that total ionization of the analyte may occur when

$$\Delta t_1 \gg (C_{2i}F)^{-1} \tag{1}$$

where Δt_1 is the irradiation time of an analyte atom by the laser light; C_{2i} is the rate of collisional ionization from the laser-populated level 2; and F is the ratio of the number density in the upper-level N_2 to the total density N_t.

Because the collisional depopulation of the upper level by the noble gas at low pressure is weak, it should be possible in most cases to saturate resonance transitions of atoms with a moderate power Ar^+-laser-pumped cw ring dye laser. With cw lasers it is easy to fulfill Eq. (1). The longer irradiation time of the cw system is a considerable advantage over pulsed laser systems with a low repetition rate.

Another important advantage of the cw system is the narrow laser line width, which permits the use of Doppler-free spectroscopy to resolve the isotopic components of analyte lines in a low-pressure noble gas atmosphere in a thermionic diode.

Using co- and counterpropagating beams of two lasers with narrow line width, Niemax et al. (47) were able to perform resonant Doppler-free two-photon spectroscopy. The overlapping beams of two single-mode cw lasers were directed through two cells. Both cells were filled with noble gas at reduced pressure. The element under investigation was evaporated continuously in one cell (the reference cell) while the sample was atomized in the other cell (the analyte cell). The lasers were first tuned to a wavelength where the analyte atoms were resonantly excited to an intermediate level; at this point the atoms were excited to a level near the ionization limit. Ionization of the analytes was achieved by thermal collisions with the buffer gas atoms. The ionization products (ions or electrons) could then be measured.

The first laser, tuned into the Doppler profile of the first transition, excited only those analyte atoms with a particular velocity component with respect to the direction of the laser beam. The second laser excited these atoms further to higher states. The detection of the second transition revealed a Doppler-free line. The authors observed Yb-isotope-selective signals and determined a detection limit of 85 pg for [173]Yb in a Eu sample. They then detected the [173]Yb content in high purity Ba (99.999%) in the parts per trillion range and [174]Yb in Sm samples.

The high-cost cw dye laser system was later replaced by easy-to-operate, low-cost semiconductor high-power diode lasers (48, 49) to measure various Ba, Ca, and Pb isotope ratios using Doppler-free laser spectroscopy with thermionic diode detection.

5.4.1. Figures of Merit

Depending on the element, typical relative standard deviations (RSDs) of 0.5–3% for isotope ratio measurement were found at continuous sample atomization. Depending on the excitation scheme, the laser intensities, and other experimental parameters (such as atomization temperature and collisions), detection limits of isotopes down into the femtogram range can be obtained.

5.4.2. Interferences

If resonant Doppler-free two-photon excitation in a collisional atmosphere is applied to the measurement of isotope ratios, systematic errors may arise due to velocity-dependent collision effects (51). This velocity-dependent broadening is unlikely to create such errors if the buffering noble gas pressure is kept low. Typical neon gas pressures used in the experiments (48) were on the order of 200 mtorr, well below the neon pressure range for which collisional broadening was observed (51).

A basic assumption of quantitative LEI spectrometry is that the analytical signal is proportional to the excited state number density. This holds true as long as the energy of the laser-excited level is approximately within the gas kT of the ionization limit[1]. However, if the excited level is far from the ionization limit, other mechanisms have to be taken into account. In such cases, it might happen that the LEI signal is no longer proportional to the laser-excited number density but dependent upon other parameters as well. Two processes, Penning ionization (49) and collisional energy pooling transfer (52–55), may also produce an LEI signal.

If Penning ionization, i.e., the collision of laser-excited atoms with excited gas atoms (e.g. in a discharge), is the major ionization process, the LEI signal depends not only on the laser-excited number density but also on the density of the excited buffer gas atoms. When Penning processes deplete the number density of excited buffer gas atoms within the interaction volume of the laser beam and analyte atoms, the LEI signal is no longer proportional to the laser-excited number density. Obrebski et al. (49) demonstrated this effect,

---

[1] Here k is Boltzmann's constant and T is the temperature.

studying the Penning ionization of low-excited barium atoms by high-excited europium atoms.

Another process that can produce LEI signals not linearly dependent on the laser-excited number density is collisional *energy pooling transfer* to high-excited levels. At these levels the ionization can take place through collisions with thermal buffer gas atoms.

Energy pooling collisions were first observed in sodium (52), where two optically excited Na* in $3p$ states collide creating high-excited sodium atoms (Na**) in states whose energy is close to twice the energy of the $3p$ resonance states. In the experiments made by Obrebski et al. (49), it was suggested that energy pooling collisions between the low-excited barium atoms with transfer to higher excited states was responsible for the unexpected ratios of the isotope components.

Narrowing of spectral lines and unexpected intensity ratios were observed by Niemax et al. (54, 55). A remarkable narrowing of the Doppler profile of the Ca $4s^2$ $^1S_0$–$4s4p$ $^3P_1$ line was measured by LEI spectrometry in a thermionic diode. Varying the laser intensity, the calcium number density, and analyzing the line profiles provided evidence that energy pooling in higher states resulting from the collision of both two and three excited calcium $4s4p$ $^3P_1$ atoms is responsible for this effect. Thus line widths as well as intensity ratios of spectral lines have to be evaluated with great care, as the effect of energy pooling can seriously distort quantitative spectroscopic measurement in LEI spectroscopy if transitions to low-excited levels are under investigation.

5.5. RESONANT LEI OF ATOMS IN AN INDUCTIVELY COUPLED PLASMA

The first measurements of LEI in an argon inductively coupled plasma (ICP) were made by Turk and Watters (56). An extended-torch ICP was used; laser excitation and ionization detection took place 10 cm above the load coil in a region of lower electron population in order to reduce radio-frequency (rf) interference from ICP and avoid arcing from plasma to the LEI electrodes. Resonant LEI was detected for Fe, Mn, Na, and Cu. The population of free atoms was too small, however, and consequently sensitivity was very poor. Limits of detection ranged from 140 ng/mL for Na to 7000 ng/mL for Cu, i.e., these limits were much poorer than the limits of detection for LEI in the flame (28, 46).

To reduce rf interference from the plasma, Turk et al. (57) used a power-modulated ICP. In this approach, the rf power to the ICP was interrupted for a period of approximately 1 ms before each pulse of the laser; the power was resumed after the laser-induced ionization was detected. During the power-off cycle, the population of background ion and electrons decayed to a level which made the detection of laser-induced ionization possible. To reduce arcing between the electrodes and the plasma, the grounding of the ICP load coil was modified so that the top of the load coil was at ground potential. A space of 3.5 mm separated the inside edges of the two water-cooled LEI electrodes bracketing the plasma. Special rf noise filters were placed between each electrode and the external electronics to filter the 40 MHz rf of the ICP from the detection electronics. Detection limits of 80 ng/mL for Fe and 20 ng/mL for Ga were achieved; these limits were still far from those that can be attained with the use of flame LEI.

Ng et al. (58) have reported significantly imporved limits of detection for LEI in an extended-torch ICP by modifying the torch and electrode designs and using continuous laser excitation; in particular, the torch–coil arrangement (59, 60) and argon flow rates were optimized for generating an extended ICP. In this case, the outer tube of the ICP torch extended 6 cm from the tip of the injector tube; a three-turn copper coil was used with a coil spacing equal to the tubing thickness. The torch was arranged such that the inner tubes were some 1–2 mm below the load coil. Samples in the form of dried aerosols were introduced into the plasma through the intermediate gas inlet of the torch. The water-cooled stainless steel tubing electrodes were slightly flattened and bent, and placed 7 mm apart on either side of the plasma. A dye laser pumped by a cw argon ion laser was used. The laser beam was directed to pass parallel, ~ 1 mm away from the collection anode. The plasma torch was enclosed in an aluminum box. This entire plasma enclosure and the impedance matching box were wrapped with a copper cloth that was subsequently grounded. This shielding and grounding greatly reduced the rf and environmental noise picked up by the LEI detection electronics.

With optimal plasma argon flows, the plasma extended approximately 27–30 cm above the load coil. The plasma tail region, where the plasma appeared homogeneous, with complete mixing of argon and air entering from the atmosphere, was used in the experiment.

The detection limit for sodium was approximately 2 orders of magnitude superior to that of Turk and Watters (56). The smaller electrode dimensions and smaller separation between electrodes used by Ng et al. (58) are probably responsible for this result. It appears that dimensionally much smaller electrodes might minimize noise pickup, improving the S/N ratio, just as smaller separation might be more efficient for analyte probing in the plasma. Limits of detection of several other elements ranged from 30 ng/mL for Ca to 810 ng/mL for Sr. Although these results represent an improvement over those of Turk and Waters (56), the figures of merit are still poor compared with other LEI systems in flames and furnaces (8, 25, 26, 28, 35). Additional progress is still needed to justify the cost of the ICP–LEI system and the complexity of the measurement.

5.5.1. ICP Plasma Diagnostics

LEI was successfully used as an informative diagnostic method for the study of fundamental processes that occur in an ICP. While the effect of easily ionizable elements on ion–electron recombination on flames has been well investigated and understood for many years, there has been no investigation of this effect in the ICP. The primary difficulty in the study of interferences caused by easily ionizable elements in the ICP is the variety of effects in addition to changing the ionization equilibrium. Direct study of recombination kinetics between the laser-produced Sr^+ ions and the electron in ICP was carried out by Turk et al. (61). Optical detection of the laser-produced Sr^+ ions allowed direct measurement of the rate of ion–electron recombination, and time decay of the strontium ionic fluorescence was observed. This rate of decay was increased by the addition of easily ionizable elements and was attributed to the recombination of Sr^+ ions with electrons. The most significant difference between Sr^+ decay in the flame and in ICP is the absence of the fast decay caused by flame chemistry. Thus in the inert argon atmosphere of the ICP there are not the complex chemical interactions that often give rise to the interferences in the flame and other excitation sources containing molecular gases. The observed recombination rate is somewhat faster in the ICP than in the flame. Although the electron number density is much higher in the ICP, comparison simply on the basis of electron number density is not justified since the detailed mechanisms of recombination are different. The absence of molecular nitrogen in the ICP as a third-body collisional partner for recombination is particularly relevant.

A spatial profile and ICP power study were also conducted as diagnostic aids for optimization of the power-modulated ICP for both fluorescence and ionization detection (62).

5.6. OTHER PLASMA ATOMIZATION SYSTEMS IN LEI SPECTROMETRY

5.6.1. Helium-Microarc-Induced Plasma

The atmospheric-pressure microarc atomizer, originally developed by Layman and Hieftje (63), was investigated by Churchwell et al. (64) as an atom reservoir for LEI spectrometry. This reservoir has a high energy density and low power requirements, affording temporal separation of the desolvation and vaporization/atomization processes; it was these features in particular that attracted the attention of the authors. Microvolumes (1–3 μL) of analyte solution were deposited on the tip of the tungsten cathode loop; prior to striking the microarc, the analyte solution was desolvated by gentle blow-drying with a heat gun. LEI measurements were performed directly in the plasma above the microarc discharge. The experimental configuration was similar to the LEI experimental scheme described by Green et al. (15) with the exception of the atom reservoir.

The operational characteristics of the microarc discharge with flowing helium serving as a primary atom source were briefly examined. The authors investigated the dual laser ionization (DLI) pulses of sputtered sodium atoms in a helium microarc-induced plasma. A detection limit of 3 ng was estimated for this element. However, an indium mass of 1 μg (1 μL of 1000 mg/L In) deposited onto the cathode loop produced a DLI signal approximately an order of magnitude lower than that obtained for an equivalent mass of sodium. Molecular recombination processes may have been occurring and depleting the indium ground-state population to a greater extent in the plasma where DLI measurements were made. Preliminary results of these initial studies have demonstrated the feasibility of direct microarc atomization for LEI spectrometry. However, additional fundamental and analytical studies are required to properly assess the full potential of this atom reservoir for LEI.

5.6.2. Microwave-Induced Active Nitrogen Plasma

Seltzer and Green (65) investigated a microwave-induced atmospheric-pressure active nitrogen plasma as a potential atom reservoir for LEI spectrometry. This active nitrogen plasma, wherein diatomic nitrogen molecules exist in a variety of electronic, vibrational, and rotational excited states, is an efficient excitation source because of its ability to collisionally transfer up to several electron volts of energy to analyte species in the plasma (66).

Discrete analyte samples were introduced into the active nitrogen plasma by a microarc atomizer. However, in the course of this investigation, a background signal was detectable in the absence of analyte in the plasma, even at low laser irradiance. With increased ultraviolet (UV) laser irradiance of the plasma, a highly structured background spectrum was revealed, corresponding to the second positive band system of nitrogen. The observation of these transitions prompted further investigation (67) of the laser-induced ionization background of the active nitrogen plasma. As the UV laser wavelength was scanned through the regions of interest (279–285 nm), the laser-induced ionization signal was monitored. It was shown that the spectral region chosen for this study overlaps several atomic lines commonly used for LEI spectrometry. Therefore transitions in other wavelength regions have to be used.

The spectral mapping obtained in this study provides a means by which optimum analytical lines can be selected and background limited lines avoided.

The further study of electrical characteristics of microwave-induced plasmas has shown that the suppression of signal detection for laser-induced ionization in the active nitrogen plasma is similar to that encountered in flames in the presence of thermally ionized Group IA elements (68).

5.6.3. Microwave Resonant Cavity in Flames

A new method for the detection of the LEI in a flame has been developed by Suzuki et al. (69). Reflected microwave power was measured for the detection of LEI inside a microwave resonant cavity when a microwave field was applied to a flame.

Suzuki et al. found that when a flame is introduced into a cavity that is filled with resonant microwave radiation of amplitude E, the absorption of the microwave field by the flame changes the resonance frequency ω and the cavity Q. When the electron density is increased by LEI, the increased density will cause changes in the plasma frequency of electrons ω_p^2, and in turn changes in ω and Q. When the resonator is adjusted such that the reflected microwave signal from the cavity is zero, i.e., at the critical coupling, the changes in ω and Q will cause an increase in the reflected microwave power. Therefore, the LEI signal in a flame was measured as a laser-induced increase in the reflected microwave power.

In a preliminary experiment the LEI signal of sodium atoms aspirated into a propane/O_2 flame was observed at a concentration below 10 ng/mL. Despite the poor detection limit obtained by Suzuki et al. (69), this method has considerable potential as a diagnostic technique for combustion systems. The advantage of microwave detection is that a microwave field does not disturb the kinetics of combustion.

5.7. LEI IN LASER-GENERATED PLUMES

One of the most common procedures in LEI spectrometry is the dissolution of a solid sample, followed by either aspiration of the solution into a flame or deposition of the solution onto the surface of an electrothermal atomizer. The sample dissolution procedure is often a time- or labor-consuming process, and the sample solution is often contaminated with impurities contained in acids or other substances used for sample dissolution.

The laser microprobe technique is an important and promising method for solid sample introduction in atomic spectrometry (70, 71) that does not require

prior dissolution of samples. Nonetheless, the use of laser atomization (ablation) with selective ionization has been almost exclusively confined to resonance ionization spectroscopy experiments carried out either in a vacuum
(72, 73) or in P-10 gas (90% argon and 10% methane by volume) at 95 torr (74).

Coche et al. (75) explored the possibility of LEI detection for analytical
purposes in a laser-created plasma at atmospheric pressure. The ablation of
a solid or liquid material with the laser was followed by resonant excitation
coming from a second dye laser, temporarily delayed and tuned to a transition
of the specific element of interest. The increase in the ionization rate caused by
resonant excitation was detected by the application of a high voltage between
two electrodes located on opposite sides of the plasma. Craters varying in
diameter from 40 to 220 μm were created in a metal target by focusing the
vaporization nitrogen laser beam on the target. The excitation laser was
a tunable dye laser pumped by a nitrogen laser; the pump laser pulse was 30 ns
to 20 μs later than the ablation laser pulse. Throughout this study, several
electrical properties of the laser-produced plasma were established. It was also
shown that during the time of existence of the laser-produced plasma, it is
impossible to obtain an electron density much smaller than that of the ions.
The charges created by resonant ionization were not detectable by electrical
means, due to the effect of the zero electric field (76). It was concluded that
apparently direct LEI detection in a laser-produced plasma at atmospheric
pressure is not feasible.

In contrast, more promising results were obtained when the experiments
were performed in a low-pressure atmosphere (77). In this case, the possibility
was examined of using LEI directly to probe the dense plasma formed by laser
vaporization (without mass spectrometry). Only one tunable dye laser in the
visible spectral region was used for both vaporization and ionization. Chen
and Yeung (78) had demonstrated earlier that measurement of the acoustic
wave can normalize the measurement of the atomic emission signal generated
from the laser plume and thus minimize the influence of laser power fluctuations and matrix effects. They demonstrated that at the pressure range of
about 50 torr of gas a linear correlation between the acoustic signal and the
amount of the evaporated material can be obtained. Therefore the correlation
of the ion signal and the acoustic signal were also investigated under different
focusing conditions and at different laser energies.

The samples, $Na_{0.7}WO_3$ and Cu were cut into disks and then mounted to
a stainless rod attached to a motor so that the sample could be rotated during
the spectroscopic experiments. All samples were polished. It was then shown
that multiphoton ionization is not the main mechanism for Na and Cu atoms.
Since the laser plasma has a relatively high density, thermal collisions occur
much frequently between the excited atoms and other atoms than in analytical
flames. There are certainly enough collisions to provide enough energy for the

ionization of the excited atoms. Enhanced ion signals were obtained when the laser wavelength was tuned in resonance with analyte transition bands generated respectively by a ground-state atom (Na), an excited-state atom (Cu), and a dimer (Na_2). Substantial broadening of the transition and wavelength shift has been observed as a result of the high plasma density. As a result, it was concluded that LEI may well be one of very few techniques that can be used to probe very dense plasmas with good spatial and temporal resolution. This is at odds with the results of Coche et al. (75), who concluded that LEI detection in a laser-produced plasma at atmospheric pressure is not feasible.

Direct laser action onto a solid sample provided a large concentration of charged particles with long lifetimes in air. As a result of this concentration, strong nonselective ionization signals were detected. Changing the distance between the sample and the LEI electrode did not diminish the strength of the nonselective signals. Therefore Novodvorsky et al. (79) proposed the laser vaporization of solid samples directly into the flame (Fig. 5.10).

Novodvorsky et al. demonstrated that with this approach the plasma produced under high-power Nd:YAG action may be rapidly recombined in the flame, with the recombination of charged particles depending upon the placement of the sample in the flame. When the sample is placed in the upper part of the flame, a large number density of charged particles of the sample

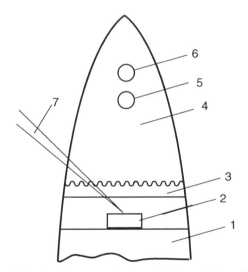

Figure 5.10. Schematic drawing of laser vaporization of solid sample into a flame: *1*, burner head; *2*, solid sample; *3*, primary combustion zone of the flame; *4*, outer cone of flame; *5*, dye laser beam for resonance excitation of analyte atoms; *6*, LEI cathode, collecting ions from a flame; *7*, laser beam for evaporation of a solid sample.

matrix causes a large nonresonant signal to be observed. If the sample is placed below the combustion zone of the flame, the nonresonant background signal decreases by a factor of 10^2. These experiments indicated that the efficiency of the recombination reactions taking place in the flame's combustion zone and in the zone's periphery was responsible for the decrease in the levels of nonresonant signals. The placement of the sample below the combustion zone of the flame also prevented the melting and evaporation of volatile samples into the flame before the application of the laser.

It was found that under particular potential differences between the cathode and the burner head, there is a spatial separation of charged and neutral components of the laser plume in the flame. The charged part of the plume rises more quickly than the cloud of neutral atoms: thus, the higher the applied potential differences, the greater the velocity of separation. It is this phenomenon that affords temporal separation of the background and analytical signals.

The distribution of neutral analyte atoms in the flame along its vertical axis was studied using laser action to vaporize samples of the metals Pt and Ni. In the case of Ni, the cloud of atoms occupied a greater volume than that in the case of Pt. This can be explained by the ejection of metal as droplets of various sizes in the flame under laser action. The heavier droplets move more slowly than free analyte atoms, which attain the upward velocity of the combustion gas very quickly. After volatilization of these droplets in the flame, the neutral Ni atoms formed the observed "tail." In the case of Pt, such an effect was not observed.

LEI combined with laser atomization of a sample into a flame was used to determine the presence of Pt in metal Ir. It was found that the Pt content did not exceed 10^{-4} wt%.

The use of such atomization systems permits the analysis of microsamples without any pretreatment or dissolution of a sample.

5.8. CONCLUSIONS

The demand for ultrasensitive methods has resulted in extensive research to develop laser-based analytical atomic spectroscopy techniques. An excellent example of these techniques is LEI. The atomization process in LEI spectrometry, i.e., the conversion of the sample containing the element to be analyzed into an atomic vapor, is closely related to the inherent qualities of the specific atoms reservoir used. LEI spectrometry applied to trace element analysis has mainly been confined to atmospheric-pressure flames with detection limits in the low picogram per milliliter range. These limits are 3–5 orders of magnitude lower than those of flame AAS.

Other potential atomic reservoirs for LEI spectrometry are currently under investigation. Electrothermal atomizers such as the graphite furnace afford a drastic increase in sensitivity. While this increase has greatly expanded the possibilities of research, a variety of problems hamper its application to LEI. The low limits of detection are themselves the most problematic, as the detection system also registers the noise of thermionic emission and noise produced by the furnace heating current.

More impressive results have been obtained with the use of hybrid rod–flame atomizers. These atomizers have attained record sensitivity in the analysis of high-purity materials. Scientists working with these atomizers have also developed effective methods for the elimination of matrix interferences. For both of these reasons, this chapter has given the most attention to applications of the hybrid rod–flame atomizers.

Many attempts have been made to use plasma reservoirs in LEI, with modest success. The use of LEI as a diagnostic tool for the plasma itself may be of great interest. There is still little research on the use of lasers for the vaporization and atomization of solid samples in LEI. This technique is a potentially rewarding one, but there are presently many problems with its use.

The continuing discovery of new atom reservoirs is essential to the development of LEI spectrometry.

REFERENCES

1. B. V. L'vov, *Inzh.-Fiz. Zh.* **2**, 44 (1959).

2. B. V. L'vov, *Atomic Absorption Spectrochemical Analysis.* American Elsevier, New York, 1970; Adam Hilger, London, 1970.

3. J. C. Travis, M. Epstein, P. K. Schenck, G. C. Turk, D. M. Sweger, and J. R. De Voe, *Proc. Colloq. Spectrosc. Int. 20th, and Int. Conf. At. Spectrosc. 7th*, Prague, 118 (1977).

4. A. S. Gonchakov, N. B. Zorov, Yu. Ya. Kuzyakov, and O. I. Matveev, *Zh. Anal. Khim.* **34**, 2312 (1979).

5. L. E. Salsedo, Torres, Ph.D. thesis, Moscow State University (1981).

6. I. V. Bykov, A. B. Skvortsov, Yu. G. Tatsii, and N. V. Chekalin, *J. Phys. (Paris), Colloq.* **44**, C7-345 (1983).

7. I. Magnusson, O. Axner, I. Lindgren, and H. Rubinsztein-Dunlop, *Appl. Spectrosc.* **40**, 968 (1986).

8. I. Magnusson, S. Sjöström, M. Lejon, and H. Rubinsztein-Dunlop, *Spectrochim. Acta* **42B**, 713 (1987).

9. I. Magnusson, *Spectrochim. Acta* **43B**, 727 (1988).

10. S. Sjöström, I. Magnusson, M. Lejon and H. Rubinsztein-Dunlop, *Anal. Chem.* **60**, 1629 (1988).

11. D. J. Butcher, R. L. Irvin, S. Sjöström, A. P. Walton, and R. G. Michel, *Spectrochim. Acta* **46B**, 9, (1991).

12. G. C. Turk, J. C. Travis, J. R. De Voe, and T. C. O'Haver, *Anal. Chem.* **50**, 817 (1978).

13. G. C. Turk, J. C. Travis, J. R. De Voe, and T. C. O'Haver, *Anal. Chem.* **51**, 1890 (1979).

14. G. J. Havrilla and R. B. Green, *Anal. Chem.* **52**, 2376 (1980).

15. R. B. Green, G. J. Havrilla, and T. O. Trask, *Appl. Spectrosc.* **34**, 561 (1980).

16. N. B. Zorov, Yu. Ya. Kuzyakov, O. I. Matveev, and V. I. Chaplygin, *Zh. Anal. Khim.* **35**, 1701 (1980).

17. T. O. Trask and R. B. Green, *Anal. Chem.* **53**, 320 (1981).

18. G. C. Turk, *Anal. Chem.* **53**, 1187 (1981).

19. V. I. Chaplygin, N. B. Zorov, and Yu. Ya. Kuzyakov, *Talanta* **30**, 505 (1983).

20. V. I. Chaplygin, Yu. Ya. Kuzyakov, O. A. Novodvorsky, and N. B. Zorov, *Talanta* **34**, 191 (1987).

21. G. J. Havrilla and C. C. Carter, *Appl. Opt.* **26**, 3510 (1987).

22. A. A. Gorbatenko, N. B. Zorov, S. Yu. Karpova, Yu. Ya. Kuzyakov, and V. I. Chaplygin, *J. Anal. At. Spectrom.* **3**, 527 (1988).

23. N. V. Chekalin, A. G. Marunkov, V. I. Pavlutskaya, and S. V. Bachin, *Spectrochim. Acta* **46B**, 551 (1991).

24. N. J. Szabo, H. W. Latz, G. A. Petrucci, and J. D. Winefordner, *Anal. Chem.* **63**, 704 (1991).

25. N. V. Chekalin and I. V. Vlasov, *J. Anal. At. Spectrom.* **7**, 225 (1992).

26. A. G. Marunkov and N. V. Chekalin, *Zh. Anal. Khim.* **42**, 638 (1987).

27. A. G. Marunkov, N. V. Chekalin, and E. I. Tihomirova, *Zh. Prikl. Spektrosk.* **48**, 542 (1988).

28. O. Axner and H. Rubinsztein-Dunlop, *Spectrochim. Acta* **44B**, 837 (1989).

29. V. P. Borzov, B. V. L'vov, and B. V. Plyushch, *Zh. Prikl. Spektrosk.* **11**, 217 (1969).

30. L. A. Pelieva, G. G. Muzykov, and I. V. Prushko, *Zh. Prikl. Spektrosk.* **20**, 771 (1974).

31. D. A. Katskov, A. P. Kruglikova, and B. V. L'vov, *Zh. Anal. Khim.* **30**, 238 (1975).

32. V. A. Razumov, *Zh. Anal. Khim.* **32**, 383 (1977).

33. F. J. Langmyhr and G. Wibetoe, *At. Spectrosc.* **8**, 193 (1985).

34. V. I. Chaplygin, N. B. Zorov, Yu. Ya. Kuzyakov, and O. I. Matveev, *Zh. Anal. Khim.* **38**, 802 (1983).

35. Yu. Ya. Kuzyakov, N. B. Zorov, V. I. Chaplygin, and O. A. Novodvorsky, *J. Phys. (Paris), Colloq.* **44**, C7-335 (1983).

36. V. I. Chaplygin, N. B. Zorov, and Yu. Ya. Kuzyakov, *Zavod. Lab.* **53**, 26 (1987).

37. S. V. Bachin, A. G. Marunkov, V. I. Pavlutskaya, I. I. Vlasov, and N. V. Chekalin, *Vysokochist. Veshchestva.* **4**, 186 (1989).

38. N. V. Chekalin, V. I. Pavlutskaya, and I. I. Vlasov, *Conf. Ser.—Inst. Phys.* **114**, Sect. 6, 283 (1990).

39. A. Miyazaki and H. Tao, *Anal. Sci.* **7**, 1053 (1991).

40. B. W. Smith, G. A. Petrucci, R. G. Badini, and J. D. Winefordner, *Anal. Chem.* **65**, 118 (1993).

41. G. A. Petrucci, R. G. Badini, and J. D. Winefordner, *J. Anal. At. Spectrom.* **7**, 481 (1992).

42. A. G. Marunkov, Ph.D. thesis, Moscow State University (1989).

43. N. V. Chekalin, V. I. Pavlutskaya, and I. I. Vlasov, *Spectrochim. Acta* **46B**, 1701 (1991).

44. N. V. Chekalin, A. Khalmanov, A. G. Marunkov, I. I. Vlasov, Y. Malmsten, O. Axner, V. S. Dorofeev, and E. Glukhan, *Spectrochim. Acta* **50B**, 753 (1995).

45. Yu. Ya. Kuzyakov and N. B. Zorov, *CRC Crit. Rev. Anal. Chem.* **20**, 221 (1988).

46. *The Guide to Techniques and Applications of Atomic Spectrometry*, p. 5. Perkin-Elmer Corp., Norwalk, CT, 1988.

47. K. Niemax, J. Lawrenz, A. Obrebski, and K.-H. Weber, *Anal. Chem.* **58**, 1566 (1986).

48. J. Lawrenz, A. Obrebski, and K. Niemax, *Anal. Chem.* **59**, 1232 (1987).

49. A. Obrebski, J. Lawrenz, and K. Niemax, *Spectrochim. Acta* **45B**, 15 (1990).

50. K. Niemax, *Appl. Phys. B.* **38**, 147 (1985).

51. J. Lawrenz, A. Obrebski, and K. Niemax, *Conf. Ser.—Inst. Phys.* **94**, 301 (1989).

52. M. Allegrini, G. Alzetta, K. Kopystyska, L. Moi, and G. Orriols, *Opt. Commun.* **19**, 96 (1976).

53. J. Huennekens and A. Gallagher, *Phys. Rev.* **A 27**, 771 (1983).

54. A. Obrebski, R. Hergenröder, and K. Niemax, *Z. Phys. D. At., Mol. Clusters* **14**, 289 (1989).

55. R. Hergenröder, A. Obrebski, and K. Niemax, *Conf. Ser.—Inst. Phys.* **114**, 65 (1991).

56. G. C. Turk and R. L. Watters, *Anal. Chem.* **57**, 1979 (1985).

57. G. C. Turk, L. Yu, R. L. Watters, and J. C. Travis, *Appl. Spectrosc.* **46**, 1217 (1992).

58. K. C. Ng, M. J. Angebrann, and J. D. Winefordner, *Anal. Chem.* **62**, 2506 (1990).

59. G. L. Long and J. D. Winefordner, *Appl. Spectrosc.* **38**, 563 (1984).

60. D. R. Demers, *Spectrochim. Acta* **40B**, 93 (1985).

61. G. C. Turk, O. Axner, and N. Omenetto, *Spectrochim. Acta* **42B**, 873 (1987).

62. G. C. Turk, *Appl. Spectrosc.* **46**, 1223 (1992).

63. L. R. Layman and G. M. Hieftje, *Anal. Chem.* **47**, 194 (1975).

64. M. E. Churchwell, T. Beeler, J. D. Messman, and R. B. Green, *Spectrosc. Lett.* **18**, 679 (1985).

65. M. D. Seltzer and R. B. Green, *Spectrosc. Lett.* **20**, 601 (1987).

66. M. D. Seltzer, R. B. Green, and E. H. Piepmeier, *ICP Inf. Newsl.* **13**, 705 (1988).

67. M. D. Seltzer, E. H. Piepmeier, and R. B. Green, *Appl. Spectrosc.* **42**, 1039 (1988).

68. M. D. Seltzer and R. B. Green, *Spectrosc. Lett.* **22**, 461 (1989).

69. T. Suzuki, T. Fukasawa, H. Sekiguchi, and T. Kasuya, *Appl. Phys. B.* **39**, 247 (1986).

70. K. Laqua, in *Analytical Laser Spectroscopy*, N. Omenetto, Ed., p. 48. Wiley (Interscience), New York, 1979.

71. K. Dittrich and R. Wennrich, *Prog. Anal. At. Spectrosc.* **7**, 139 (1984).

72. D. W. Beekman and T. A. Calcott, in *Resonance Ionization Spectroscopy 1984* (G. S. Hurst and M. G. Payne, Eds.), p. 143. IOP Publishing, Bristol, 1984.

73. M. W. Wiliams, D. W. Beekman, J. B. Swan, and E. T. Arakawa, *Anal. Chem.* **56**, 1348 (1984).

74. S. Mayo, T. B. Lukatorto, and G. G. Luther, *Anal. Chem.* **54**, 553 (1982).

75. M. Coche, T. Berthould, P. Manchien, and P. Camus, *Appl. Spectrosc.* **43**, 646 (1989).

76. J. C. Travis, G. C. Turk, J. R. De Voe, and P. K. Schenck, *Prog. Anal. At. Spectrosc.* **7**, 199 (1984).

77. H. Pang and E. S. Yeung, *Anal. Chem.* **61**, 2546 (1989).

78. G. Chen and E. S. Yeung, *Anal. Chem.* **60**, 2258 (1988).

79. O. A. Novodvorsky, A. B. Ilyuhin, N. B. Zorov, and Yu. Ya. Kuzyakov, *Vestn. Mosk. Univ., Ser. 2: Khim.* **30**, 99 (1989).

CHAPTER

6

IONS AND PHOTONS: INTERPLAY OF LASER-INDUCED IONIZATION AND FLUORESCENCE TECHNIQUES IN DIFFERENT ATOMIC AND MOLECULAR RESERVOIRS

NICOLÒ OMENETTO

Environment Institute,
Commission of the European Communities, Joint Research Centre—Ispra Site,
Ispra (Varese), Italy

PAUL B. FARNSWORTH

Department of Chemistry, Brigham Young University,
Provo, Utah 84602

6.1. INTRODUCTION

The choice of the title of this chapter has been prompted by an article by A. L. Gray (1) in which the relative merits of the two techniques of optical emission spectrometry and mass spectrometry, both using as excitation–ionization source an inductively coupled plasma, i.e., ICP–OES and ICP–MS, were assessed. Differences in the operational behavior as well as in analytical performances were considered and conclusions drawn on the relevance to be expected in the future in analytical laboratories for both techniques. Clearly, in that paper, ions (mass spectrometry) were contrasted with photons (emission spectrometry), whence the title: "Ions or Photons...." Along the same lines, the analytical performances of the techniques of laser-induced fluorescence (LIF) and laser-enhanced ionization (LEI) could be compared in the same atom reservoir, e.g., in a flame, an ICP, or a graphite furnace, and again, ion detection would be contrasted with photon detection. This chapter, however, is based on a different approach, in the sense that the techniques of LIF and LEI are in most cases used simultaneously in order to gain some insight on several important physical processes occurring in atmospheric pressure atomizers. In addition,

Laser-Enhanced Ionization Spectrometry, edited by John C. Travis and Gregory C. Turk.
Chemical Analysis Series, Vol. 136.
ISBN 0-471-57684-0 © 1996 John Wiley & Sons, Inc.

several examples will be illustrated in which a fluorescence measurement is capable of providing a parameter pertinent to the ionization process and vice versa, an ionization measurement is useful to interpret a fluorescence process. The title of this chapter, "Ions *and* Photons..." is therefore meant to emphasize the combined use of both techniques as well as their interactions, the understanding of which is essential for the interpretation of the experimental data.

When dealing with LIF and LEI techniques in flames, it is customary to refer separately to the analytical and diagnostic aspects of the experiment. This distinction, which appears to be in most cases unnecessary when each technique is used alone, is worthy of somewhat more elaborate consideration in the cases considered here. For example, if a fluorescence experiment is devised to provide the total number of emitting atoms in the excitation volume, a simultaneous measurement of the ionization signal will allow evaluation of the number of atoms undergoing a collisional ionization process and will therefore account for this loss. As a second example, an efficient multistep ionization scheme could be devised so that the first step in the excitation ladder is provided by fluorescence photons of that excitation energy. In this case, one would use the ionization technique as a means of detecting fluorescence photons from a sample. These two examples clearly indicate the analytical exploitation of simultaneous fluorescence and ionization measurements. On the other hand, there is a remarkable variety of experiments, which will be described in detail in Section 6.5., in which the fluorescence and/or the ionization data are used to extract information about the ionization cross sections, the chemical kinetics and dynamics of the ions formed, the type of excitation–ionization mechanism responsible for the observed signal, and other typical diagnostic aspects of the system under investigation. Many experiments have been performed in atmospheric pressure flames and plasmas, i.e., in collision-dominated systems, where LEI is the dominant process compared to direct photoionization. This would be in the line with the overall approach of this book, which specifically addresses LEI methodologies. However, several interesting applications have been described in which the experiment was carried out in vapor cells, operated at reduced pressure. One important difference between a flame and such a cell is that in the latter, number densities on the order of 10^{14}–10^{15} atoms cm^{-3} are reached. Such a concentration could be obtained in a typical flame or plasma system only by aspirating a solution containing approximately 100,000 μg/mL of the species investigated. Such abnormally high concentration would preclude the use of a conventional burner and premix nebulizing chamber due to clogging and other undesirable effects. Nevertheless, experiments performed in vapor cells will be reported below since they fit perfectly into the scope of this chapter, with the understanding that it would be difficult, if not impossible, to reproduce the same results in flames and plasmas.

Four sections follow these introductory remarks. In Section 6.2, a theoretical background is presented, based upon the derivation of selected analytical relations obtained from the use of the rate equation approach and a multilevel atomic system. Several experimental considerations relevant to the simultaneous monitoring of the fluorescence and ionization signals are discussed in Section 6.3. Section 6.4 illustrates the analytical implications of combined ionization and fluorescence measurements, with specific reference to saturated single-step and two-step excitation processes, and to the concept of a resonance ionization detector. Section 6.5 is devoted to the discussion of spectroscopic studies carried out with both techniques, either simultaneously or independently. In this context, the modulation of the fluorescence signal, caused by ionization, has recently played a significant role in many different experiments. The technique associated with such modulation is generally referred to as *fluorescence-dip spectroscopy* (or other semantic variations on the theme). The theoretical foundations for the topics treated here have been mainly laid down in Chapter 1 of this book; therefore, lengthy derivations of signal expressions will either be omitted or presented only in a tabular form.

Finally, as is always the case when a review of a particular topic is attempted, the coverage of the literature is certainly not complete. It then follows that no priorities, either in the inception of an idea or in its application, are implied here.

6.2. GENERAL THEORETICAL CONSIDERATIONS

As stated previously, only the theory pertinent to the concepts and experiments reported in this chapter will be discussed below, since the necessary background information concerning the interaction of atoms with laser light has already been extensively presented in Chapter 1. When dealing with LEI or LIF experiments carried out in atmospheric pressure flames and plasmas, it is customary to present a model in which the atoms are ideally schematized by a few energy levels connected by one or two lasers, which are also ideally characterized in terms of uniform intensity and temporal shape. This "wishful modeling" (2), while detracting from the *direct* applicability of the theory to the experimental situation on hand, is nevertheless useful, since it provides an insight into the general behavior of the system and usually correctly predicts the essential outcome of an experiment in flames and plasmas. In diagnostic experiments, however, aimed at the absolute evaluation of a physical parameter, special care must be exercized in assessing the temporal, spectral, as well as spatial characteristics of the lasers used for the excitation of the atomic

Table 6.1. Commonly Adopted Simplifications and Assumptions Made in Modeling the Interaction of Laser Radiation and the Atomic System in LEI and LIF Experiments in Flames and Plasmas

The Atomic System

(a) Two-level or three-level systems are considered, radiatively and/or collisionally coupled, with well-defined parities. The total atomic population is distributed among the levels, and the ionization continuum is described as an additional level.

(b) Thermal excitation–ionization from the ground state to higher levels and to the continuum is neglected under laser irradiation. Thermal population of excited states as well as ion–electron recombination and fast chemical reactions involving ions are also neglected.

(c) Radiation trapping is absent, i.e., the system is optically thin.

The Laser Source

(a) Two limiting cases are considered: (i) the laser is broadband, with a spectral full width at half-maximum (fwhm) much larger than the absorption profile; or (ii) the laser is monochromatic with a spectral fwhm much narrower than the absorption profile. In the former, a high number of equally spaced longitudinal modes is uniformly distributed over the absorption width.

(b) The laser has a uniform spatial irradiance distribution.

(c) The temporal profile of the laser can be fairly well approximated by a rectangle.

(d) The laser is linearly or circularly polarized.

Interaction

(a) The rate equation approach is valid. This means that atomic coherences between the level populations are neglected and that power broadening and splitting of the atomic levels are negligible.

(b) The fluorescence is spectrally integrated and unpolarized. The atoms are neither oriented nor aligned by the laser radiation, so that the Einstein B coefficients (defined for an isotropic, unpolarized radiation field) can still be used.

(c) Cooperative effects such as superfluorescence and superradiance are not important at the atomic densities used in the experiment.

---

species. In addition, the direct relation between the excited-state population and the fluorescence signal might lose its validity if polarization or anisotropy effects are present. In Table 6.1, an attempt has been made to collect the assumptions and simplifications usually given when describing the interaction of laser light with atomic systems. Some of these assumptions will be discussed in a more detailed manner below. It is clear, however, that the adaptation of a theoretical model to an experiment should always be taken *cum grano salis*.

The only correct way of describing the interaction of laser light with atomic and molecular systems is to solve the time-dependent Schrödinger equation in the rotating wave approximation, using the density-matrix approach. This time-dependent perturbation treatment leads to the well-known Rabi oscillations in the amplitude describing the quantum mechanical state of the atom in the radiation field. Details of this treatment can be found in many books on lasers and quantum mechanics (see, e.g., refs. 3–5) as well as monographs (e.g., ref. 6) or specific reports (e.g., 7–9), and the reader is again referred to Chapter 1 of this book. The vast majority of papers dealing with laser fluorescence and ionization techniques in flames and plasmas, however, resort to the more conventional rate equations approach for the modeling of the interaction. The density-matrix approach reverts to the rate equation approach if coherence effects can be disregarded. Coherence effects can be disregarded if two conditions are met: (i) the laser bandwidth, δv_l, is much larger than the effective width, δv_{eff}, of the absorption profile of atoms, and remains larger even when power broadening of the transition occurs; and (ii) the rate of dephasing collisions in the atom reservoir is much higher than the rate of coherent interaction with the monochromatic laser field. The two conditions together imply that the coherence time of the laser radiation ($\propto \delta v_l^{-1}$) is much shorter than the residence time of the atoms in the excitation volume, so that any "memory" of the coherent excitation process is quickly lost. As summarized by Alkemade (2), for a broadband laser, the rate equation approach is valid as long as the laser bandwidth is much greater than the excitation rate (Hz), given by the product of the Einstein coefficient of induced absorption, $B\,(J^{-1}\,cm^3\,s^{-1}\,Hz)$, and the spectral volume energy density, $\rho(v)\,(J\,cm^{-3})$. By using the classical definitions and relationships, in the case of a two-level system with equal statistical weights, the above condition can be formulated as

$$(\delta v_l)^2 \gg A \frac{\lambda_0^3}{8\pi hc} \frac{Q_l}{T_l S_l} \tag{1}$$

Here, $A\,(s^{-1})$ represents the Einstein coefficient for spontaneous emission of the transition whose center wavelength is λ_0 (cm); $h\,(J\,s)$ is Planck's constant; $c\,(cm\,s^{-1})$, the velocity of light in vacuum; $Q_l\,(J)$, the energy of the laser pulse; $T_l\,(s)$, the (rectangular) duration of the laser pulse; and $S_l\,(cm^2)$, its geometric cross section (assuming uniform irradiation). The quantity (Q_l/T_l) is the laser power P_l or $I_l\,(W)$, while (P_l/S_l) represents the laser irradiance, $E_l\,(W\,cm^{-2})$. Therefore, $\rho_l = (E_l/c)$. Often in the literature the foregoing definitions are used with different terminologies, e.g., the irradiance is indicated with the symbol I and called intensity. Also, the intensity, I, of a monochromatic laser at

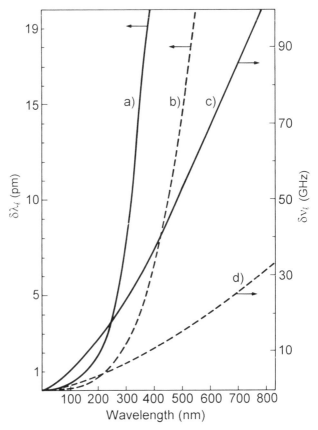

Figure 6.1. Wavelength dependence of the limiting lower values of the laser bandwidth necessary to justify the use of the rate equations approach when a broadband laser is used as an excitation source in fluorescence and ionization experiments. Laser irradiance for cases *a* and *c* is $100 \, KW/cm^2$ and cases *b* and *d* is $10 \, KW/cm^2$.

frequency v is given in terms of *photon flux*, or photons per unit area per unit time (P_l/hvS_l); *photon fluence*, or photons per unit area integrated over the entire laser pulse (Q_l/hvS_l), is also used.

Equation (1) is shown in Fig. 6.1, where both δv_l and $\delta \lambda_l$ are plotted as a function of wavelength for a strong atomic transition ($A = 10^8 \, s^{-1}$) and for two values of laser irradiance, i.e., 10 and $100 \, kW/cm^2$. These values are typical of many analytical and diagnostic experiments with pulsed lasers. Figure 6.1 will be useful in the discussion of selected experiments presented in Section 6.5.

For a monochromatic laser and a two-level system of equal statistical weights, the calculation of the Rabi frequency on resonance, when the irradiance is $100 \, kW/cm^2$, results in values ranging from $4 \, GHz$ ($0.53 \, pm$) at $200 \, nm$ to $32 \, GHz$ ($68.3 \, pm$) at $800 \, nm$, if the same strong ($A = 10^8 \, s^{-1}$) transition is considered. These values would also correspond to the energy splitting of the levels (ac Stark shift or Autler–Townes doublet). The dephasing collisional rate, in addition to the spontaneous emission rate, would then have to be on the order of 10–$100 \, GHz$ for the rate equation approach to retain its validity.

The abundant variety of mathematical software available nowadays with personal computers would allow the numerical solution of a many-level system of rate equations directly, while providing the temporal behavior of the population of each level considered. However, as pointed out earlier, the advantage resulting from the simple two-level model is that one can immediately see the general behavior of the interaction from the analytical expression obtained, which shows the role played by the different parameters. In addition, a multilevel atomic scheme can often be reduced to a simple three-level scheme if saturation occurs between the levels coupled by the laser radiation. Therefore, the relations presented here also have a tutorial value. Figure 6.2 shows a general three-level scheme and four variants, which include the ionization continuum. The analytical solution of the general case (Fig. 6.2a) is given in Table 6.2, which also gives the definitions of the various parameters involved. These equations can be considerably simplified when applied to a particular case in which the rates between the various levels are either zero or negligible compared to other rates in the expressions. In Fig. 6.2, the levels are given in increasing order of energy, level 1 being the ground state. Intermediate level 2 can be metastable (and therefore not radiatively coupled with level 1) or part of an excited multiplet (and therefore not radiatively coupled with level 3). Two lasers can simultaneously be tuned at v_{12} and v_{23} (in which case $R_{13} = 0$), and so on.

Most cases considered in this chapter can be represented by the schemes shown in Fig. 6.2b–e. It can easily be seen that all these cases are amenable to the general three-level scheme (Fig. 6.2a). In fact, the ionization continuum and the metastable level are considered as traps in which the ionic and atomic population will accumulate during the short (10–20 ns) interaction time with the laser excitation. In saturation, the laser will keep the population of the levels connected by radiation locked (in the ratio of their respective degeneracies) for the entire interaction. This locked population will then decay away toward the ionization continuum and/or the metastable level, depending upon the magnitude of the pertinent rate coefficients. The four- and five-level schemes of Fig. 6.2c–e can thus be simplified by considering levels 1–2, 2–3, and/or 1–3 as single levels for the entire duration of the saturating laser pulse.

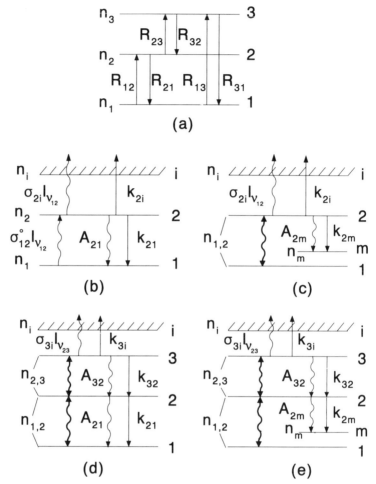

Figure 6.2. Excitation–ionization schemes pertinent to several simultaneous LEI and LIF experiments: (a) general three-level system, where the expressions for $n_2(t)$ and $n_3(t)$ are given in Table 6.2; (b) single-step excitation, followed by collisional (k_{2i}) and/or photoionization ($\sigma_{2i}I_{v_{12}}$); (c) same as in scheme (b), with saturation of the transition and the presence of a metastable level m; (d) saturated two-step excitation followed by collisional (k_{3i}) and/or photoionization ($\sigma_{2i}I_{v_{23}}$); (e) same as in scheme (d), in the presence of a metastable level. In saturation, the locked population of levels j and k is indicated as $n_{j,k}$. The ionization continuum is represented by level i. Downward collisional (k_{21}, k_{32}, k_{2m}) and spontaneous radiative (A_{21}, A_{2m}, A_{32}) rates are also shown. The thicker double-arrows in schemes (c)–(e) are meant to indicate saturation.

Table 6.2. Temporal Dependence of the Population of a General Three-Level System in a Collisional and Radiative Environment[a]

$$n_3(t) = D^{-1}\left\{ n_3^{ss} E\left[1 + \beta\exp(-\alpha_2 t) - \gamma\exp(-\alpha_3 t) + \frac{\beta\alpha_2}{E}(\exp(-\alpha_3 t) - \exp(-\alpha_2 t)) \right] \right.$$

$$+ \frac{R_{12} n_T E}{\alpha_2 - \alpha_3}\left[\frac{\alpha_2(\exp(-\alpha_2 t) - 1)}{E} + \frac{\alpha_3(1 - \exp(-\alpha_3 t))}{E} \right.$$

$$\left.\left. + \exp(-\alpha_3 t) - \exp(-\alpha_2 t) \right] \right\}$$

$$n_2(t) = n_2^{ss}\left\{ 1 + \beta\exp(-\alpha_2 t) - \gamma\exp(-\alpha_3 t) + \frac{R_{12} n_T}{n_2^{ss}(\alpha_2 - \alpha_3)}(\exp(-\alpha_3 t) - \exp(-\alpha_2 t)) \right\}$$

$$D \equiv R_{32} - R_{12} \qquad n_3^{ss} = C^{-1} n_T[R_{12}R_{23} + R_{13}(R_{21} + R_{23})]$$

$$E \equiv R_{12} + R_{21} + R_{23} \qquad n_2^{ss} = C^{-1} n_T[R_{12}(R_{31} + R_{32}) + R_{13}R_{32}]$$

$$\beta \equiv \frac{\alpha_3}{\alpha_2 - \alpha_3} \qquad X \equiv R_{12} + R_{21} + R_{13} + R_{31} + R_{23} + R_{32}$$

$$\gamma \equiv \frac{\alpha_2}{\alpha_2 - \alpha_3} \qquad C \equiv (R_{31} + R_{32})(R_{12} + R_{21}) + R_{13}(R_{21} + R_{23} + R_{32})$$

$$+ R_{23}(R_{12} + R_{31})$$

$$\alpha_2, \alpha_3 = \tfrac{1}{2}[X \mp (X^2 - 4C)^{1/2}]$$

[a] Here R_{jk} represents the total (radiative plus collisional) rate coefficient (s^{-1}) coupling levels j and k; the total atomic population $(n_1 + n_2 + n_3)$ is indicated by n_T (cm^3); n_2^{ss} and n_3^{ss} are the steady-state values obtained when the duration of the unit step excitation function approaches infinity. For $t = 0$, both $n_2(t)$ and $n_3(t)$ are zero.

This procedure, as well as the general solution given in Table 6.2, can be found in several papers dealing with fluorescence and ionization experiments (e.g., see refs. 2 and 10–13, and older references cited·therein, as well as Chapter 1 of this book).

The theoretical expressions derived for cases (b) through (e) in Fig. 6.2 are given below.

Case (b):

$$[Y_i]_{\text{lin}} = 1 - \exp\left[-\sigma^\circ_{12} I_{v_{12}} \left(\frac{R_{2i}}{R_{2i} + R_{21}} \right) T_p \right] \tag{2}$$

$$[Y_i]_{\text{sat}} = 1 - \exp\left[-\left(\frac{g_2}{g_1 + g_2} \right) R_{2i} T_p \right] \tag{3}$$

$$[I_F(t)]_{\text{lin}} = \text{Cn}_T \frac{\sigma^\circ_{12} I_{v_{12}}}{R_{2i} + R_{21}} \exp\left[-\frac{\sigma^\circ_{12} I_{v_{12}} R_{2i}}{R_{2i} + R_{21}} t \right] \tag{4}$$

$$[I_F(t)]_{\text{sat}} = \text{Cn}_T \left(\frac{g_2}{g_1 + g_2} \right) \exp\left[-\left(\frac{g_2}{g_1 + g_2} \right) R_{2i} t \right] \tag{5}$$

$$[Q_F]_{\text{lin}} = \text{Cn}_T \left(\frac{1}{R_{2i}} \right) \left\{ 1 - \exp\left[-\frac{\sigma^\circ_{12} I_{v_{12}} R_{2i}}{R_{2i} + R_{21}} T_p \right] \right\} \tag{6}$$

$$[Q_F]_{\text{sar}} = \text{Cn}_T \left(\frac{1}{R_{2i}} \right) \left\{ 1 - \exp\left[-\left(\frac{g_2}{g_1 + g_2} \right) R_{2i} T_p \right] \right\} \tag{7}$$

Case (c):

$$[Y_i]_{\text{sat}} = \frac{R_{2i}}{R_{2i} + R_{2m}} \left\{ 1 - \exp\left[-\left(\frac{g_2}{g_1 + g_2} \right) (R_{2i} + R_{2m}) T_p \right] \right\} \tag{8}$$

$$[I_F(t)]_{\text{sat}} = \text{Cn}_T \left(\frac{g_2}{g_1 + g_2} \right) \exp\left[-\left(\frac{g_2}{g_1 + g_2} \right) (R_{2i} + R_{2m}) t \right] \tag{9}$$

$$[Q_F]_{\text{sat}} = \text{Cn}_T \left(\frac{1}{R_{2i} + R_{2m}} \right) \left\{ 1 - \exp\left[-\left(\frac{g_2}{g_1 + g_2} \right) (R_{2i} + R_{2m}) T_p \right] \right\} \tag{10}$$

Case (d):

$$[Y_i]_{\text{sat}} = 1 - \exp\left[-\left(\frac{g_3}{g_1 + g_2 + g_3} \right) R_{3i} T_p \right] \tag{11}$$

$$[I_F(t)]_{\text{sat}} = \text{Cn}_T \left(\frac{g_2}{g_1 + g_2 + g_3} \right) \exp\left[-\left(\frac{g_2}{g_1 + g_2 + g_3} \right) R_{3i} t \right] \tag{12}$$

$$[Q_F]_{\text{sat}} = \text{Cn}_T \left(\frac{1}{R_{3i}} \right) \left\{ 1 - \exp\left[-\left(\frac{g_2}{g_1 + g_2 + g_3} \right) R_{3i} T_p \right] \right\} \tag{13}$$

Case (e):

$$[Y_i]_{sat} = \frac{g_3 R_{3i}}{g_3 R_{3i} + g_2 R_{2m}} \{1 - \exp[-(g_3 R_{3i} + g_2 R_{2m})T_p]\} \qquad (14)$$

$$[I_F(t)]_{sat} = Cn_T \left(\frac{g_2}{g_1 + g_2 + g_3}\right) \exp\left[-\left(\frac{g_3 R_{3i} + g_2 R_{2m}}{g_1 + g_2 + g_3}\right)t\right] \qquad (15)$$

$$[Q_F]_{sat} = Cn_T \left(\frac{g_2}{g_3 R_{3i} + g_2 R_{2m}}\right) \{1 - \exp[-(g_3 R_{3i} + g_2 R_{2m})T_p]\} \qquad (16)$$

In all these expressions, $g_j (j = 1, 2, 3)$ is the statistical weight of level j; Y_j represents the ion yield (n_i/n_T); T_p, the laser probing time; I_F, the spectrally integrated fluorescence power; Q_F, the fluorescence energy $[\int I_F(t)\,dt]$. The fluorescence is monitored at v_{21}, and $C \equiv A_{21} hv_{21} V_F$, where A_{21} (s^{-1}) is the Einstein coefficient for the transition $2 \rightarrow 1$; hv_{21} is the photon energy (J) and V_F (cm^3) is the fluorescence volume seen by the detector. The subscripts "lin" and "sat" signify that the expressions in Table 6.2 have been simplified for *linear* and *saturated* interaction. In the former case quenching collisions and spontaneous emission dominate the laser-induced transition rates, whereas the contrary holds in saturation. The radiative absorption rate [in the linear case (b)] and the photoionization rate are given by the product of the cross section (cm^2) and the laser photon irradiance, I_v (s^{-1} cm^{-2}). For broadband excitation, these rates can be estimated by (14)

$$B_{12}\rho(v) = \sigma_{12}^{\circ} I_{v_{12}} \frac{\delta v_{eff}}{\delta v_l} \qquad (17a)$$

where

$$I_{v_{12}} \equiv \int_{\delta v_l} I(v)\,dv \qquad (17b)$$

$$\delta v_{eff} \equiv \frac{\int \sigma_{12}(v)\,dv}{\sigma_{12}^{\circ}} \qquad (17c)$$

where σ_{12}° is the absorption cross section at the center frequency of the transition, and δv_{eff} is the effective width of the adsorption line. The other rates, R's, have the following definitions:

$$R_{2i} \equiv k_{2i} + \sigma_{2i} I_{v_{12}} \tag{18a}$$

$$R_{21} \equiv A_{21} + k_{21} \tag{18b}$$

$$R_{2m} \equiv A_{2m} + k_{2m} \tag{18c}$$

$$R_{3i} \equiv k_{3i} + \sigma_{3i} I_{v_{12}} \tag{18d}$$

These expressions have been intentionally selected to form the basis of the discussion of the experiments described later. However, a few remarks can already be made here:

1. All equations are obtained by simplifying the general expressions for $n_2(t)$ and $n_3(t)$ given in Table 6.2. The immediate consequence of considering the ionization continuum and the metastable level as traps is that steady-state values will not be reached during the excitation pulse T_p. Also, when saturation is achieved, it is tacitly assumed that the relaxation time for setting up a balance in the populations of the levels connected by the laser radiation (sometimes referred to as the "pumping time") is much shorter than the laser pulse duration and the other decay rates out of the pumped level. This explains why $I_F(t)$, which is always proportional to $n_2(t)$, is different from zero for $t = 0$ but reaches its maximum value under saturation.

2. All expression are limiting cases, i.e., valid for linear interaction or saturation. The dependence of the ion yield and the time-integrated fluorescence upon the laser photon irradiance will be discussed in Section 6.5.1.

3. The decay of the fluorescence power and the decrease in the fluorescence energy when the processes of collisional ionization and/or photoionization become important are clearly evident in the foregoing expressions. Fluorescence dip spectroscopy will be discussed in Section 6.5 in connection with the potential diagnostic value of time resolving the fluorescence and the ionization waveforms during the laser pulse.

4. In Eq. (8), the ratio outside the braces stresses the negative influence of a metastable level on the efficiency of ionization when the system is used as a resonance ionization detector. In fact, if R_{2m} is greater than R_{2i}, the ion yield will be low, irrespective of the product $R_{2i} T_p$, in clear contrast with the yield given by Eq. (3). This matter will be discussed in Section 6.4.2.

5. Finally, the analytical use of atomic fluorescence under saturation conditions, especially when an absolute calibration of the signal is

attempted, will benefit from the knowledge of the fraction of excited atoms that are ionized during the excitation pulse (see Section 6.4.1).

6.3. GENERAL EXPERIMENTAL CONSIDERATIONS

Most analytical and diagnostic studies described in this chapter are performed with two (or even three) laser beams, whose spatial and temporal overlap in the interaction volume can vary from one experiment to another, depending upon the particular outcome sought. A general setup, simple and yet versatile, that allows fluorescence and ionization measurements in a flame is shown in Fig. 6.3. Here, both LEI and photoionization studies can be performed while monitoring the modulation of the fluorescence signal resulting from the ionization process. The two dye lasers are pumped by the same excimer laser, which limits a practical delay in a pump-and-probe experiment to several tens of nanoseconds. Both dye lasers can be scanned in wavelength throughout the UV–visible–near–IR region of the spectrum, thus allowing the choice of several double-resonance excitation schemes on many atomic species present in an air/acetylene flame. In Fig. 6.3, a common ionization detection approach

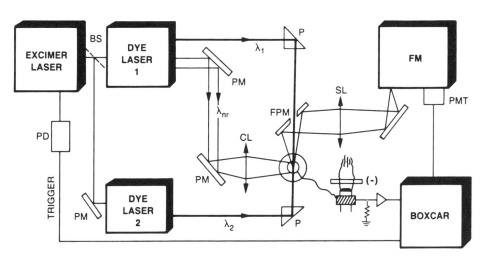

Figure 6.3. Typical experimental setup for simultaneous fluorescence and ionization measurements in flames. Two dye lasers assure tunability at the atomic transitions λ_1 and λ_2, while part of the excimer beam pumping the amplifier of the first dye laser is diverted and directed into the flame to cause direct nonresonant (λ_{nr}) photoionization. The fluorescence radiation is collected longitudinally with a pierced plane mirror. *Key:* PD, photodiode; BS, beam splitter; PM, plane mirror; P, prism; CL, cylindrical lens; FPM, fluorescence (pierced) plane mirror; SL, spherical lens; FM, fluorescence monochromator; PMT, photomultiplier tube. Reproduced from Omenetto (71) by permission of IOP Publishing Ltd., Bristol, United Kingdom.

utilizing an electrode "immersed" in the flame (15) is shown. The fluorescence is collected longitudinally with another commonly used detection system, i.e., a plane mirror having a hole in its center to allow the passage of the laser beams. This scheme is especially useful in laser-excited fluorescence with graphite furnace atomization (16).

The assumptions usually made in the modeling of the interaction between the laser light and the atoms concerning the temporal and spatial characteristics of the beam have been collected in Table 6.1. Some more detailed considerations can now be discussed.

   a. Undoubtedly, the laser beam cannot be considered as spatially uniform, with the obvious consequence that the irradiance over the sampled volume will be different at different points. This will lead to different probabilities of interaction between the atoms and the radiation if this interaction is linear. Fortunately, in the cases where the fluorescence is highly saturated, such as in an argon plasma, the spatial variations of the irradiance do not influence the signal. Nevertheless, some spatial filtering is always desirable, since this will greatly improve the beam uniformity by reducing hot spots and inhomogeneities in the output. An interesting experimental approach to circumvent the aforementioned problem, suggested recently in the literature (17), consists of forcing the laser beam to assume a "speckle distribution" by sending it, for example, through a random phase plate. The rationale behind this idea is that the laser beam is already somewhat speckled due to the index inhomogeneities in the dye and in the optical steering elements used, and that the statistics of the speckle are well known. In other words, the most natural fluence probability density function of the laser output is that for speckle, and therefore the signal at a given fluence can be multiplied by this distribution function and then integrated over all fluences (17).

   b. In connection with the above point, care must be exercized when the laser beams need to be focused in the interaction volume in order to increase the irradiance. Indeed, a large part of the observed signal (ionization and also fluorescence) can be attributed to scattered laser light that lies totally outside of the interaction volume (18; see also Chapter 1 of this book). Such an experimental artifact, which is especially severe when the optical transition investigated is strongly saturated, significantly complicates the interpretation of the experiment.

   c. The bandwidth of the dye laser in several of the experiments described in the following sections is much larger than a typical absorption line width of atomic species in atmospheric pressure flames and plasmas. For example, a bandwidth of 44 GHz at 400 nm (corresponding to $1.5\,cm^{-1}$ or to 24 pm) is typical of an excimer-pumped dye laser with a special eight-cell configuration (19), used in many of the experiments described later. For a cavity length of

18 cm, this would correspond to 53 longitudinal modes, of which 9 would be approximately uniformly distributed underneath an absorption profile of about 4 pm. When a Fabry–Perot etalon is inserted in the oscillator cavity, a considerable reduction (up to 10-fold) in the bandwidth occurs: this brings the laser profile down to the same width of the absorption profile, thus increasing the importance of effects such as mode instabilities and frequency hopping. Other dye laser configurations are already characterized by a spectral bandwidth similar to that of the atomic profile, and such widths can be made even narrower by the use of an intracavity etalon. It is therefore essential to characterize the spectral output of the laser in order to correctly calculate the atomic excitation rates. Moreover, and more important, a detailed knowledge of the spectral fluctuations is needed in theoretically modeling the effects of noise in nonlinear spectroscopic measurements (20).

   d. The temporal profile of a pulsed dye laser is far from the idealized rectangular pulse assumed in the solution of the equations (see Section 6.2 and Table 6.2). In most instances, a clearly structured output is observed, with periodic oscillations of considerable amplitude. This could be ascribed to either mode beating (21) or, as in our case, to transient effects in the early development of the lasing action (22). A striking example of such behavior is shown in Fig. 6.4, which shows a time-resolved pulse profile for a frequency-doubled, excimer-pumped pulsed dye laser. As in the discussion concerning item *a*, above, if the transition is optically saturated, the fluorescence waveform will not be affected by such oscillations. On the other hand, the *rise time* of the laser pulse should be very fast, so that one can take full advantage of resolving the time behavior of the fluorescence within the excitation pulse. It is also essential to note that the time profile of the laser pulse changes considerably when the gain of the lasing medium changes, e.g., when the dye output is spectrally scanned from the region of maximum efficiency to both higher and lower wavelength limits.

   e. Finally, in connection with the last item above, it is essential to check the time coincidence of the laser pulses used in double-resonance experiments for each wavelength setting chosen. In particular, in experiments where one laser is fixed at one transition and a second laser is scanned, the time coincidence between the two pulses can easily be lost during the scan.

   As concluding general remark, we even should emphasize that, even though it increases the complexity of the experimental setup, it should become common practice to characterize both the spatial and temporal distributions of the irradiance of the laser beams during the experiment. This can be done with the use of a suitable CCD (charge-coupled device) camera for the spatial profile and a fast microchannel plate photomultiplier for the temporal profile. Such a setup has the significant advantage that one can immediately visualize

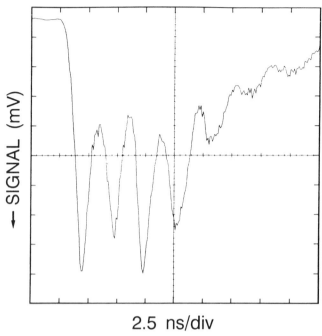

2.5 ns/div

Figure 6.4. Time-resolved profile of the frequency-doubled output of an excimer- pumped dye laser. The excimer laser was a model LPX-200 (Lambda Physik, Göttingen, Germany), and the dye laser was a special eight-cuvette system (Jobin Yvon, Longjumeau, France) (19). A solution of coumarin 540 in methanol and a KDP (potassium dihydrogen phosphate) crystal were used. The laser beam was diffused into a monochromator, set at 276 nm, and monitored with a fast microchannel plate photomultiplier (Model R1546U, Hamamatsu, Hamamatsu City, Japan) whose output was fed into a fast digitizer (Model D602A, Tektronix, Beaverton, Oregon). The waveform shown is the result of 128 averages. From N. Omenetto, unpublished results (1993).

the effect obtained on the output characteristics of the laser when a given parameter in the oscillator cavity or in the amplifier is changed.

6.4. ANALYTICAL STUDIES

6.4.1. Evaluation of the Total Number Density by Means of Saturated Fluorescence

The possibility of evaluating the number density of the emitting species in a combustion system by a saturated fluorescence measurement was realized in several early studies of LIF in flames (14, 23). Even though the initial excitement generated by such experiments was later mitigated by a more detailed theoretical and experimental understanding of the parameters involved in the process, it is clear that, if the conditions of Table 6.1 are met, an absolute measurement of the fluorescence signal emitted by a two-level atom under optically saturated conditions allows the direct evaluation of the number density if the transition probability of spontaneous emission is known. This can be seen from an inspection of Eq. (7), in conditions of *negligible ionization losses* from the excited level. This last requirement, which was not taken into account in many early papers, needs to be considered in particular in cases of strong UV transitions or high laser irradiances. In fact, in the former case the decrease in the "energy defect," i.e., in the energy difference between the level reached by the laser and the ionization continuum, enhances the occurrence of collisional ionization, whereas in the latter case the simultaneous absorption of a second photon from the same laser can result in direct photoionization from the excited level.

The loss of excited atoms due to the aforementioned ionization processes can be quantified with the help of Table 6.2 and Fig. 6.2. In the case of single-step excitation by one laser tuned at v_{12} [case (b) in Fig. 6.2], the general expressions given in Table 6.2 can be manipulated and simplified. In this particular example, level 3 becomes level i (the ionization continuum), and since no steady state will be reached during the laser pulse because ion recombination processes or fast ion chemistry are not considered, C_0 as well as R_{31} and R_{32} will be equal to zero. In addition, R_{13} is also neglected. In this case, the time evolution of the excited atom density, $n_2(t)$, and of the ion density, $n_i(t)$, will be given by the following expressions (where $n_T = n_1 + n_2 + n_i$):

$$\frac{n_2(t)}{n_T} = \frac{\alpha_2 \alpha_3 [\exp(-\alpha_3 t) - \exp(-\alpha_2 t)]}{R_{2i}(\alpha_2 - \alpha_3)} \tag{19}$$

$$\frac{n_i(t)}{n_T} = \frac{\alpha_2 [1 - \exp(-\alpha_3 t)] - \alpha_3 [1 - \exp(-\alpha_2 t)]}{(\alpha_2 - \alpha_3)} \tag{20}$$

Note that Eq. (19), if saturation is achieved, simply reverts to Eq. (5). In Fig. 6.5 the ratio $R(t)$, defined for $t > 0$, between Eqs. (19) and (20), is plotted for several ratios of the parameter R_{2i} (which includes both collisional ionization and photoionization) to an assumed spontaneous emission rate of $A_{21} = 10^8 \, \text{s}^{-1}$. From the curves shown, one can conclude that, for short ($\sim 10 \, \text{ns}$) excitation pulses, ion losses will be negligible whenever R_{2i} is 100-fold smaller than the radiative (and quenching) rates; this is expected to hold in flames for transitions where the energy defect is greater than $\sim 3 \, \text{eV}$. In these cases, the peak of the time-resolved, saturated fluorescence signal will accurately reflect the total number of atoms (12).

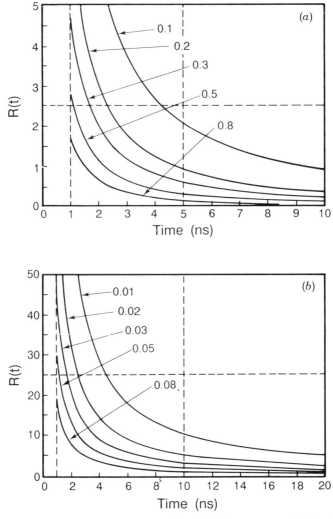

Figure 6.5. Calculated time behavior of the ratio of the fluorescence and ionization signals (proportional to the excited atom density and to the ion density) for several different values of the ionization rate coefficient, R_{2i}. These values are normalized to a radiative rate of $A_{21} = 10^8\,\mathrm{s}^{-1}$. Calculations start from $t = 1$ ns: (a) $R_{2i}/A_{21} = 0.8$, 0.5, 0.3, 0.2, and 0.1; (b) $R_{2i}/A_{21} = 0.08$, 0.05, 0.03, 0.02, and 0.01.

It must be emphasized, however, that the foregoing conclusions are not valid for the fluorescence signals resulting from double-resonance or two-step excitation processes (24). Here, the rate coefficients for collisional ionization can be of the same order of magnitude, if not larger, than the radiative relaxation rates. Any attempt to derive the total number density from a saturated double-resonance fluorescence experiment cannot leave out of consideration a simultaneous evaluation of the ion yield pertinent to the excitation process. This is especially important in two-step excitation, UV fluorescence experiments in a graphite furnace, where a highly excited atomic level (or manifold) is reached. Here, the aim is to monitor the fluorescence radiation in the low UV region, where furnace emission noise would be negligible (25, 26). The knowledge of the interaction process between the laser radiation and the atomic system is essential if an attempt is made to evaluate the fluorescence technique in a graphite furnace as an "absolute" method of analysis (27).

6.4.2. Resonance Ionization Detection of Photons

The concept of "resonance detection" or a "resonance monochromator" has been familiar to atomic spectroscopists for a long time. In one of its earliest versions, the atomic vapor generated in a sputtering device was selectively excited by a hollow cathode lamp and the resulting fluorescence measured, i.e., the device acted as a very-high-resolution monochromator (28). Since then, this concept has been repeatedly applied, with many variants [for a review, see Matveev (29)]. To the present authors' knowledge, the detection of the fluorescence radiation with a multistep excitation/ionization scheme involving the fluorescence photons as one step was first proposed by Matveev et al. (30). The idea was subsequently discussed and extended to different forms of radiation, including Raman, and to different atom reservoirs, from flames to discharge cells, by other authors as well with widely different applications (31–39).

The principle of operation of the ionization detector is schematized in Fig. 6.6, which refers specifically to the detection of fluorescence and Raman photons by means of a two-step excitation, collisional ionization scheme (e.g., magnesium atoms in an air/acetylene flame). In ideal conditions, no signal is observed in the detector when only λ_{23} is present; this is due to the negligible population of level 2 in the absence of the primary absorption step. The important difference between the two schemes shown is as follows: in the fluorescence case, the detector needs to be filled with pure vapor of the same analyte emitting the fluorescence signal, so one detector must be devised for each element of interest; on the other hand; in the Raman case, only one detector is needed for different sample solutions, since here the laser excitation

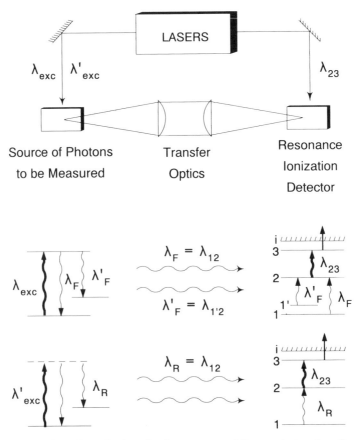

Figure 6.6. Simplified scheme for detecting fluorescence and Raman photons by a flame resonance ionization detector. In the detector, always illuminated by λ_{23}, the first excitation step to level 2 is accomplished by the absorption of either resonance and direct-line fluorescence or Raman scattering.

wavelength can be scanned until the scattered photons appear at the ground-state absorption transition ($\lambda_R = \lambda_{12}$). In both approaches, a large signal will be observed when the laser excitation wavelength (λ_{exc} or λ'_{exc}) will be resonant with λ_{12}, i.e., in the presence of spurious scattering in the fluorescence sample and of Rayleigh scattering in the Raman sample. As indicated in Fig. 6.6, scattering in the fluorescence detection can be overcome by taking advantage of a direct-line fluorescence transition, which occurs at a wavelength, $\lambda_{1'2}$, different from λ_{exc}, which in this case is filtered out. This scheme assumes that level 1' is significantly populated by thermal collisions in the flame.

As an example of application of the flame ionization detector, Fig. 6.7 shows three Raman spectra [(a) carbon tetrachloride; (b) dimethyl sulfoxide; (c) chloroform] detected in a miniature air/acetylene flame fed with a magnesium solution. The laser illuminating the sample solution was tuned until the Raman photons emitted and transferred into the flame matched the resonance absorption of magnesium atoms at 285.213 nm. Another laser, tuned at

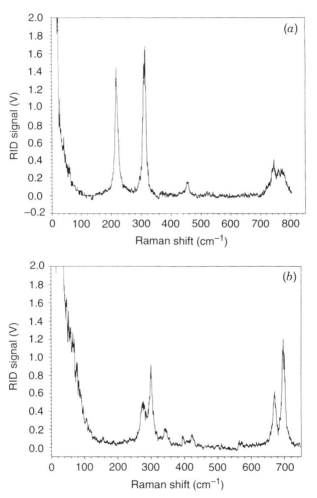

Figure 6.7. [pp. 285–286] Raman spectra of (a) carbon tetrachloride, (b) dimethyl sulfoxide, and (c) chloroform obtained with the magnesium flame resonance ionization detector (RID). The spectral response and the rejection of Rayleigh scattering is dictated by the Lorentzian absorption profile in the flame. Reproduced from Petrucci et al. (38) by permission of the Royal Society of Chemistry.

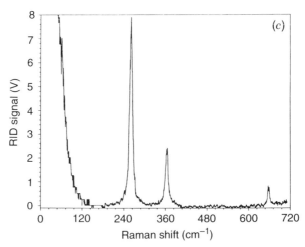

Figure 6.7. (*Contd.*)

435.191 nm, provided the second excitation step, which was followed by fast collisional ionization (38). The minimum detectable number of photons was experimentally determined to be 1×10^3 (7×10^{-16} J).

It is interesting to note that, while Fig. 6.6 shows two *independent* systems for the generation and detection of the fluorescence and Raman photons, in principle these processes can occur in a *single* atomizer because of a fortuitous spectral coincidence of radiation emitted by flame radicals or gases with λ_{12}. For example, an interesting way of detecting OH species in a capillary hydrocarbon flame was reported in 1994 by Petrucci et al. (40) in which a rotational fluorescence transition of the native OH radical was coincident with a resonance absorption line of indium atoms, simultaneously present in the flame and further excited by another laser beam to a level easily depopulated by ionizing collisions. Fluorescence could also be partially responsible for the curious anomalies observed in some ionization experiments with focused laser beams, even though scattering effects outside the laser volume offer a more plausible explanation (18, 41; see also Chapter 1, Section 1.8, in this volume).

From the analytical point of view, the detection limits achievable when the fluorescence photons are detected with an ionization scheme have been calculated for a two-flame system (30, 37) and found to be in the picograms per milliliter range. It was also anticipated that absolute amounts close to attograms would be approached if a graphite furnace is used for atomizing the sample. The limiting noise considered in these calculations was the statistical shot-noise fluctuations in the flame background ionization current (42). However, scattering of laser photons at λ_{12} constitutes by far the dominant noise source in the overall detection system. In the presence of scattering, this approach of measuring the fluorescence radiation by an ionization detector may well lose its attractiveness.

An important application foreseen with the resonance ionization detector is its operation as a suitable *atomic line filter*, such as those amply discussed in the literature in the context of atmospheric lidar applications (43, 44), and based upon the detection of fluorescence radiation with a photomultiplier. The ionization detector would offer the advantage of extremely fast response time and an inherently higher overall quantum efficiency than fluorescence-based filters (33, 35).

The concept of *quantum efficiency* is essential for this detector and should be clearly distinguished from that of *ion yield* (36, 45). Both definitions and their mutual relationship are shown in Fig. 6.8 and 6.9 (see also Section 6.5.2). These figures are mostly self-explanatory, as the level schemes reported strictly follow those of Fig. 6.2b, c, together with the expressions for Y_i, which hold in conditions of linear interaction for R_{12} and with the substitution of Δt_i (the interaction time) for T_p. A simple inspection of these figures enables us to summarize the following considerations (36).

1. If the ionization scheme shown is used an as analytical tool, the ion yield should be unity irrespective of the number of primary photons absorbed and needed to form an ion pair. For example, a near-unity ionization yield for sodium in a continuous-wave (cw) laser excitation experiment was obtained, despite the fact that about 10^4 excitation and quenching collisions occurred before an ionizing collision produced an ion pair (45). On the other hand, if the above scheme is used as a photon detector, its quantum efficiency should be unity, i.e., an ion pair must be created for every primary photon absorbed;

2. In the short timescale of the interaction (10–20 ns), the presence of an additional trap for the excited atoms, such as a long-lived metastable level, will have the effect of deteriorating both the ion yield and the quantum efficiency. The quantum efficiency can never exceed the *ion-branching ratio*, which is defined as the relative proportion of atoms that are lost in the ionization process with respect to the total number of atoms lost in the two traps.

Several remarks (derived from the preceding discussion) concerning the essential requisites of an ideal ionization detector are collected in Table 6.3.

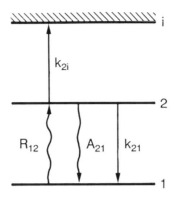

Ionization Quantum Efficiency

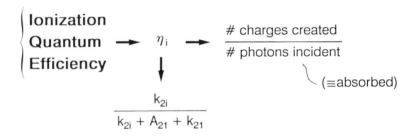

Ion yield

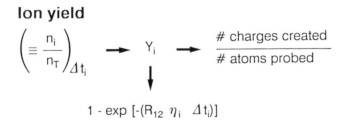

➤ Y_i can approach unity even if η_i is (much) less than unity

Figure 6.8. Definition of quantum efficiency, η_i, and ion yield, Y_i, for the level scheme shown, which is similar to that of Fig. 6.2b in the absence of photoionization: k_{2i} is the effective collisional ionization rate coefficient, while k_{21} and A_{21} are the quenching and spontaneous emission rate coefficients, respectively. The equation for Y_i shown is derived from the general expressions reported in Table 6.2 [see the discussion in the text and Eq. (3)].

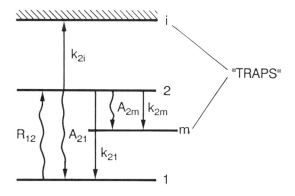

$$\eta_i = \frac{k_{2i}}{k_{2i} + A_{21} + k_{21} + A_{2m} + k_{2m}}$$

$$\xi_i = \frac{k_{2i}}{k_{2i} + A_{2m} + k_{2m}}$$

$$\eta_i = \frac{\xi_i}{1 + \xi_i \dfrac{A_{21} + k_{21}}{k_{2i}}}$$

$$Y_i = \xi_i \left[1 - \exp\left(-R_{12} \frac{\eta_i}{\xi_i} \Delta t_i\right)\right]$$

● η_i and Y_i will never exceed ξ_i

Figure 6.9. Modification of the expressions for η_i and ion yield, Y_i, as defined in Fig. 6.8, when a metastable level, m, acting as a trap for the atoms excited in level 2, is present. The parameter ξ_i is defined as the *ion-branching ratio*. The expressions indicate how this parameter limits the maximum value of η_i and Y_i achievable with the collisional ionization scheme shown.

Table 6.3. Requisites of an Ideal Photon Detector Based on LEI

- Existence of a suitable atomic excitation scheme, characterized by unity quantum efficiency and ion yield
- Absence of ionization background when only the ionizing laser is present in the detector
- Presence of a sufficient number density of atoms in the ground state or in the state originating the primary absorption step to ensure total absorption
- Good atomization efficiency and long residence time of the atoms in the interaction volume
- Large acceptance solid angle for the incoming fluorescence and Raman photons
- Availability of a method of charge detection characterized by unity collection efficiency
- Gaussian-dominated absorption profile for maximum stray light rejection in Raman experiments

6.5. SPECTROSCOPIC STUDIES

The versatility of the ionization technique when applied to the study of fundamental atomic parameters, or, as in the case of flames and plasmas, to the characterization of the combustion process itself, is well documented in some examples collected in Table 6.4. Since a similar versatility is also offered by the fluorescence technique, it seems logical to imagine that the combined use of both methods will provide an important diagnostic tool in atomic and molecular spectroscopy. As stated earlier in Section 6.1, both simultaneous and independent measurements can be devised and have indeed been exploited experimentally, as reported in Tables 6.5 and 6.6. Because of the practical impossibility of discussing all the applications here, a minimum of descriptive information has been provided for each experiment, together with the reference to the original literature, so that a reader can easily find the detailed treatment of a particular topic of interest.

In the following subsections, we discuss some concepts and experiments whose underlying theoretical principles and pertinent mathematical expressions were given in Section 6.2. The reader should also consult a few pertinent references (2, 69, 70).

6.5.1. Simultaneous Fluorescence and Ionization Measurements: Power Dependence of the Signals

The double-logarithmic plot of the fluorescence signal vs. laser power represents the well-known *saturation curve*. This curve has been used as a test of the

Table 6.4. Some Examples of Spectroscopic Studies Performed by LEI Spectrometry and Resonance Ionization Spectroscopy in Flames and Other Atom Reservoirs

Methodology and Laser Source	Spectroscopic Parameter and Information Provided	Reference
Delayed, two-step, disconnected excitation–collisional ionization; excimer-pumped pulsed dye lasers	*Lifetime of metastable states* of Au, Bi, Cd, Mg, Pb, Sr, and Tl in an air C_2H_2 flame	46, 47
Single-step LEI with measurement of ion arrival times; N_2-pumped pulsed dye laser	*Mobility of atomic ions* (Li, Na, K, Ca, Sr, Ba, Fe, In, Tl, U) C_2H_2 and CO/O_2 flames	48
Multiphoton laser-induced heating and ionization; N_2-pumped pulsed dye laser	*Mobility of very small particles* in an air C_2H_2 flame at the sooting limit	49
Temporal and spatial probing, by absorption, of the evolution of the depletion of the neutral atom sodium density caused by a cw laser	*Flow velocity* of flame gases; air/H_2 and air/C_2H_2 flames	50
Single-step LEI and photoionization; excimer-pumped pulsed dye laser	*Ionization efficiency* of excited Na and Li atoms in air/C_2H_2 flames	51
Two-step excitation, collisional and photoionization; Nd:YAG-pumped pulsed dye lasers	*Collisional ionization* yield of excited In and Pb atoms in an air/C_2H_2 flame	52
Two-step excitation, field ionization of Rydberg states, performed in an atomic beam produced by laser vaporization; excimer-pumped pulsed dye lasers	*First ionization potential* of Ru, Rh, and Pd	53
Two-step photoionization of atoms, maintained in a heat-pipe oven with helium buffer gas; Nd-glass-pumped pulsed dye lasers	*Photoionization cross section* of a Mg autoionizing transition	54
Single-step excitation, collisional ionization of gallium in an air/C_2H_2 flame; Nd:YAG-pumped pulsed dye laser	*Flame temperature*	55

(Contd.)

Table 6.4. (*Contd.*)

Methodology and Laser Source	Spectroscopic Parameter and Information Provided	Reference
Single-step excitation, nonresonant photo ionization; N_2-laser-pumped pulsed dye laser	*Ionic diffusion and mobility coefficients* of Na and Li in an $H_2/O_2/Ar$ flame; *flame temperature*	56, 57
Single-step excitation, collisional ionization; Nd:YAG-pumped pulsed dye laser	*Atomization efficiency* of Na and Li in an air/C_2H_2 flame	58
Single-step excitation, collisional ionization of Li Rydberg states; excimer-pumped pulsed dye laser	*Electrical field distribution* in an air C_2H_2 flame	59
Two-step excitation, collisional ionization of magnesium atoms in an air/acetylene flame; excimer-pumped pulsed dye lasers	Detection of *forbidden* Mg *transition* at 457.1 nm, with very low oscillator strength	60

Table 6.5. Simultaneous Laser-Induced Fluorescence and Ionization Measurements in Flames and Other Atom Reservoirs: Some Selected Studies

Methodology and Laser Source	Spectroscopic Parameter and Information Provided	Reference
Fluorescence-dip and ion-dip spectrometry of single rovibronic states of benzene; excitation–ionization in a molecular beam is provided by pulsed amplification of a single-mode cw dye laser and by an excimer-pumped dye laser	Determination of the *harmonic frequencies* of the vibrational modes, their *anharmonic constants*, and the coupling between states	61
Two-step excitation–ionization and multiphoton ionization of magnesium atoms in a heat-pipe oven; Nd:YAG-pumped pulsed dye lasers	*Single-* and *Double-ionization mechanism*	62
Resonance-enhanced multiphoton ionization (REMPI) and two-photon excited fluorescence of O atoms in a discharge plasma; Nd:YAG-pumped pulsed dye laser	Determination of *absolute number densities*	63

Table 6.5. (*Contd.*)

Methodology and Laser Source	Spectroscopic Parameter and Information Provided	Reference
REMPI and two-photon excited fluorescence of CO in a CH_4/air diffusion flame; Nd:YAG-pumped pulsed dye laser	Signal (electron or ion) *detection efficiency* as a function of flame position	64
Two-step ionization-dip and fluorescence-dip spectrometry of jet-cooled molecules; excimer-pumped, pulsed dye lasers	Spectroscopy of *highly excited* molecular states; *energetics* and *dynamics* of vibrational states of molecules in their ground electronic state and of the electronic excited state of molecular ions	65
Single-step excitation, collisional ionization, and fluorescence of Li atoms in the air/C_2H_2 flame; excimer-pumped pulsed dye lasers	*Ionization yield*	66
Two-step excitation, collisional ionization, or photoionization and resonance fluorescence dip of Mg atoms in an air/C_2H_2 flame; excimer-pumped pulsed dye lasers	*Ionization efficiency* and *Ionization yield*	36–38
Two-step photoionization and delayed ionic fluorescence of Sr ions in an air/C_2H_2 flame an in an inductively coupled argon plasma	*Kinetic behavior* of ions	67, 68
Single-step excitation, collisional ionization, and/or photoionization and resonance fluorescence in and air/C_2H_2 flame; excimer-pumped pulsed dye lasers	*Ionization mechanisms* in flames	69, 70

approach of optical saturation in the interaction between the laser radiation and the atomic system since the very beginning of the analytical atomic fluorescence technique in flames. Several experimental as well as more subtle theoretical considerations affecting the shape of these curves, which were overlooked in the past, have now been properly addressed (2). In the ideal case, the saturation curve shows two asymptotes, one linear at low laser intensities and a final horizontal asymptote at high laser intensities, showing that the

Table 6.6. Fluorescence Measurements Yielding Ionization Information in Flames, Plasmas, and Other Atom Reservoirs: Some Selected Studies

Methodology and Laser Source	Spectroscopic Parameter or Application	Reference
Resonance fluorescence dip of Mg atoms in a plasma caused by single-step excitation, non-resonant photoionization; excimer-pumped pulsed dye laser	*Photoionization cross section*	71
Resonance fluorescence dip of Mg atoms in an $O_2/Ar/H_2$ flame caused by two-step ionization; Nd:YAG-pumped pulsed dye lasers	*Autoionization cross section* at 300.9 nm	72
Two-photon excited, direct-line fluorescence dip of Cs atoms in an evacuated gas cell caused by non resonant photoionization	*Photoionization cross section* of the $7s\,D_{3/2}$ level	17
Decrease of X–UV fluorescence signal originating from a quasi metastable core-excited level of Rb in a heat pipe, caused by transfer into an autoionizing manifold; Nd:YAG-pumped pulsed dye laser	*Line widths* and *autoionizing times* of core-excited levels and transition	73, 74
Decrease of the X–UV emission of core-excited levels of Li in a hollow cathode discharge; Nd:YAG-pumped pulsed dye laser	*Autoionizing line widths*	75
Direct monitoring of the ionic fluorescence signals resulting from multiphoton absorption process in Ca and Sr vapors; Nd:YAG-pumped pulsed dye laser	*Multiphoton ionization mechanism*	97
Double resonance fluorescence-dip Rydberg spectroscopy of Ti, V, Fe, Co, and Ni in an rf glow-discharge sputtering cell; excimer-pumped pulsed dye laser	*Ionization potentials*	77

Table 6.6. (*Contd.*)

Methodology and Laser Source	Spectroscopic Parameter or Application	Reference
Two-step excitation, collisional ionization and/or photoionization in flames and vapor cells; excimer-pumped and Nd:YAG-pumped pulsed dye laser	*Resonant detection* of fluorescence and Raman photons	29–38
Two-step excitation to a Rydberg level of Rb in a vapor cell, followed by field ionization; diode laser and excimer-pumped pulsed dye laser	*Atomic resonance filters*	35
Modulation of the emission signal of Mg ion in a plasma caused by an abrupt depletion of ground-state atoms by LEI; excimer-pumped pulsed dye lasers	Mechanism of *charge transfer*	91, 98
Photoionization controlled-loss spectroscopy of H atoms in $H_2/O_2/N_2$ flames; Nd:YAG-pumped pulsed dye laser	*Quenching* of fluorescence radiation in flames	94, 95

transition has been optically saturated. The intersection between the two asymptotes, which marks the onset of saturation, corresponds to the *saturation parameter*, which depends on the intrinsic characteristics of the atom as well as on the spectral characteristics of the laser. For the simple three-level system shown in Fig. 6.2b, the occurrence of collisional ionization *modifies the fluorescence saturation curve*. This can be seen in Eqs. (7) and (8), arranged to extract explicitly the saturation parameter, called here I_v^s, i.e., the laser saturation photon irradiance $(s^{-1} cm^{-2})$. For the transition $1 \rightarrow 2$, we use the following definition:

$$I_v^s \equiv \frac{1}{[(g_1 + g_2)/g_2]\sigma_{12}^\circ \tau^*} \tag{21}$$

where

$$\tau^* \equiv \frac{1}{A_{21} + k_{21} + k_{2i}} \tag{22}$$

The resulting relations between the integrated (spectrally and temporally) fluorescence signal, Q_F, and I_v, in the absence and in the presence of collisional ionization, are given in Fig. 6.10, where for the sake of convenience the

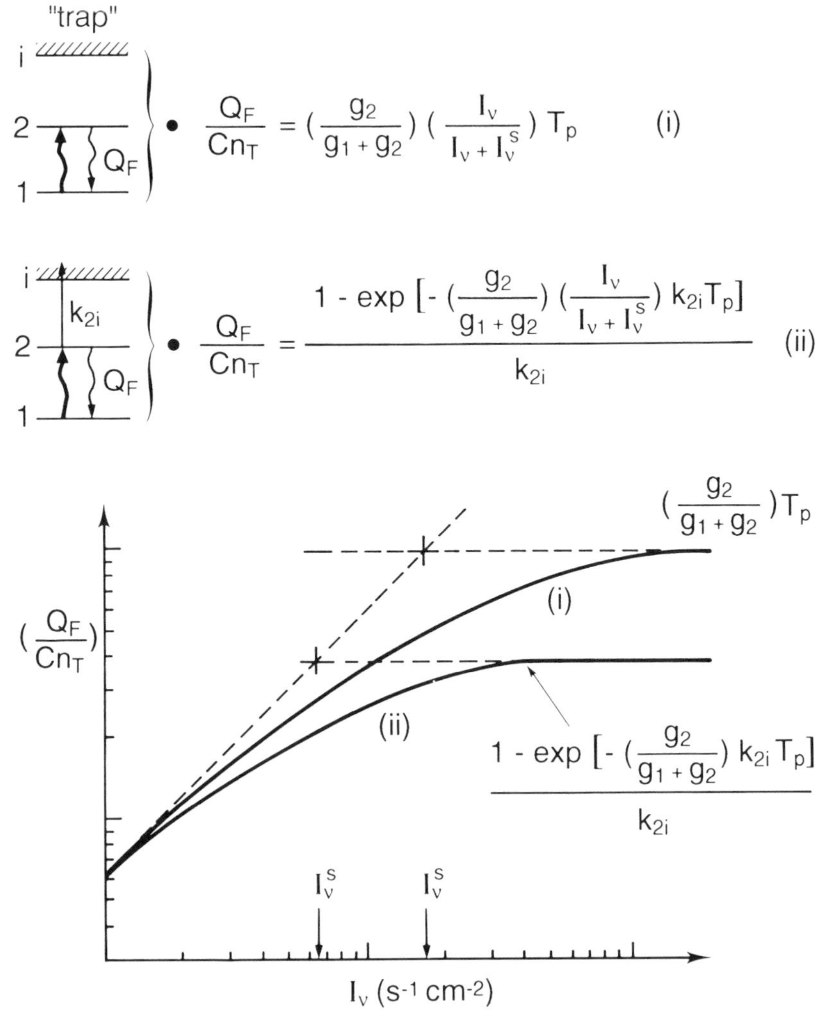
"trap"
i
2
1
Q_F
$\frac{Q_F}{Cn_T} = (\frac{g_2}{g_1 + g_2})(\frac{I_\nu}{I_\nu + I_\nu^S}) T_p$
(i)
i
k_{2i}
2
1
Q_F
$\frac{Q_F}{Cn_T} = \frac{1 - \exp[-(\frac{g_2}{g_1+g_2})(\frac{I_\nu}{I_\nu + I_\nu^S}) k_{2i} T_p]}{k_{2i}}$
(ii)
$(\frac{g_2}{g_1 + g_2}) T_p$
(i)
$(\frac{Q_F}{Cn_T})$
(ii)
$\frac{1 - \exp[-(\frac{g_2}{g_1+g_2}) k_{2i} T_p]}{k_{2i}}$
I_ν^S
I_ν^S
I_ν (s-1 cm-2)

Figure 6.10. Qualitative behavior of the fluorescence energy as a function of the laser photon irradiance at transition $1 \rightarrow 2$. Case (i) refers to the absence of collisional ionization, while case (ii) implies that atoms excited into level 2 are ionized at a rate given by the collisional rate coefficient k_{2i}. See the text for further discussion.

corresponding three-level schemes are also reported. The qualitative behavior of the curves is emphasized to illustrate the fact that the ionization process, at a given photon irradiance, *lowers* the fluorescence plateau, with a consequent similar effect on the value of the saturation parameter. The higher the value of k_{2i}, the lower will be the fluorescence plateau. If photoionization is also

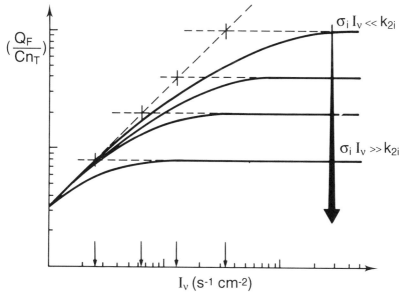

Figure 6.11. Qualitative behavior of the fluorescence energy as a function of the laser photon irradiance at the transition $1 \rightarrow 2$. Both collisional ionization and photoionization occur. The various shapes shown indicate the increasing importance of the photoionization rate over the collisional rate. The four arrows point to four different values of the saturation parameter obtained from the intersection of the asymptotes. See the text for further discussion.

possible (see Fig. 6.2b) because the photon energy at the transition $1 \rightarrow 2$ is sufficient to exceed the gap between level 3 and level i, then the photoionization rate, $\sigma_i I_{v_{12}}$, adds to the collisional rate by worsening the effect on the fluorescence plateau and the corresponding saturation parameter. This is again qualitatively illustrated for increasing values of $\sigma_i I_{v_{12}}$ compared to k_{2i} in Fig. 6.11. In a completely analogous manner, one can derive an *ionization*

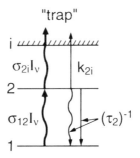

$$\frac{n_i}{n_T} = 1 - \exp\left\{-\left[\left(\frac{g_2}{g_1 + g_2}\right)\left(\frac{I_v}{I_v + I_v^s}\right)(k_{2i} + \sigma_{2i}I_v)T_p\right]\right\}$$

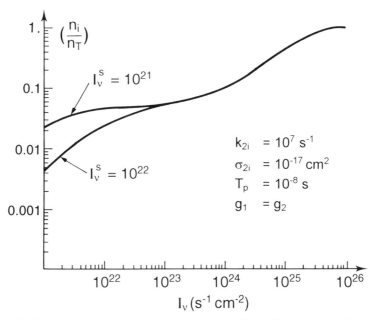

Figure 6.12. Simulated ionization saturation curve. Both collisional ionization and photoionization from level 2 occur. The curves are calculated with the numerical values shown in the illustration and for two values of the saturation parameter. The lifetime of level 2, in the absence of ionization, is indicated as τ_2. A photon irradiance of $10^{24}\,\mathrm{s}^{-1}\,\mathrm{cm}^{-2}$ corresponds to a peak irradiance of $400\,\mathrm{kW/cm^2}$ at $500\,\mathrm{nm}$.

saturation curve, i.e., the behavior of the ionization signal vs. laser power, obtained at a given value of k_{2i} from Eqs. (3) and (4). Such behavior is illustrated in Fig. 6.12, where the equation shown is plotted quantitatively for selected values of the ionization rate coefficient, pulse duration, and saturation parameters (the value chosen for k_{2i} is purposely exaggerated, to emphasize the particular shape shown). For the lower of the two values of I_v^s, a plateau is reached; this makes the ionization curve isomorphous with the integrated fluorescence curve (see the equations in Figs. 6.11 and 6.12). For higher values of I_v, however, the ionization signal increases toward its limiting plateau, where all the atoms within the laser volume have become ions. It necessarily follows that Q_F, at these high powers, decreases and eventually vanishes. It is interesting to note that, when I_v^s is much less than I_v and in absence of collisional ionization, the ionization signal grows as the *square* of the laser power. Moreover, a plateau obtained in both the ionization and fluorescence curves reflects the negligible occurrence of photoionization as compared to collisional ionization.

In conclusion, even if the ionization curve *alone* contains the information needed to distinguish between collisional and photoionization processes, the fluorescence curve obtained simultaneously from the same flame volume makes the interpretation of these mechanisms easier. On the other hand, the evaluation of the saturation parameter from the intersection between the linear and the horizontal asymptotes obtained in the fluorescence saturation curve cannot exclude the possibility of ionization losses. From the experimental point of view, it is important to be aware of and check the relevance of artifacts such as those attributed to scattered laser light outside the interaction volume (18), since they can affect the shape of both saturation curves.

As a final remark, which we will again consider in Section 6.5.4, we note that the direct evaluation of the ionization rate is in principle feasible by measuring $I_F(t)$ rather than Q_F, i.e., by time resolving the fluorescence waveform under optically saturated conditions [see Eq. (5)].

6.5.2. Optical Detection of LEI and Multiphoton Ionization

In this context, *optical detection* of ionization means that the ions are probed by their fluorescence emission rather than by applying an electric field to the atomic system and measuring the charges produced in whatever ionization process has occurred. The ion *production* aspects of one technique are therefore combined here with the ion *detection* aspects of the other technique. In principle, laser excitation can be arranged in such a way that, with two or more steps, a bound energy level of the ion can be *directly* reached from the ground state of the atom. While this possibility has never been exploited in flames, it has been successfully adopted in low-pressure atomic vapors (discussed below).

In flame and plasma work (67), the ions formed are excited with a laser tuned to an allowed ionic transition and the resulting resonance or non-resonance fluorescence is measured by a monochromator–photomultiplier combination. The obvious advantage of using one independent laser system for the ionization and another for the excitation of the ionic fluorescence is that the second can be delayed from the first, thus allowing the direct observation of the decay of the enhanced ion population back to the original level, i.e., to the thermal ion density, As pointed out in the literature (67), the optical detection of ions, although less sensitive than the usual direct electrical detection, has the advantage of being specific and free from perturbations such as nonuniformity of the electric field in the flame, which are especially severe in the presence of easily ionizable elements. Moreover, this technique addresses specifically the ions created rather than monitoring a generic atom depletion process, as was done in a somewhat similar experiment in which the population of sodium atoms in a flame was monitored by absorption (50). Finally, there may be analytical reservoirs, such as the inductively coupled argon plasma (ICP), in which a suitable means of electrical detection of ionization is difficult to develop due to arcing problems and radio frequency (rf) interference.

A versatile setup that was used to detect and follow the decay of strontium ions in an air/acetylene flame and in an argon plasma is shown in Fig. 6.13 (67, 68). An excimer-pumped dye laser, here called the *ionizing laser*, was tuned to the resonance Sr atomic transition, and nonresonant photoionization was accomplished with the excimer beam, which was partially deflected from the amplifier section of the dye and directed into the flame or plasma together with the dye output. A second excimer-pumped dye laser, here called the *fluorescence laser*, was used to probe the fluorescence of the ions created by the first laser combination. The time delay between the two lasers could be varied continuously from zero (when the ionizing and fluorescence laser were temporally coincident) to several hundred microseconds.

The decay time observed in the flame and in the ICP showed striking differences (which could have been expected) and similarities (which were unexpected). A major difference was the absence in the argon plasma of the fast decay (58 ns) observed in the flame, which consumes 85% of the laser-produced ions before reaching an equilibrium. This fast decay was therefore attributed to ion chemistry, whose mechanism was not specified because of insufficient characterization of the combustion process. On the other hand, as illustrated in Fig. 6.14a,b, in both atomizers a long decay (tens of microseconds), attributed to electron–ion recombination, was significantly affected (shortened) by the concomitant presence of an excess of easily ionizable

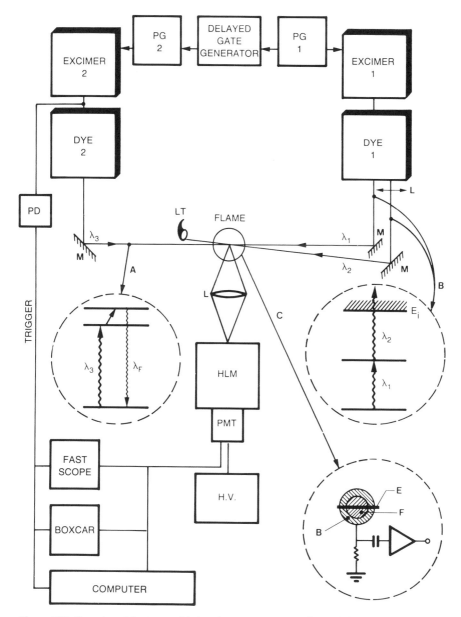

Figure 6.13. Experimental setup used in ion decay measurements in an air/acetylene flame. *Key*: PG, pulse generator; PD, photodiode; Ms: mirrors; LT; light trap; L; lens; HLM, high luminosity monochromator; PMT, photomultiplier tube; HV: high voltage. *Insert A*: excitation–detection scheme of the strontium ionic fluorescence ($\lambda_3 = 421.552$ nm; $\lambda_F = 407.771$ nm). *Insert B*: two-step ionization scheme of the strontium atom ($\lambda_1 = 460.733$ nm; $\lambda_2 = 308.2$ nm; E_i indicates the ionization continuum). *Insert C*: detection scheme of the laser-induced ionization current (B, burner; F, flame; E, water-cooled electrode). Reproduced from Turk and Omenetto (67) by permission of the Society of Applied Spectroscopy.

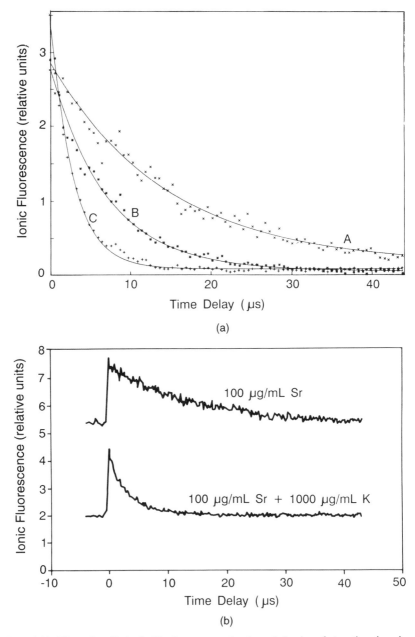

Figure 6.14. Effect of easily ionizable elements on the decay behavior of strontium ions in an air/acetylene flame and in an inductively coupled argon plasma. (a) Flame results: *A*, Cs = 100 μg/mL; *B*; Cs = 300 μg/mL; *C*, Cs = 1000 μg/mL. The strontium concentration is 10 μg/mL. (b) ICP results. Flame results have been reproduced from Turk and Omenetto (67) by permission of the Society for Applied Spectroscopy.

elements (Cs, and Li, and K). While this outcome seems logical in the case of the flame, it was unexpected in the argon plasma, since the native electron number density of the plasma is several orders of magnitude larger than in the flame.

In both original publications (67, 68), the decay time observed experimentally was associated with the "recombination time constant," thus neglecting the inverse ionization process and the initial pool of thermal ions present in the flame or the ICP. This can be justified in the flame, especially in the presence of an excess of cesium, but much less in the ICP, where the "thermal" ionization rate for alkaline earth metals is very high. In any case, as correctly pointed out in a paper using the "statistical moment approach" to estimate the rate constants (81), a rigorous treatment of the decay behavior cannot neglect the ionization time constant. The following considerations are therefore meant to improve the theoretical description of the experiment as well as to show that the conclusions and speculations given in the original papers are still valid.

The number densities (cm^{-3}) involved in the process are defined as follows:

$[Sr_T]_{ss}^{th}$ = total steady-state number density of strontium atoms and ions due to thermal processes, i.e., without lasers;

$[Sr^+(t)]_l$ = number density of strontium ions, at any time t, in the presence of the ionizing laser;

$[Sr^+]_{ss}^{th}$ = steady-state number density of strontium ions thermally present in the flame or plasma in the absence of the ionizing laser;

$[Sr(t)]_l$ = number density of neutral strontium atoms in the presence of the ionizing laser;

$[Sr]_{ss}^{th}$ = steady-state number density of strontium atoms thermally present in the flame or plasma in the absence of the ionizing laser.

With the above definitions, the following equalities hold:

$$[Sr_T]_{ss}^{th} = [Sr^+]_{ss}^{th} + [Sr]_{ss}^{th} \tag{23}$$

and

$$[Sr^+(t)]_l - [Sr^+]_{ss}^{th} = -\{[Sr(t)]_l - [Sr]_{ss}^{th}\} \tag{24}$$

In the absence of laser excitation and ionization, the steady-state concentrations of strontium atoms and ions are related by the expression

$$\frac{[Sr^+]_{ss}^{th}}{[Sr]_{ss}^{th}} = \frac{k_i}{k_r[e^-]} = \frac{k_i}{k_r'} k_i \tau_r \tag{25}$$

where $k'_r \equiv k_r[e^-] \equiv (\tau_r)^{-1}$ is treated as a first-order recombination rate coefficient (s^{-1}), since the electron number density, $[e^-]$, is much larger than the concentration of strontium ions; τ_r is the recombination time constant (s), and $k_i(s^{-1})$ is defined, like k'_r, as the ionization rate coefficient. The foregoing definitions are commonly used in discussions of flame kinetics (14). The ratio on the left-hand side of Eq. (25) can be evaluated experimentally by measuring the ionic as well as the atomic fluorescence signals. The decay of the laser-created strontium ions, for times *longer* than the duration of the ionizing laser pulse, T_l, can now be described by

$$\frac{d[Sr^+(t)]_l}{dt} = k_i[Sr(t)]_l - k'_r[Sr^+(t)]_l \tag{26}$$

By using Eqs. (23) and (24), we find the following solution:

$$[Sr^+(t)]_l = Sr^+(t = T_l)\exp[-(k_i + k'_r)(t - T_l)]$$

$$+ \frac{k_i}{k_i + k'_r}[Sr_T]_{ss}^{th}\{1 - \exp[-(k_i + k'_r)(t - T_l)]\} \tag{27}$$

An inspection of Eq. (27) shows that, for $t = T_l$ the concentration of strontium ions is that found at the end of the ionizing laser pulse; for $(t - T_l)$ approaching infinity, the steady-state value of the thermal concentration of strontium ions is restored. The important point is that the *measured* value of the time decay, τ_m, is given by the sum of both ionization and recombination terms, i.e., by the relationship

$$\tau_m \equiv \frac{1}{k_i + k'_r} \tag{28}$$

By combining Eqs. (25) and (28), it is easy to show that

$$\tau_m(1 + y) = \frac{1}{k'_r} = \tau_r \tag{29a}$$

and

$$k_i = \frac{y}{\tau_r} \tag{29b}$$

where

$$y = \frac{[Sr^+]_{ss}^{th}}{[Sr]_{ss}^{th}} \tag{29c}$$

The experimentally measured values of y and τ_m, as done in the study by Turk et al. (68), can then still allow evaluation of the recombination as well as the ionization rate coefficients. Reconsidering the experimental data provided in that paper, we find that the authors' conclusions that a *single* ionization rate coefficient holds from 7 to 19 mm above the load coil and that *longer* recombination times are observed in higher regions of the plasma are still valid. The absolute value of k_i is recalculated as 0.05 μs^{-1}, while τ_r varies from 2 to 7 μs when the height above the load coil increases from 6 to 20 mm. The corresponding decrease of k'_r range from 0.5 to 0.14 μs^{-1}. The addition of lithium is not expected to change the number density of the electrons in the plasma. However, it could have a significant effect on the electron energy distribution. These conclusions were, and still remain, of speculative value: many more experimental data are needed for these and other ionic species and plasma operating conditions. The experimental approach used, however, has proved to be a desirable addition to the existing methods of plasma diagnostics.

An interesting corollary to the strontium experiment just described is the evaluation of its "thermal degree of ionization," β_i, in the flame or plasma, i.e., of the ratio between the number of ions and the total number of strontium-containing species. The setup shown in Fig. 6.13 allows the sequential measurement of the ionic fluorescence (λ_F in insert A) as well as of the atomic resonance fluorescence (λ_1 in insert B). As schematized in Fig. 6.15, in coincidence with the laser enhancement of the ionic fluorescence signal (LE–IF), shown in part a, there will be a corresponding decrease in atomic fluorescence (AFD: atomic fluorescence dip) caused by the depletion of excited atoms, shown in part b, where both effects result from photoionization with the excimer laser. Under the assumption that neutral atoms and ions are the major species present, it is easy to show that the ratio between the enhancement in the ionic fluorescence signal (LE–IF/IF) and the dip in atomic fluorescence (AFD/AF) can be used to calculate β_i.

Another "all-optical" approach to the study of multiphoton ionization of calcium and strontium vapors was reported by Haugen and Stapelfeldt (76). As shown in Fig. 6.16, in the calcium case (part a) the simultaneous absorption of four photons and six photons leads to the population of the $Ca^+(4p)$ and $Ca^+(5s, 4d)$ levels, respectively, whereas in the strontium case (part b) a four-photon absorption and a five-photon absorption populate the $Sr^+(5p)$ and $Sr^+(6s)$ levels, respectively. These ionization pathways can be directly monitored by the resulting ionic fluorescence signals at 393.5 and 397.0 nm for $Ca^+(4p)$; 370.7 and 373.8 nm for $Ca^+(5s)$; 316.0, 318.2, and 318.0 nm for $Ca^+(4d)$; 407.7 and 421.5 nm for $Sr^+(5p)$; and 416.3 and 430.7 nm for $Sr^+(6s)$. The multiphoton process could be identified by the slope of the plot of the ionic fluorescence signal vs. the energy of the laser. The atomic vapors were

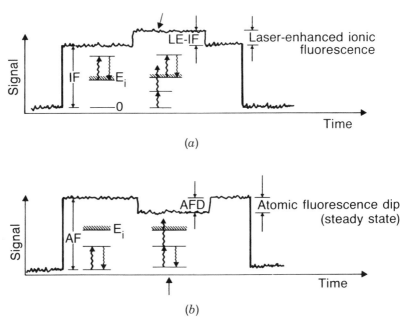

Figure 6.15. Schematic modeling of the behavior of (a) ionic and (b) atomic fluorescence signals when a strontium solution is aspirated into a plasma. Steady-state conditions are assumed. Atomic fluorescence and ionic fluorescence signals are indicated as AF and IF, respectively; E_i is the ionization energy. The arrows indicate the presence of the excimer beam, which results in the enhancement of IF and in the decrease of AF.

maintained in a oven (~ 1 mtorr) in the presence of $\leqslant 100$ mtorr of argon buffer gas. This technique was shown to be sensitive and particularly useful as a complementary approach to electron spectroscopy for low electron energies ($\leqslant 100$ meV) or where some overlap existed in the electron energies for different ionization channels.

6.5.3. Evaluation of Fundamental Ionization Parameters by Fluorescence-Dip Spectroscopy

This subsection discusses several experiments in which ionization information is derived from fluorescence techniques (82). The common denominator in all these experiments is the modulation of the fluorescence signal from a selected excited level caused by its depletion due to an ionization process. For example, in Fig. 6.2d,e, resonance fluorescence (A_{21}) or direct-line fluorescence (A_{2m}) can be monitored instead of two-step resonance fluorescence (A_{32}). The monitoring of the fluorescence signal from an *intermediate* level rather than the

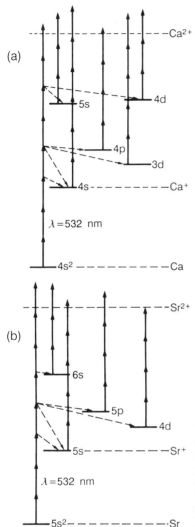

Figure 6.16. Simplified energy level diagrams and relevant multiphoton ionization channels for neutral and singly ionized Ca (*a*) and Sr (*b*). The excitation was provided by the frequency-doubled output of a Q-switched (300 mJ, 12 ns) or mode-locked (12 mJ, 35 ps) Nd:YAG laser at 10 Hz. Reproduced from Haugen and Stapelfeldt (76) by permission of the American Physical Society.

direct monitoring of the final level reached in a multistep process presents several advantages: (i) the last transition can be weak (e.g., a Rydberg transition) and therefore difficult to measure by absorption; (ii) the lifetime of the final level can be short (e.g., because of fast ionization), which then also results in a weak fluorescence signal from that level; and (iii) when the last transition is spectrally scanned, the resolution obtained is given by the line width of the laser used. Since the fluorescence strength is usually reduced, the technique has been variously called fluorescence-dip spectroscopy (69, 77, 83), modulated fluorescence (84), fluorescence reduction (17), laser depletion spectroscopy (73, 74), and extinction spectroscopy (75). Several applications are reported in Table 6.6, and some examples are discussed below.

The *ionization potentials* of five transition elements have been evaluated by double-resonance fluorescence-dip Rydberg spectroscopy (77). The principle of the technique can be seen in Fig. 6.17a, b, which shows the experimental results obtained for the iron atom. The usual method of evaluating the ionization potential is by extrapolation of Rydberg series in absorption experiments. Since the oscillator strengths of Rydberg transitions are generally weak (scaling as the inverse cube of the principal quantum number), a substantial vapor pressure may be required to achieve a sufficient signal-to-noise ratio. For refractory materials, a high temperature is therefore needed, which results in overlapping absorption spectra. In the technique proposed, the atoms are excited by a laser beam to a level from which fluorescence is observed, e.g., at 248.3 nm for iron in Fig. 6.17a, and forms the baseline level. A second tunable beam is scanned through a region that includes several Rydberg states. Since these states do not fluoresce at 248.3 nm, the baseline fluorescence level is reduced. A complete fluorescence-dip spectrum will therefore be generated, as shown in Fig. 6.17b, which is greatly simplified compared with an ordinary linear absorption spectrum. The analysis of the spectrum in terms of the usual hydrogenic formula allows calculation of the series limit and therefore the ionization potential (77).

Another attractive use of the fluorescence-dip method has been demonstrated in the study of *core-excited levels* of atoms (73, 74). These levels result from the excitation of an inner-shell electron to an outer valence orbital, a configuration that lies above the first continuum, and the knowledge of their location and autoionizing times is relevant to the understanding of many physical processes including dielectronic recombination, multiphoton and multielectron ionization, and harmonic generation (74). The technique has been applied to the study of core-excited levels of neutral rubidium. A *quasi-metastable level*, which radiates strongly in the X–UV region, is impulsively excited by a flash of soft X-rays from a plasma created by focusing a 150 mJ, 7 ns pulse of 1064 nm radiation onto a rotating tantalum target. A tunable dye laser is then used to transfer the population of this level into a manifold of autoionizing levels; this transfer causes a depletion of the X–UV fluorescence emitted from the quasi-metastable level. The position and shape of the depleted fluorescence is used to determine the position and width of the autoionizing level. In some cases, the core-excited level reached by the dye laser is stable enough against autoionization to have a significant branching ratio to radiation in the X–UV (see Fig. 6.18a–c). In this case, both laser-

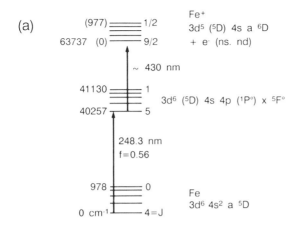

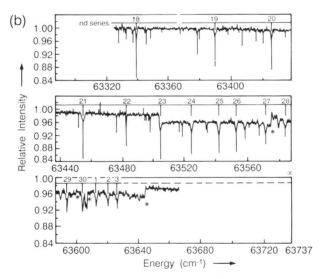

Figure 6.17. Energy levels of the iron atom pertinent to the fluorescence-dip Rydberg spectroscopy experiment. (*a*) Ground, intermediate, and ion states are labeled. The zero scale of the ion levels (in parentheses) is shown as the Rydberg series convergence limit. The resonance absorption transition (248.3 nm) has an oscillator strength $f = 0.56$. The range of wavelengths around 430 nm, obtained by scanning the second laser, reaches the series $4s4p \rightarrow 4snd$ and $4s4p \rightarrow 4sns$. (*b*) Fluorescence-dip spectrum obtained. The *nd* series has been assigned with principal quantum number up to $n = 33$. Sudden changes in the baseline levels, marked by an asterisk, are due to laser frequency adjustments during the scan. Reproduced from Page and Gudeman (77) by permission of the Optical Society of America.

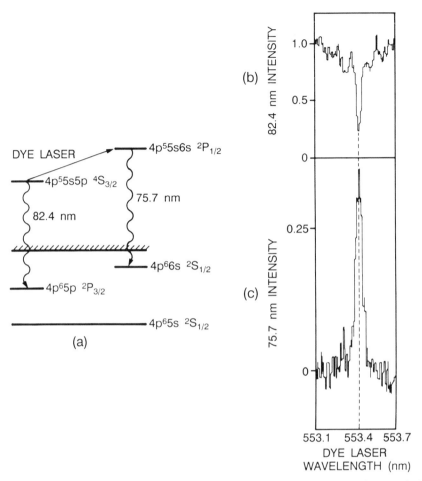

Figure 6.18. Energy level diagram and fluorescence signals pertinent to the study of core-excited levels of the Rb atom. (*a*) The quasi-metastable level, $^4S_{3/2}$, pumped by a flash of soft X-rays from which fluorescence at 82.4 nm is monitored, is connected to a core-excited level, $^2P_{1/2}$, by a dye laser tuned at 553.4 nm. As a result of this coupling, the fluorescence at 82.4 nm decreases (*b*), while the fluorescence at 75.7 nm is now observed (*c*). Reproduced from Spong et al. (74) by permission of the American Physical Society.

induced fluorescence and laser-depleted fluorescence can be observed and are used to identify the emitting level (74).

Photoionization cross section and *ion yields* can also be evaluated by fluorescence measurements alone, and a few experimental approaches on Mg and Cs atoms can be found in the literature (17, 36, 71, 72). The possibility of evaluating both parameters in flames can be understood with reference to the expressions given in Section 6.2 and to Figures 6.2 b, c (see also Figs. 6.10 and 6.11). The cases considered here are those in which collisional ionization and photoionization with $I_{v_{12}}$ are negligible compared to the photoionization achieved with a second laser, either tuned to an autoionizing level or reaching the continuum in a nonresonant ionization process. In Eqs. (8) and (10), R_{2i} is therefore identified with the product of the ionization cross section and the photon irradiance of the second laser. The product $I_v T_p$ is the photon fluence. By using the above expressions and defining the "relative integrated fluorescence dip," Δ', as the normalized difference between the signals obtained without and with the ionizing laser, we have, for $I_v \gg I_v^s$ and $k_{2m} \ll R_{2i}$, the very simple relation:

$$\Delta' \equiv \frac{(Q_F)_{\text{off}} - (Q_F)_{\text{on}}}{(Q_F)_{\text{off}}} = 1 + \frac{Y_i}{\ln (1 - Y_i)} \tag{30}$$

which we can also write, using Eq. (8), as

$$\Delta' = 1 - \frac{1 - \exp\{-[g_2/(g_1 + g_2)] R_{2i} T_p\}}{[g_2/(g_1 + g_2)] R_{2i} T_p} \tag{31}$$

The above expressions have been used to calculate the photoionization cross section of the Mg $^1P_1$ level in an argon plasma (71) and the autoionization cross section of the Mg 300.9 nm transition in an oxygen/argon/hydrogen flame (72). In both cases, the ground-state transition at 285.213 nm was saturated. In the former experiment, a value of 2.03×10^{-17} cm^2 was reported, while the autoionizing cross section was evaluated as 3×10^{-16} cm^{-2}. It should be stressed that many requirements must be met in order to justify the use of the equations as simplified above: (i) the photoionizing laser must be characterized in terms of its photon irradiance, i.e., its energy per pulse, pulse duration, and geometric cross section must be accurately measured; (ii) the intensity of the first laser must be sufficient to saturate the fluorescence but not high enough to induce photoionization (this might require the construction of a full saturation curve); (iii) the condition imposed by Eq. (1) concerning the Stark splitting of level 2 must be satisified; and (iv) the presence of a metastable level must be fully accounted for, which implies that the parameter ξ_i in Fig. 6.9 must be known (see also Section 6.4.2). The last three requirements were satisfied in both magnesium photoionization experiments: in one case (71) the laser irradiance was 57 kW/cm^2 with a bandwidth of 110 GHz; in the other (72) the laser irradiance was 20 kW/cm^2 with a bandwidth of 64 GHz. For an A value of 5.3×10^8 s^{-1}, Eq. (1) is satisfied in both examples. The weakest point of the aforementioned experiments is the accurate characterization of the spatial and temporal behavior of the laser (see Section 6.3).

Two points still need to be considered. First, the precision of the fluo-rescence-dip method is limited to a range of ionization cross sections such that $\sigma_i \leqslant (I_v T_p)^{-1}$. In fact, as seen from Eq. (31), at high values of the product $\sigma_i I_v T_p$, the relative fluorescence dip is not sensitive to even significant varia-tions of this product. Second, since the atoms are present in a collisional environment, any other level, 2′, whose energy difference from level 2 matches the energy overshoot, E_{os}, of the ionizing laser with respect to the continuum, i.e., any level 2′ for which $|E_2 - E_2'| = E_{os}$, might also be involved in the nonresonant ionization process, being populated very rapidly by collisions. If this is the case, the σ_i measured cannot be assigned to a single level.

When the experiment is performed at reduced pressure or in vacuum rather than in a flame, the foregoing simplifications of the theory cannot hold. A clear example is the use of the fluorescence dip to measure the photoionization cross section of the cesium $7s D_{3/2}$ level by Bonin et al. (17). Here, fluorescence was monitored at 672 nm after two-photon excitation at 767 nm. The cross section was measured at several different photoionizing photon energies. Since the atoms were produced in a cell evacuated to 10^{-6} torr and then heated at 70 °C, the magnetic quantum number sublevels and the polarization of the laser had to be taken into account in order to give an exact meaning to the measured cross section (17). As previously mentioned in Section 6.3, the spatial distribu-tion of the laser fluence was purposely made a "speckle distribution" in order to account for the different ionization probabilities in different parts of the beam.

As a final example in this subsection, it is worth remembering that fluorescence-dip spectroscopy can also be used to evaluate the *quantum efficiency* of a resonance ionization detector (see Section 6.4.2 and Figs. 6.8 and 6.9). Whether the fluorescence technique can be used alone or in conjunction with ionization measurements depends essentially on the scheme chosen. These considerations have been amply discussed in the literature for the magnesium flame detector (36).

6.5.4. Simultaneous Molecular Ionization and Fluorescence Spectroscopy

Although the main purpose of this chapter is to describe the interaction between fluorescence and ionization in the case of neutral atoms and atomic ions, one should be aware of the abundant panorama of experiments involving molecules that can be found in the literature [see, e.g., the review by Ito (65)]. Indeed, *optical–optical double-resonance spectroscopy* is a well-known tech-nique for studying excitation dynamics and structural conformation of ground and excited molecular states. As in the atomic counterpart described in the previous subsections, the double-resonance excitation process from an initial ground state to a final excited state can be studied by monitoring the dip in the signal associated with an intermediate state. To report just one example, this principle was used to study the UV–visible excitation spectrum of CO and the technique called *laser-reduced fluorescence* (85). As in the case of autoionizing transitions of atomic species, the advantage of monitoring an intermediate state here lies in the fact that the final transition can reach a predissociating state, which is difficult to analyze by conventional absorption and fluorescence methods.

As discussed by Ito (65), when double-resonance spectroscopy is applied to low-pressure gaseous molecules, one can directly probe the ions created from the final state of autoionization or photoionization, as in the resonance enhanced multiphoton ionization (REMPI) technique, or probe the modulation of the *fluorescence or ion signal* associated with an intermediate state and caused by the presence of the second excitation step. The approach has therefore been called "two-color dip spectroscopy" since both fluorescence dips and ion dips can be monitored. When large molecules are studied at room temperature, the ground-state population is distributed over many low-lying vibrational levels, which are extremely congested. Supersonic jet expansion is therefore used to produce ultracold molecules in the gas phase, allowing high selectivity in addressing a particular level.

An interesting way of studying the ground-state configuration of a molecule that involves the simultaneous use of stimulated emission and photoionization was called "two-color ionization dip spectroscopy using stimulated emission." Here, the molecule in the electronic state S_0 is excited to a particular level in S_1 with a laser tuned to frequency v_1. A second laser at v_2 photoionizes the molecule. If v_2 is now made resonant with a vibrational level in S_0, stimulated emission occurs and the population of S_1 decreases. This decrease is reflected in a dip in the ion signal. The novelty of this approach is that v_2 plays here the dual role of ionizing S_1 while stimulating emission into a particular vibrational level of S_0. Compared to the fluorescence-dip scheme, where any *spatial* mismatching of v_1 and v_2 in the excitation volume effects of signal, in this case both ionization and stimulated emission can only occur in the overlapping volume of v_1 and v_2 and therefore the ion dip intensity is not seriously influenced by the unavoidable difference in size between the two lasers. As an example of the utility of the technique, the ion dip as well as the fluorescence spectrum of jet-cooled aniline is shown in Fig. 6.19. As seen in the figure, the number of vibrational levels detected in the ion dip spectrum is much larger than in the corresponding fluorescence spectrum, with better spectral resolution.

Simultaneous ionization and fluorescence measurements of molecular species in atmospheric pressure flames have not been extensively performed, to the present authors' knowledge, apart from some isolated applications, as reported in Table 6.5 and Section 6.5.6.

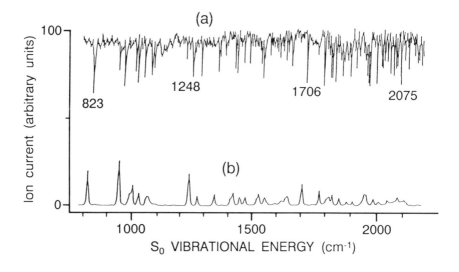

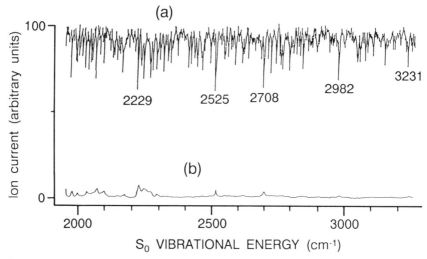

Figure 6.19. Two-color ionization dip spectrum of jet-cooled aniline (*a*) and corresponding fluorescence spectrum (*b*). The ionization spectrum is due to stimulated emission from the zero-point level of S_1 to the ground-state vibrational levels in the frequency region from 800 to 3200 cm^{-1}. Reproduced from Ito (65) by permission of Taylor & Francis Ltd.

6.5.5. Time-Resolved Studies

The pump-and-probe experiments on the ion decay described in Section 6.5.2 are a clear example of time-resolved studies, which are the natural consequence (and advantage) of using multistep excitation schemes where, in general, the first step is provided by one laser and the second by another independent laser. A noticeable exception to this is the possibility of using two pulses of identical frequency from the same laser, one delayed with respect to the other, such as in the technique of *picosecond fluorescence depletion spectroscopy*, in which pure dephasing rather than energy relaxation processes in molecules are studied (86). In this subsection, two other aspects of time-resolved measurement are discussed: (i) the atomic or ionic fluorescence signal is measured from a level *not directly pumped* by the laser radiation, and (ii) the temporal behavior of the fluorescence is investigated *during* the excitation pulse. In the former case, the time lag between the onset of the radiation pumping and attainment of the maximum population of the fluorescent level is dictated by the collisional environment of the excited atoms and by their energy level scheme. In the last case, if the optically pumped transition is saturated, any decay of the fluorescence signal while the pump is still on indicates the occurrence of a depletion mechanism operating on the population of the levels locked by the laser excitation. Time-resolved fluorescence with *long* laser excitation pulses is therefore advocated, in contrast with the conventional approach of measuring fluorescence decay after a *short* exciting pulse has subsided.

The attractive use of time resolving the fluorescence signal during the long excitation pulse provided by a flashlamp-pumped dye laser in order to evaluate atomic parameters of a helium plasma was suggested more than 20 years ago (87). Some time later, while studying the temporal behavior of the saturation of sodium atoms in an $O_2/Ar/H_2$ flame with a flashlamp-pumped dye laser characterized by a pulse width of 460 ns, Van Calcar et al. (78) were the first to report on the decay of the saturated fluorescence signal during the excitation pulse. Such a decay, attributed to collisional ionization, was found to be dependent on the oxygen concentration in the flame. Indeed, in an oxygen-rich mixture, the peak of the fluorescence waveform decreased practically to zero before the end of the excitation pulse (78). Along the same lines, several papers reported on laser-induced depletion effects on alkali atoms in flames, which were attributed to laser-enhanced chemical reactions involving the excited atoms (88–90). Due to the relatively long (microseconds) timescale involved here, the significance of these last processes is marginal when short excitation pulses are considered and therefore they are neglected in out treatment. A thorough discussion of the relevance of all these effects on the saturation behavior of atomic species has been given by Alkemade (2).

On the timescale of the experiments reported for flames and plasmas in the previous sections, it is clear that the experiment constraints on the apparatus are much more stringent, since the laser pulses seldom exceed an average duration of 15 ns and can be highly structured (see Fig. 6.4). However, under saturation, the fluorescence should remain constant in time until the laser irradiance has fallen below the threshold of saturation. Indeed, for strongly saturated waveforms, the fluorescence maximum can be so prolonged in time that the natural decay of the level can still be measured after the disappearance of the excitation pulse.

For the level schemes of Fig. 6.2, the corresponding time behavior of the fluorescence signal is predicted by Eqs. (5), (9), (12), and (15). If a decay is observed after a fast rise time of the signal (due to saturation), then either R_{2i}, R_{2m}, or their sum can be evaluated (36, 37, 82). Such an experimental outcome would allow the direct measurement of many parameters related to these rates (ξ_i, η_i, σ_i), whose evaluation by time-integrated fluorescence was already discussed in Section 6.5.3. Until now, however, the experimental demonstration of a time-resolved fluorescence decay during the laser pulse that can be unequivocally attributed to photoionization is still lacking, despite several attempts made in flames and plasmas (36, 82). Is is still worth emphasizing here that the theoretical expressions derived in Section 6.2 do consider ion formation as a complete loss process (no recombination reactions are considered). However, especially in flames, ion chemistry might occur on a surprisingly fast timescale (67). Its relevance to the process of restoring the population of neutral atoms therefore needs to be assessed.

The difference in the time behavior of the fluorescence of magnesium ions in an argon plasma, after single-step and two-step excitation by two frequency-doubled dye lasers, is shown in Fig. 6.20. In this experiment, the kinetics of *charge exchange* between the argon ion and the magnesium ground-state atom was investigated (91). If the inverse reaction occurs, i.e., energy transfer between the quasi-resonant level of the magnesium ion, Mg_{qr}^+, and the argon neutral, Ar_0, to regenerate Ar^+ and Mg_0, the population of Mg_{qr}^+ should decay during the excitation pulse, since no fast replenishment is expected from the reaction between the regenerated species. This is shown in Fig. 6.20 for the two-step excited fluorescence signal. If the second excitation step is absent, the resonance fluorescence is indeed constant for about 10 ns. From these experiments, an effective rate constant for the transfer of charge from Mg^+ to Ar_0 was calculated to be $1.1 \times 10^8 \, s^{-1}$.

Finally, the evolution of the population of several atomic and ionic levels of an atomic system illuminated by a short excitation pulse has been successfully used to interpret the ionization mechanism in dense vapors of alkaline-earth metals (see e.g., refs. 79, 80, 92). Although the atomic densities are of the order of 10^{15} atoms cm^{-3}, which would be impractical for flame and plasma work,

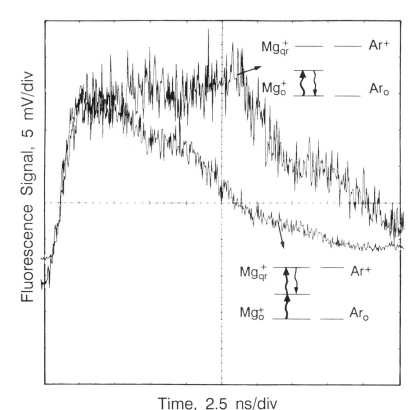

Figure 6.20. Time-resolved single-step and two-step excited fluorescence waveforms of the magnesium ion in an argon plasma. The corresponding (oversimplified) energy level scheme is also shown: Ar and Mg_0^+ indicate the ground-state population of argon atoms and magnesium ions, while Ar^+ and Mg_{qr}^+ refer to the argon ions and the excited level of magnesium that is "quasi-resonant" with the ionization energy of argon, thus allowing an efficient inverse charge transfer reaction. Adapted from Farnsworth and Omenetto (91).

the principle of the method could be advantageously applied also to these atom/ion reservoirs. Figure 6.21a–d reports some time-resolved fluorescence measurements of Sr(I) and Sr(II) that show the role played by the metastable population and by the superelastic collisions, i.e., collisions between seed electrons and excited atoms, in which electrons gain energy until final ionization is achieved. A dye laser, pumped by a 7 ns N_2 laser, was used to excite the strontium atoms either to the $5s\,5p$ level (single photon absorption) or to the $5s\,10s$ level (two-photon absorption). In this last case, superelastic collisions with the excited atoms do not occur and the fluorescence temporal shape is drastically changed (92). As seen in Fig. 6.21, the timescale of the experiment

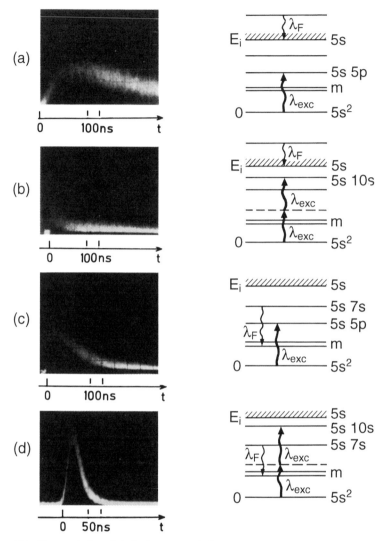

Figure 6.21. Time evolution of the ionic and atomic fluorescence signals (λ_F) in a dense strontium vapor. The pertinent energy level scheme, with the excitation and fluorescence wavelength, is shown to the right of each photo. In parts (a) and (b), the ionic fluorescence was monitored at 407.771 nm while laser excitation was set at 460.733 nm (a) and 459.51 nm (b). In parts (c) and (d), atomic fluorescence was measured at 443.804 nm, with laser excitation set at 460.733 nm (c) and 459.5 nm (d). In the energy level scheme, E_i is the ionization energy and m represents a manifold of metastable levels acting as energy reservoirs in the collisional ionization process. The left part of the figure was reproduced from Bréchignac et al. (92) by permission of the American Physical Society.

was several hundreds of nanoseconds, which is very much longer than the collisional times expected in an air/acetylene flame or in an argon plasma. Nevertheless, time resolution in the nanosecond range can easily be achieved: this would offer a systematic way of studying the difference in the rise time between resonance and nonresonance fluorescence waveforms. In the former case, the rise time would be essentially due to the laser pumping time, whereas in the latter case it would indicate the collisional transfer rates in and out of the excited level. The time behavior of a strong *ionic* fluorescence transition monitored during and after the excitation of an *atomic* level lying a few electron volts below the ionization potential might then help distinguish between collisional ionization and direct laser photoionization.

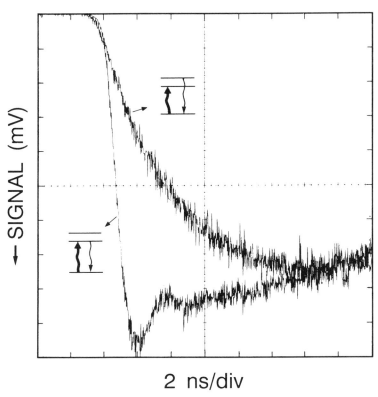

2 ns/div

Figure 6.22. Time-resolved ionic fluorescence waveforms of barium in an argon plasma. Laser excitation is set at the transition $^2S_{1/2} \rightarrow {}^2P_{1/2}$ (493.409 nm). Fluorescence is observed at the same transition and at the transition $^2P_{3/2} \rightarrow {}^2S_{1/2}$ (455.403 nm). The measurements were obtained with a fast photomultiplier (Model H3376, Hamamatsu, Hamamatsu City, Japan) and a fast digitizer (Model DG02A, Tektronix, Beaverton, Oregon). From N. Omenetto, unpublished results (1992).

This type of *systematic* investigation has not yet been reported in the LIF and LEI literature, apart from some scattered experiments (36, 82, 93). One of these is illustrated in Fig. 6.22, which shows the difference in the rise time of the resonance fluorescence and thermally assisted fluorescence signals of barium ions in an argon plasma. Here, laser excitation was set at 493.409 nm and fluorescence was measured at 493.409 and 455.403 nm. The difference in energy between the two excited levels ($^2P_{3/2}$ and $^2P_{1/2}$) is only 0.21 eV, and therefore fast collisional mixing between the two levels is expected at a plasma temperature of ~ 6000 K. The use of the complete expressions given in Table 6.2 would allow comparison of the experimental behavior with theoretical predictions and evaluation of the mixing rates between the levels, if thermal equilibrium conditions prevail.

6.5.6. Miscellaneous Applications

An intriguing example of the use of an ionization process to interpret a fluorescence experiment is found in the technique called photoionization controlled-loss spectroscopy (PICLS) (94, 95). The fundamental objective of PICLS is to make the ionization rate purposely higher than the radiative and collisional rates out of the fluorescent level; the fluorescence signal will thus be independent of quenching. This can be seen with reference to the energy level scheme of Fig. 6.8, where k_{2i} is now substituted by R_{2i}, i.e., by the photoionizing rate due to a strong second laser. Under *linear* interaction, the resonance fluorescence signal (A_{21}) is affected by quenching (k_{21}). A simple balance equation can be written:

$$\frac{dn_2}{dt} = R_{12}n_1 - n_2(A_{21} + k_{21} + R_{2i}) \tag{32}$$

If the peak of the fluorescence pulse is measured, $dn_2 dt = 0$, and since for linear interaction $R_{12} \ll A_{21} + k_{21}$,

$$n_2 = n_T \frac{R_{12}}{A_{21} + k_{21} + R_{2i}} \tag{33}$$

This expression shows that, if R_{2i} is made much greater than ($A_{21} + k_{21}$), the fluorescence signal, even if reduced, is virtually independent of quenching. However, when the maximum power in the second laser is not sufficient to make R_{2i} dominate the other rates, the technique still allows specific accounting for the influence of quenching. This can be seen in the following way. In the absence of the ionizing laser ($R_{2i} = 0$), Eq. (33) reduces to

$$(n_2)_{\text{off}} = n_T \frac{R_{12}}{A_{21} + k_{21}} \tag{34}$$

By ratioing the last two equations, we obtain

$$n_2^* \equiv \frac{n_2}{(n_2)_{\text{off}}} = \frac{1}{1 + R_{2i}/(A_{21} + k_{21})} \tag{35}$$

It can therefore be seen that the ratio of two fluorescence signals, with and without the photoionizing laser, directly yields the ratio between the ionization rate and the total rate of deexcitation, which in a flame is a highly localized parameter because of the difference in flame composition and therefore in the quenching coefficient. By manipulating Eqs. (33) and (35), it can be shown that

$$n_T = \frac{n_2}{1 - n_2^*} \frac{R_{2i}}{R_{12}} \tag{36}$$

This equation, which is valid at any point in the flame and for any flame conditions, is the keystone of PICLS, since it can be used even when R_{2i} is of the same order of magnitude as $(A_{21} + k_{21})$ (94). The technique has been applied in a flat, premixed $H_2/O_2/N_2$ flame at a pressure of 20 torr, in which the fluorescence of atomic hydrogen at the Balmer-α line (656 nm) was excited by a two-photon absorption transition at 205.1 nm (95).

Fluorescence-dip spectroscopy can also be used to study *stimulated relaxations* such as amplified spontaneous emission (ASE) (96). This is a variant of the ion dip spectroscopy technique described in Section 6.5.4. The principle of the method and its relevant level scheme is reported in Fig. 6.23. Two lasers, tuned at λ_1 and λ_2, excite the fluorescence radiation emitted at λ_{3m}. ASE can occur at λ_{2m} because the strong pumping of level 2 by λ_1 generates a population inversion between the levels 2 and m. However, because of the metastable character of level m, this population inversion quickly decreases, with a corresponding decrease in the ASE signal. If a third laser, tuned at λ_3 now depletes level m, the population inversion is restored with two effects: (i) the population of level 2 will decrease as a result of an increased ASE; and (ii) because of effect (i), the population of level 3, caused by λ_2, will decrease. The final outcome of this population shuffling will be a dip in the fluorescence signal at λ_{3m}. Although the scheme of Fig. 6.23 can be typical of several elements, the original study was made with atomic barium (96).

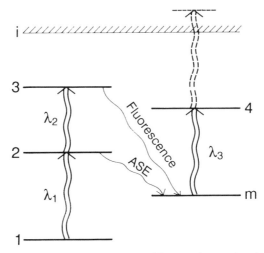

Figure 6.23. Possible energy level scheme and transitions used to study stimulated relaxation processes by fluorescence-dip spectroscopy. Three lasers are used, tuned at λ_1, λ_2, and λ_3. Level 4 can also serve as an intermediate level to an autoionizing level above the continuum, i. ASE means amplified spontaneous emission; m is a metastable level. Adapted from Xu et al. (96).

As a final example, the simultaneous monitoring of the fluorescence and ionization signals of a given flame radical can be used as a means of calibrating the relative *electron detection efficiency* (64). This last quantity is known to vary in a flame whenever the electron–ion recombination rates vary with the local ion density or when the native ion density of the flame is surpassed by that created in the laser ionization process. If the flame is optically thin, the relative *photon* detection efficiency can be assumed to be unity throughout the entire excitation volume. The fluorescence signal can therefore be used for each flame position to normalize the ionization signal. The method has been successfully applied to a laminar CH_4/air diffusion flame in which the $B^1 \Sigma^+$ state of CO was excited by a two-photon absorption transition at 230 nm and both ionization and fluorescence from this electronic state were then measured. The ratio of these signals directly yields the relative electron detection efficiency. By extrapolation of the above principle, if the detection of charges is accurately calibrated, by ratioing the ionization and fluorescence signals originating from the same volume one could also calibrate the optical chain and the detector used for measuring the fluorescence.

6.6. CONCLUSIONS

The usefulness of the interaction between the ionization and fluorescence methods has been documented by the various applications described in this chapter. Some peculiar examples—more pertinent to the field of atomic and nuclear physics than to analytical chemistry and flame or plasma diagnostics—have been reported to illustrate the significant versatility of the approach described. Still, it appears that there is ample room for further development and studies.

The analytical application of the resonance ionization detector is expected to grow, not only as a Raman photon detector, which could be exploited in surface enhanced Raman experiments, but also for high-sensitivity fluorescence measurements. For example, pulsed fluorescence emitted from a graphite furnace could be focused into a corresponding detector. Although inflexible and plagued with scattering problems, this combination is capable of extremely high sensitivity. In this respect, other detectors, in addition to the flame, should be analytically evaluated.

Ion decay experiments in flames should be pursued, especially in view of the fast "ion chemistry" observed in the strontium case. Such experiments are improved when matched with molecular ionization and fluorescence measurements.

The most promising—even though experimentally challenging—approach is that of temporally resolving the fluorescence and ionization waveforms, both during the laser excitation and after it. These measurements, which have often been carried out in pure atomic vapors, are still lacking in atmospheric pressure flames and plasmas.

The most important need, which is the common denominator of all the experiments described, is the accurate experimental characterization of the spatial and temporal profile of the laser. This need is intensified by the frequent use of two and even three beams simultaneously. The lack of temporal beam profiles hampers evaluation of the assumptions made in theoretical modeling.

REFERENCES

1. A. L. Gray, *Frescenius' Z. Anal. Chem.* **324**, 561 (1986).
2. C. Th. J. Alkemade, *Spectrochim. Acta* **40B**, 1331 (1985).
3. R. H. Pantell and H. E. Puthoff, *Fundamental of Quantum Electronics* Wiley, New York, 1969.
4. A. Yariv, *Quantum Electronics*, 3rd ed. Wiley, New York, 1989.
5. B. W. Shore, *The Theory of Coherent Atomic Excitation*, Vols. 1 and 2. Wiley, New York, 1990.
6. P. L. Knight and P. W. Milonni, *Phys. Rep.* **2**, 21 (1980).
7. T. J. McIlrath and J. L. Carlsten, *Phys. Rev. A* **6**, 1091 (1972).
8. J. W. Daily, *Appl. Opt.* **16**, 2322 (1977).

9. O. Axner and S. Sjöström, *Spectrochim. Acta* **47B**, 245 (1992).

10. G. Zizak, J. D. Bradshaw, and J. D. Winefordner, *Appl. Opt.* **19**, 3631 (1980).

11. G. S. Hurst, M. H. Nayfeh, and J. P. Young, *Phys. Rev. A* **15**, 2283 (1977).

12. N. Omenetto, B. W. Smith, and L. P. Hart, *Fresenius' Z. Anal. Chem.* **324**, 683 (1986).

13. C. M. Miller and N. S. Nogar, *Anal. Chem.* **55**, 481 (1983).

14. C. Th. J. Alkemade, Tj. Hollander, W. Snelleman, and P. J. Th. Zeegers, *Metal Vapours in Flames.* Pergamon, Oxford, 1982.

15. G. C. Turk, *Anal. Chem.* **53**, 1187 (1981).

16. D. J. Butcher, J. P. Dougherty, F. R. Preli, A. P. Walton, G. T. Wei, R. L. Irwin, and R. G. Michel, *J. Anal. At. Spectrom.* **3**, 1059 (1988).

17. K. D. Bonin, M. Gatzke, C. L. Collins, and M. A. Kadar-Kallen, *Phys. Rev. A* **39**, 5624 (1989).

18. O. Axner and S. Sjöström, *Appl. Spectrosc.* **44**, 864 (1990).

19. F. Bos, *Appl. Opt.* **20**, 3553 (1981).

20. T. T. Kajava, H. M. Lauranto and R. R. E. Salomaa, *Appl. Opt.* **31**, 6987 (1992).

21. A. Tomaselli, P. G. Gobbi, and P. Benetti, *Proc. Int. Conf. Lasers '86*, p. 420 (1987).

22. R. Wyatt, *Appl. Phys.* **21**, 353 (1980).

23. N. Omenetto and J. D. Winefordner, *Prog. Anal. At. Spectrosc.* **2**, 1 (1979).

24. N. Omenetto, B. W. Smith, L. P. Hart, P. Cavalli, and G. Rossi, *Spectrochim. Acta* **40B**, 1411 (1985).

25. N. Omenetto, *Spectrochim. Acta* **44B**, 131 (1989).

26. J. A. Vera, C. L. Stevenson, B. W. Smith, N. Omenetto, and J. D. Winefordner, *J. Anal. At. Spectrosc.* **4**, 619 (1989).

27. N. Omenetto, *Mikrochim. Acta* **II**, 277 (1991).

28. J. V. Sullivan and A. Walsh, *Spectrochim. Acta* **21**, 727 (1965).

29. O. I. Matveev, *J. Appl. Spectrosc.* (*Engl. Transl.*) **46** (3), 217 (1987).

30. O. I. Matveev, N. B. Zorov, and Y. Y. Kuzyakov, *J. Anal. Chem. USSR* (*Engl. Transl.*) **34**, 654 (1979).

31. N. Omenetto, B. W. Smith, and J. D. Winefordner, *Spectrochim. Acta, Part B, Spec. Suppl.*, p. 101 (1089).

32. B. W. Smith, N. Omenetto, and J. D. Winefordner, *Spectrochim. Acta, Part B, Spec. suppl.* p. 91 (1989).

33. T. Okada, H. Andou, U. Moriyama, and M. Maeda, *Opt. Lett.* **14**, 987 (1989).

34. B. W. Smith, P. B. Farnsworth, J. D. Winefordner, and N. Omenetto, *Opt. Lett.* **15**, 823 (1990).

35. S. H. Bloom, E. Korevaar, M. Rivers, and C. S. Lin, *Opt. Lett.* **15**, 294 (1990).

36. N. Omenetto, B. W. Smith, P. B. Farnsworth, and J. D. Winefordner, *J. Anal. At. Spectrosc.* **7**, 89 (1992).

37. N. Omenetto, B. W. Smith, and P. B. Farnsworth, *Conf. Ser.*—**114**: Sect. 9, 369 (1991).

38. G. A. Petrucci, R. G. Badini, and J. D. Winefordner, *J. Anal. At. Spectrosc.* **7**, 481 (1992).

39. G. A. Petrucci and J. D. Winefordner, *Spectrochim. Acta* **47B**, 437 (1992).

40. G. A. Petrucci, D. Imbroisi, B. W. Smith, and J. D. Winefordner, *Spectrochim. Acta* **49B**, 1569 (1994).

41. G. C. Turk, *Anal. Chem.* **64**, 1836 (1992).

42. J. C. Travis, G. C. Turk, J. R. De Voe, P. K. Schenck, and C. A. Van Dijk, *Prog. Anal. At. Spectrosc.* **7**, 199 (1984).

43. J. A. Gelbwachs, *IEEE J. Quantum Electron.* **QE-24**, 1266 (1988).

44. G. Mageri, B. P. Oehry, and W. Ehrlich-Schupita, *Study of Atomic Resonance Narrow Band Filters*, Final Report, ESTEC/Contract No. 8488/89/NL/PM(SC). 1991.

45. G. J. Havrilla, S. J. Weeks, and J. C. Travis, *Anal. Chem.* **54**, 2566 (1982).

46. N. Omenetto, T. Berthoud, P. Cavalli, and G. Rossi, *Appl. Spectrosc.* **39**, 500 (1985).

47. O. Axner, P. Ljungberg, and Y. Malmsten, *Appl. Phys. B* **54**, 144 (1992).

48. W. G. Mallard and K. C. Smyth, *Combust. Flame* **44**, 61 (1982).

49. K. C. Smyth and W. G. Mallard, *Combust. Sci. Technol.* **26**, 35 (1981).

50. P. K. Schenck, J. C. Travis, G. C. Turk, and T. C. O'Haver, *Appl. Spectrosc.* **36**, 168 (1982).

51. O. Axner and T. Berglind, *Appl. Spectrosc.* **43**, 940 (1989).

52. A. G. Marunkov and N. V. Chekalin, *Opt. Spectrosc.* (*Engl. Tranal.*) **61**(4), 461 (1986).

53. C. L. Callender, P. A. Hackett, and D. M. Rayner, *J. Opt. Soc. Am. B* **5**, 614 (1988).

54. D. J. Bradley, C. H. Dugan, P. Ewart, and A. F. Purdie, *Phys. Rev. A* **13**, 1416 (1976).

55. K. D. Su, C. Y. Chen, K. C. Lin, and W. T. Luh, *Appl. Spectrosc.* **45**, 1340 (1991).

56. K. C. Lin, P. M. Hunt, and S. R. Crouch, *Chem. Phys. Lett.* **90**, 111 (1982).

57. O. I. Matveev, Ph.D. dissertation, Moscow State University (1979).

58. K. D. Su, K. C. Lin, and W. T. Luh, *Appl. Spectrosc.* **46**, 1370 (1992).

59. O. Axner and T. Berglind, *Appl. Spectrosc.* **40**, 1224 (1986).

60. B. W. Smith, P. B. Farnsworth, and N. Omenetto, *Spectrochim. Acta* **45B**, 1085 (1990).

61. Th. Weber, E. Riedle, and H. J. Neusser, *J. Opt. Soc. Am. B* **7**, 1875 (1990).

62. Z. Jingyuan, L. Qiongru, Y. Jian, Z. Lizeng, and N. Yuxin, *J. Phys. B* **19**, L75 (1986).

63. G. Sultan, G. Baravian, and J. Jolly, *Chem. Phys. Lett.* **175**, 37 (1990).

64. K. C. Smyth and P. J. H. Tjossem, *Appl. Opt.* **29**, 4891 (1990).

65. M. Ito, *Int. Rev. Phys. Chem.* **8**, 147 (1989).

66. B. W. Smith, L. P. Hart, and N. Omenetto, *Anal. Chem.* **58**, 2147 (1986).

67. G. C. Turk and N. Omenetto, *Appl. Spectrosc.* **40**, 1085 (1986).

68. G. C. Turk, O. Axner, and N. Omenetto, *Spectrochim. Acta* **42B**, 873 (1987).

69. O. Axner, M. Norberg, and H. Rubinstzein-Dunlop, *Spectrochim. Acta* **44B**, 693 (1989).

70. N. Omenetto, B. W. Smith, B. T. Jones, and J. D. Winefordner, *Appl. Spectrosc.* **43**, 595 (1989).

71. N. Omenetto, *Conf. Ser.—Inst. Phys.* **94**, Sect. 3, 141 (1989).

72. G. A. Petrucci, C. L. Stevenson, B. W. Smith, J. D. Winefordner, and N. Omenetto, *Spectrochim. Acta* **46B**, 975 (1991).

73. J. K. Spong, J. D. Kmetec, S. C. Wallace, J. F. Young, and S. E. Harris, *Phys. Rev. Lett.* **58**, 2631 (1987).

74. J. K. Spong, A. Imamoglu, R. Buffa, and S. E. Harris, *Phys. Rev. A* **38**, 5617 (1988).

75. K. D. Pedrotti, *Opt. Commun.* **62**, 250 (1987).

76. H. K. Haugen and H. Stapelfeldt, *Phys. Rev. A* **45**, 1847 (1992).

77. R. H. Page and C. S. Gudeman, *J. Opt. Soc. Am. B* **7**, 1761 (1990).

78. R. A. Van Calcar, M. J. M. Van de Ven, B. K. Van Uitert, K. J. Biewenga, Tj. Hollander and C. Th. J. Alkemade, *J. Quant. Spectrosc. Radat Transfer* **21**, 11 (1979).

79. T. J. McIlrath and T. B. Lucatorto, *Phys. Rev. Lett.* **38**, 1390 (1977).

80. R. M. Measures and P. G. Cardinal, *Phys. Rev. A* **23**, 801 (1981).

81. K. P. Li, T. Yu, J. D. Hwang, K. S. Yeah, and J. D. Winefordner, *Anal. Chem.* **60**, 1599 (1988).

82. N. Omenetto, *Conf. Ser.-Inst. Phys.* **128**, Sect. 4, 151 (1992).

83. N. Omenetto, G. C. Turk, M. Rutledge, and J. D. Winefordner, *Spectrochim. Acta* **42B**, 807 (1987).

84. S. L. Gilbert, M. C. Noecker, and C. E. Wieman, *Phys. Rev. A* **29**, 3150 (1984).

85. P. Klopotek and C. R. Vidal, *J. Opt. Soc. Am. B* **2**, 869 (1985).

86. M. J. Côté, J. F. Kauffman, P. G. Smith, and J. D. McDonald, *J. Chem. Phys.* **90**, 2865 (1989).

87. D. D. Burgess and C. H. Skinner, *J. Phys. B* **7**, L297 (1974).

88. C. H. Muller, III, K. Schofield, and M. Steinberg, *Chem. Phys. Lett.* **57**, 365 (1978).

89. C. H. Muller, III, K. Schofield, and M. Steinberg, *J. Chem. Phys.* **72**, 6620 (1980).

90. M. Lino, H. Yano. Y. Takubo, and M. Shimazu, *J. Appl. Phys.* **52**, 6025 (1981).

91. P. B. Farnsworth and N. Omenetto, *Spectrochim. Acta* **48B**, 809 (1993).

92. C. Bréchignac, Ph. Cahuzac, and A. Débarre, *Phys. Rev. A* **31**, 2950 (1985).

93. O. Axner, personal communication (1990).

94. J. T. Salmon and N. M. Laurendeau, *Appl. Opt.* **26**, 2881 (1987).

95. J. T. Salmon and N. M. Laurendeau, *J. Quant. Spectrosc. Radiat. Transfer* **43**, 155 (1990).

96. L. Xu, Y. Zhao, G. Wang, M. He, and Z. Wang, *J. Opt. Soc. Am. B* **9**, 1017 (1992).

97. J. O. Gaardstedt, T. Andersen, H. K. Haugen, J. E. Hansen, and N. Vaeck, *J. Phys. B.* **24**, 4363 (1991).

98. P. B. Farnsworth, B. W. Smith, and N. Omenetto, *Spectrochim. Acta* **46B**, 843 (1991).

INDEX

327